Taxonomy, Evolution, and Biostratigraphy of Conodonts from the Kechika Formation, Skoki Formation, and Road River Group (Upper Cambrian to Lower Silurian), Northeastern British Columbia

NRC Monograph Publishing Program

Inquiries: Monograph Publishing Program, NRC Research Press, National Research Council of Canada, Ottawa, Ontario K1A 0R6, Canada. Web site: www.monographs.nrc.ca

Correct citation for this publication: Pyle, L.J., and Barnes, C.R. 2002. *Taxonomy, Evolution, and Biostratigraphy of Conodonts from the Kechika Formation, Skoki Formation, and Road River Group (Upper Cambrian to Lower Silurian), Northeastern British Columbia*. NRC Research Press, Ottawa, Ontario, Canada. 227 pp.

A Publication of the
National Research Council of Canada
Monograph Publishing Program

Taxonomy, Evolution, and Biostratigraphy of Conodonts from the Kechika Formation, Skoki Formation, and Road River Group (Upper Cambrian to Lower Silurian), Northeastern British Columbia

Leanne J. Pyle and Christopher R. Barnes

School of Earth and Ocean Sciences
University of Victoria
P. O. Box 3055
Victoria, BC V8W 3P6

NRC·CNRC
NRC Research Press
Ottawa 2002

Printed in Canada on acid-free paper. ♾

ISBN 0-660-18521-0
NRC No. 44461

Canadian Cataloguing in Publication Data

Pyle, Leanne J.

Taxonomy, Evolution, and Biostratigraphy of Conodonts from the Kechika Formation, Skoki Formation, and Road River Group (Upper Cambrian to Lower Silurian), Northeastern British Columbia

Issued by National Research Council of Canada.
Includes bibliographical references.
ISBN-0-660-18521-0

1. Conodonts — British Columbia — Classification.
2. Paleontology — British Columbia.
3. Paleontology — Paleozoic.
I. Barnes, C.R. (Christopher R.), 1940– .
II. National Research Council Canada.
III. Title.

QE899.2.C65P94 2002 562'.2'097118 C2002-980243-9

Contents

Abstract / Résumé

This study establishes the stratigraphic framework and conodont biostratigraphy of Lower Paleozoic strata of the Northern Canadian Cordilleran Miogeocline. Only a few reconnaissance stratigraphic studies have been conducted previously in the study area. Nine key sections span an east–west transect from the Macdonald Platform to the Kechika Trough (platform–miogeocline–basin). Over 9000 m of strata from the Kechika and Skoki formations and Road River Group in northeastern British Columbia were measured and described from which a total of 320 conodont samples (4–5 kg each) were taken.

Conodonts of Late Cambrian to Early Silurian age are described taxonomically from the Kechika, Skoki, and Road River Group (Ospika and Pesika formations) across the transect and have been used to refine the conodont biostratigraphy across the transect. A total of 38 600 identifiable conodonts indicate high species diversity and abundance in shallow-water facies and less diversity and abundance within deeper water facies. Elements are moderately to well preserved, typically with a colour alteration index (CAI) of 3–5.

A total of 185 species, representing 69 genera are identified and illustrated among which 7 new genera and 39 new species are established. Fifteen of the 39 new species were represented by few specimens and are described in open nomenclature. The new genera are *Graciloconus*, *Kallidontus*, *Planusodus*, *Ventricodus*, and three new genera (A, B, C) treated in open nomenclature. The new species are *Acodus kechikaensis*, *Acodus neodeltatus*, *A. quentinensis*, *A. warenesis*, *Cordylodus delicatus*, *Colaptoconus greypeakensis*, ?*Diaphorodus* n. sp., *Drepanoistodus minutus*, *Graciloconus concinnus*, *Kallidontus nodosus*, *K. princeps*, *K. serratus*, *Laurentoscandodus sinuosus*, *Macerodus*? n. sp. A and B, *Microzarkodina* n. sp., *Oepikodus* n. sp., *Oistodus* n. sp., *Paroistodus* n. sp., *Planusodus gradus*, ?*Prioniodus* n. sp., *Protoprioniodus* n. sp., *Rossodus kwadachaensis*, *R. muskwaensis*, *R. sheffieldensis*, *R. subtilis*, *Scolopodus amplus*, *Striatodontus strigatus*, *Triangulodus akiensis*, *Tricostatus infundibulum*, *T. terilinguis*, also three unnamed new species of three unnamed new genera and five new species of *Drepanoistodus* (A, B, C, D, E).

The conodont zonation for Upper Cambrian to Lower Silurian strata is refined, using Sections 4, 5, 13, and Grey Peak as reference sections. It allows close dating of stratigraphic boundaries. The oldest zones in the Kechika are cosmopolitan and include the *Eoconodontus* Zone (upper Cambrian), *Cordylodus proavus* (uppermost Cambrian), and *Iapetognathus* Zone (base of Tremadocian). Nine higher zones are recognized and redefined for shallow-water platform facies containing faunas of the Midcontinent Realm. Four of these are new (*Polycostatus falsioneotensis*, *Rossodus tenuis*, *Scolopodus subrex*, and *Acodus emanualensis* zones), and 10 new subzones are established. Those for the Kechika Formation include, in ascending order, the *Polycostatus falsioneotensis* Zone (lower Tremadocian), *Rossodus tenuis* Zone (lower Tremadocian), *Rossodus manitouensis* Zone with *R. muskwaenesis* and *R. sheffieldensis* subzones (middle Tremadocian), Low Diversity Interval (upper Tremadocian), *Scolopodus subrex* Zone with *Graciloconus concinnus* and *Colaptoconus bolites* subzones (lower Arenigianian), and *Acodus kechikaensis* Zone with *Kallidontus serratus*, *Diaphorodus russoi*, and *Kallidontus nodosus* subzones (lower Arenigian). Those for the Skoki Formation include the *Oepikodus communis* Zone with *Tropodus sweeti*, *Bergstroemognathus extensus*, and *Juanognathus variabilis* subzones (middle Arenigian). The *O. communis* Zone spans the Kechika–Skoki boundary, and the uppermost Kechika lies within the lowermost part of the *O. communis* Zone underlying the *T. sweeti* Subzone. The Skoki Formation also contains the *Jumudontus gananda* Zone (middle Arenigian) and *Tripodus laevis* Zone (upper Arenigian).

Eleven deep-water zones are recognized for basinal facies containing faunas of predominantly the Atlantic Realm. Five new zones are established (*Drepanoistodus nowlani*, *Acodus deltatus*, *Paracordylodus gracilis*, *Paroistodus horridus*, and *Dzikodus tablepointensis* zones), and one new subzone within the *P. gracilis* Zone is proposed. Those within the Kechika Formation include *Cordylodus angulatus* Zone (lower Tremadocian), *Paltodus deltifer* Zone (middle Tremadocian), *Drepanoistodus nowlani* Zone (middle Tremadocian), *Acodus deltatus* Zone (middle Tremadocian), *Paroistodus proteus* Zone (upper Tremadocian), *Paracordylodus gracilis* Zone with *Oelandodus elongatus* Subzone (upper Tremadocian), and *Prioniodus elegans* Zone (base of Arenigian). Those within the Skoki and Ospika formations include *Oepikodus evae* Zone (Skoki Formation, middle Arenigian), *Paroistodus originalis* Zone (Skoki and Ospika formations, upper Arenigian), *Paroistodus horridus* and *Dzikodus tablepointensis* zones (both within the Ospika Formation, lower Llanvirn). The *Distomodus staurognathoides* Zone lies within the Pesika Formation (middle Llandovery).

The conodont faunas provide detailed temporal constraints for the stratigraphic framework. Some evolutionary remarks are made for selected species involved in radiations, especially in the Tremadocian and Arenigian, which are useful in further refining the standard Midcontinent Realm zonation.

Cette étude établit le cadre stratigraphique de la biostratigraphie des Conodontes des strates du Paléozoïque inférieur, du miogéocline de la Cordillère canadienne septentrionale. Seules quelques études stratigraphiques de reconnaissance avaient été effectuées antérieurement dans cette région. Neuf sections clefs couvent un transect est–ouest, de la plate-forme Macdonald à la fosse Kechika (plate-forme–miogéocline–bassin). Plus de 9000 m de strates des formations Kechika et Skoki et du Groupe de Road River, dans le nord-est de la Colombie-Britannique, ont été mesurées et décrites, on y a de plus récolté 320 échantillons de Conodontes (4 à 5 kg/échantillon).

Les Conodontes du Cambrien tardif au Silurien précoce sont décrits selon leur taxonomie; ils proviennent des formations Kechika et Skoki et du Groupe de Road River (formations de Ospika et Pesika), ils recoupent le transect et ont été utilisés pour raffiner la biostratigraphie des Conodontes à travers le transect. D'un total de 38 600 Conodontes identifiables, on déduit une grande diversité d'espèces et une abondance dans les faciès d'eau peu profonde ainsi que moins de diversité et d'abondance dans les faciès d'eau à plus grande profondeur. Les éléments sont de modérément à bien préservés et ont typiquement un index d'altération des couleurs (CAI) de 3 à 5.

Cent quatre-vingt-cinq espèces, représentant 69 genres, sont identifiées et illustrées parmi lesquelles on établit 7 nouveaux genres et 39 nouvelles espèces. Quinze des 39 nouvelles espèces sont représentées par peu de spécimens et sont décrites selon une nomenclature ouverte. Les nouveaux genres sont *Graciloconus*, *Kallidontus*, *Planusodus*, *Ventricodus* et 3 nouveaux genres (A, B, C) sont traités selon une nomenclature ouverte. Les nouvelles espèces sont *Acodus kechikaensis*, *Acodus neodeltatus*, *A. quentinensis*, *A. warenesis*, *Cordylodus delicatus*, *Colaptoconus greypeakensis*, ?*Diaphorodus* n. sp., *Drepanoistodus minutus*, *Graciloconus concinnus*, *Kallidontus nodosus*, *K. princeps*, *K. serratus*, *Laurentoscandodus sinuosus*, *Macerodus*? n. sp. A et B, *Microzarkodina* n. sp., *Oepikodus* n. sp., *Oistodus* n. sp., *Paroistodus* n. sp., *Planusodus gradus*, ?*Prioniodus* n. sp., *Protoprioniodus* n. sp., *Rossodus kwadachaensis*, *R. muskwaensis*, *R. sheffieldensis*, *R. subtilis*, *Scolopodus amplus*, *Striatodontus strigatus*, *Triangulodus akiensis*, *Tricostatus infundibulum*, *T. terilinguis*, ainsi que 3 nouvelles espèces sans nom de 3 nouveaux genres sans nom et 5 nouvelles espèces de *Drepanoistodus* (A, B, C, D, E).

La zonation des Conodontes des strates du Cambrien supérieur au Silurien inférieur est précisé en utilisant les sections 4, 5, 13 et Grey Peak comme sections de référence. Cela permet une datation serrée des limites stratigraphiques. Les plus vieilles zones dans le Kechika sont cosmopolites et comprennent la zone *Eoconodontus* (Cambrien supérieur), *Cordylodus proavus* (Cambrien terminal) et la zone *Iapetognathus* Zone (base du Trémadocien). Neuf zones supérieures sont reconnues et redéfinies pour un faciès de plate-forme d'eau peu profonde contenant des faunes du domaine du centre du continent. Quatre de celles-ci sont nouvelles (les zones *Polycostatus falsioneotensis*, *Rossodus tenuis*, *Scolopodus subrex* et *Acodus emanualensis*) et 10 nouvelles sous-zones sont établies. Celles de la Formation de Kechika comprennent, selon un ordre ascendant, la zone *Polycostatus falsioneotensis* (Trémadocien inférieur), la zone *Rossodus tenuis* (Trémadocien); la zone *Rossodus manitouensis* avec les sous-zones *R. muskwaenesis* et *R. sheffieldensis* (Trémadocien moyen), un intervalle de faible diversité (Trémadocien supérieur), la zone *Scolopodus subrex* avec les sous-zones *Graciloconus concinnus* et *Colaptoconus bolites* (Arénigien inférieur) et la zone *Acodus kechikaensis* avec les sous-zones *Kallidontus serratus*, *Diaphorodus russoi* et *Kallidontus nodosus* (Arénigien inférieur). Celles de la Formation de Skoki comprennent la zone *Oepikodus communis* avec les sous-zones *Tropodus sweeti*, *Bergstroemognathus extensus* et *Juanognathus variabilis* (Arénigien moyen). La zone *O. communis* chevauche la limite Kechika–Skoki et le Kechika terminal se retrouve à l'intérieur de la zone *O. communis* sous-jacente à la sous-zone *T. sweeti*. La Formation de Skoki comprend aussi la zone *Jumudontus gananda* (Arénigien moyen) et la zone *Tripodus laevis* (Arénigien supérieur).

Onze zones d'eau profonde sont reconnues comme faciès de basin contenant surtout des faunes du domaine de l'Atlantique. Cinq nouvelles zones sont établies (les zones *Drepanoistodus nowlani*, *Acodus deltatus*, *Paracordylodus gracilis*, *Paroistodus horridus* et *Dzikodus tablepointensis*) et on propose une nouvelle sous-zone à l'intérieur de la zone *P. gracilis*. Les zones à l'intérieur de la Formation de Kechika comprennent la zone *Cordylodus angulatus* (Trémadocien inférieur), la zone *Paltodus deltifer* (Trémadocien moyen), la zone *Drepanoistodus nowlani* (Trémadocien moyen), la zone *Acodus deltatus*, (Trémadocien moyen), la zone *Paroistodus proteus* (Trémadocien supérieur), la zone *Paracordylodus gracilis* avec la sous-zone *Oelandodus elongatus* (Trémadocien supérieur) et la zone *Prioniodus elegans* (base de l'Arénigien). Celles à l'intérieur des formations Skoki et Ospika comprennent la zone *Oepikodus evae* (Formation de Skoki, Arénigien moyen), la zone *Paroistodus originalis* (formations Skoki et Ospika, Arénigien supérieur), les zones *Paroistodus horridus* et *Dzikodus tablepointensis* (toutes deux à l'intérieur de la Formation Ospika, Llanvirnien inférieur). La zone *Distomodus staurognathoides* se trouve à l'intérieur de la Formation de Pesika (Llandovérien moyen).

La faune Conodonte détaille les contraintes de temps pour le cadre stratigraphique. Quelques remarques évolutives sont faites pour des espèces choisies impliquées dans des radiations, surtout au Trémadocien et à l'Arénigien, qui sont utiles pour la mise au point de la zonation standard du domaine du milieu du continent.

Acknowledgements

L.J. Pyle gratefully acknowledges receiving financial support from the Natural Sciences and Engineering Research Council of Canada (NSERC) and additional support from the Northern Scientific Research Training Program and a Geological Society of America grant. C.R. Barnes acknowledges primary financial support for the project provided by LITHOPROBE/SNORCLE and NSERC. Logistical support for field work was generously provided by the British Columbia Geological Survey and Inmet Mining Corp.

Valuable advice on stratigraphic sections was provided by M.P. Cecile, F. Ferri, H. Gabrielse, J.L. Nelson, and B.S. Norford. L.J. Pyle thanks G.S. Nowlan for significant discussion on conodont taxonomy and G.L. Albanesi, S. Zhang, J. Zhang, and R.L. Ethington for additional critical taxonomic discussions. We thank the reviewers of the manuscript, R.L. Ethington and A. Löfgren, for their time and many helpful suggestions. We thank C. Sullivan and D. Pellerin for cheerful and able field assistance and M. Landry, M. Deagle, and E. Delahaye for assistance in preparation of figures. We appreciate the time and assistance in operation of the scanning electron microscope by C. Singla.

Introduction

Conodonts are phosphatic microfossils that are found in marine strata ranging from Upper Cambrian to Upper Triassic. Conodont elements represent the teeth, which formed a feeding apparatus, of an extinct, soft-bodied chordate. Their biostratigraphic usefulness was noted long before an understanding of the conodont animal developed. Based on morphological characteristics known from fossils, soft-bodied impressions of the conodont animal, conodonts are most likely a group of gnathostome jawless fish (Donoghue et al. 2000).

This monograph examines the conodont taxonomy, evolution, and biostratigraphy of Lower Paleozoic strata of remote northeastern British Columbia. The stratigraphic framework of the study area has been established by Pyle and Barnes (2000) and includes a platformal succession, Kechika and Skoki formations (Upper Cambrian to Lower Ordovician), and the shelf-break to basinal equivalents of the upper Kechika and Skoki, the Road River Group (Lower Ordovician to Lower Silurian). Over 38 000 conodont elements were recovered from 320 samples (4–5 kg each) from a total of nine sections (over 9000 m of strata) comprising the transect. Detailed taxonomic study of the abundant and well-preserved conodont fauna has resulted in the identification and description of a total of 185 species, representing 69 genera. Several important evolutionary lineages have been established or clarified. The large taxonomic database from platform to basin strata of late Cambrian to early Silurian age has resulted in the establishment of a refined conodont biostratigraphy consisting of shallow, warm water Midcontinent Realm faunas and deeper, cooler water Atlantic Realm faunas. The biostratigraphy is detailed for the continuous Lower to Middle Ordovician succession and also provides age constraints on strata ranging into the Lower Silurian. The zones recognized in this study can be correlated to the standard zonation proposed for the Ordovician. This refined biozonation can serve as a reference for the Canadian Cordillera and likely much of the Canadian Innuitian Orogen and adjoining platform.

Location and regional geology

A total of nine sections have been measured and described in detail across an east–west, platform–miogeocline–basin transect located within the Ware (94F), Tuchodi Lakes (94K), and Kechika (94L) map areas (Fig. 1). Exposed strata in the Northern Rocky Mountains, within the Canadian Cordilleran Foreland Belt, comprise shallow-water carbonates of the Macdonald Platform bounded to the west by equivalent deep-water slope and basin facies of the Kechika Trough (Fig. 2). The western boundary of the Foreland Belt and Kechika Trough is the Northern Rocky Mountain Fault/Trench (NRMT); the eastern boundary is the erosional edge of preserved Lower Paleozoic strata above the regional sub-Devonian unconformity.

The Lower Paleozoic succession of the Macdonald Platform comprises platform carbonates of the Kechika Formation, a laterally extensive unit which thickens westward into the Kechika Trough. The Kechika is overlain by platform carbonates of the Skoki Formation, which abruptly changes facies westward to the Road River Group which overlies the Kechika in the Kechika Trough.

Materials and methods

Field work was completed over three field seasons (1994, 1995, and 1996). The stratigraphic sections in the study area are accessible only by helicopter. Fly camps near remote sections were established, either within cols upon ridges or within cirques above the treeline. Logistic and helicopter support from the British Columbia Geologic Survey (BCGS) was received in each field season and from Inmet Mining Corp. in 1996. Sections were measured across strike using a Jacob's staff. Measurements were comparable to thicknesses of previously measured sections (Sections 1, 4, 5, and 13 of Cecile and Norford 1979). Conodont samples were collected at an average interval of 20 m, or greater in expanded sections or less than 20 m in condensed sections. Carbonate beds were sampled to give a total of 320 (4–5 kg each) samples from over 9000 m of strata.

The technique for processing carbonate rocks to extract phosphatic conodont elements first involves the dissolution of a 3 kg sample in dilute (10%) acetic acid, following a standard conodont analysis technique (Austin 1987). The resulting residue is wet sieved into a coarse (16 mesh) fraction and a fine (200 mesh) fraction. The concentration of conodonts from the fine fraction is accomplished by heavy-liquid separation in which the conodont elements with a specific gravity of 2.84–3.10

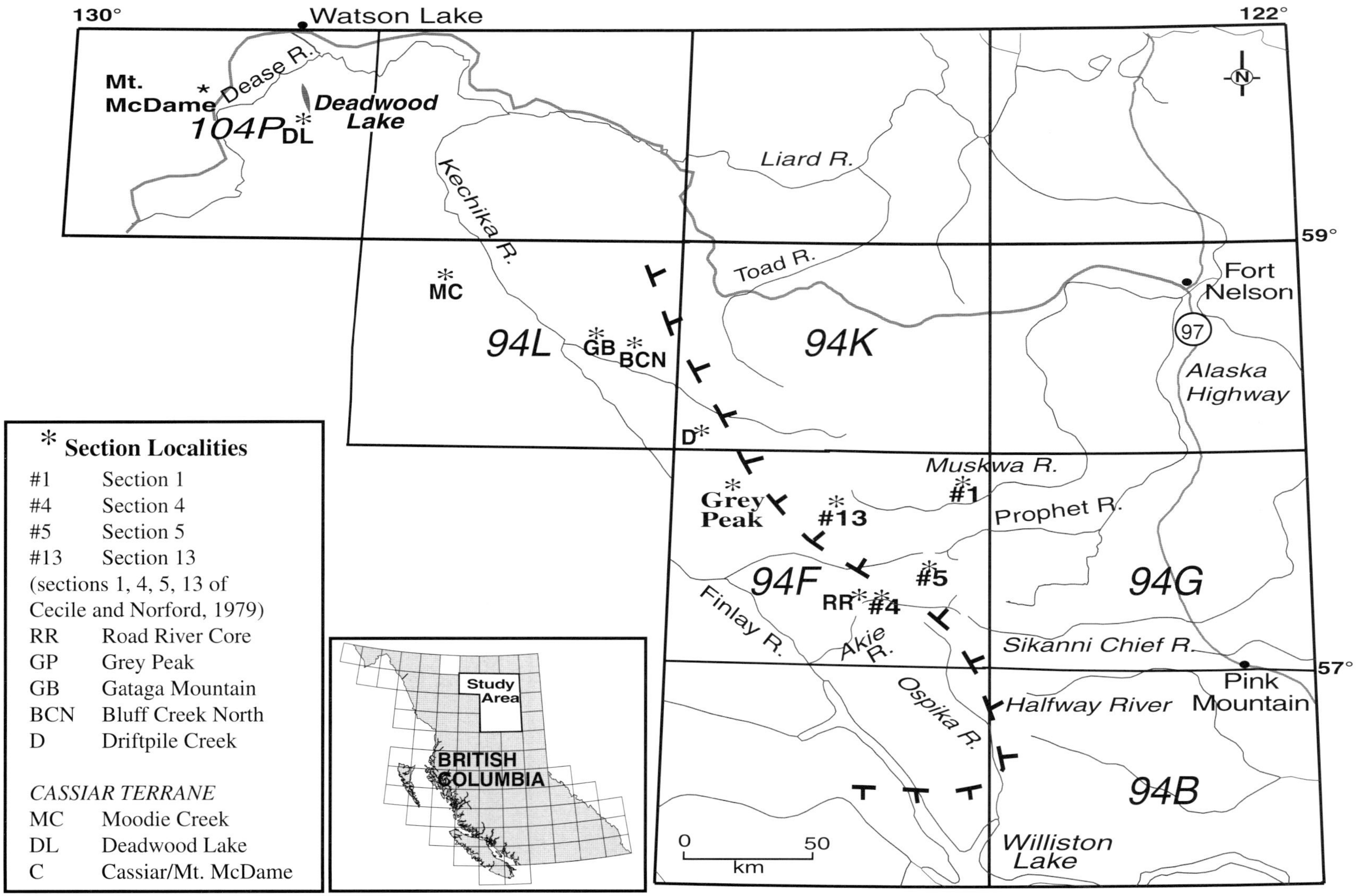

Fig. 1. Map showing geographic localities, section localities, position of shelf to off-shelf transition (TT teeth point toward shelf, after Cecile and Norford 1979; Thompson 1989).

Fig. 2. Map showing Lower Paleozoic paleogeography and occurrences of Lower Paleozoic volcanic rocks modified from Cecile et al. 1997).

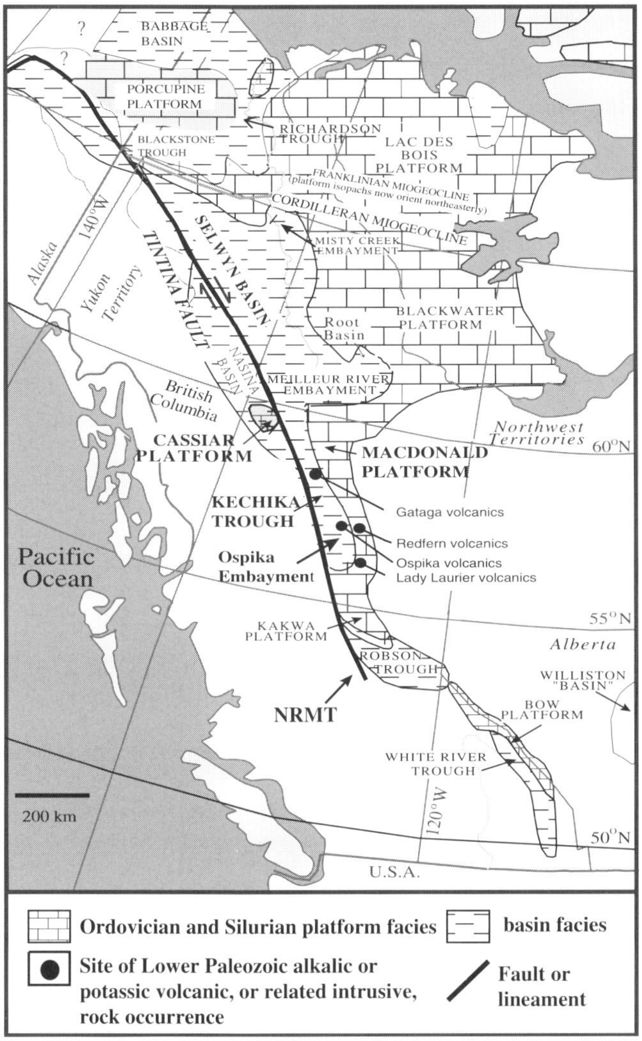

are separated from the "light" fraction using sodium polytungstate (Stone 1987). Electromagnetic separation, using a Carpco Inc. magnetic separator, is used on samples with a high proportion of magnetic minerals.

The final extraction of conodonts is accomplished by handpicking under a binocular microscope. All conodonts illustrated were photographed by a Hitachi S-3500N scanning electron microscope.

Previous biostratigraphic studies

Few detailed conodont biostratigraphic studies have been initiated for the Lower Paleozoic of the Canadian Cordillera (Barnes et al. 1991). In the northern Canadian Cordillera, previous work on Lower Paleozoic conodont biostratigraphy has been based on some small Cambrian to Devonian collections made in the study area during regional mapping projects. Pohler and Orchard (1990) reported on 59 Ordovician conodonts assigned to 17 species from the Kechika Trough. Reports on conodont collections from Road River strata in the Yukon and Northwest Territories, the northern equivalents of the Kechika Formation, are reported by McCracken and Lenz (1987), McCracken (1989*a*, *b*, 1991), and Pohler and Orchard (1990).

Unpublished paleontological reports of the Geological Survey of Canada contain information on conodont and graptolite collections from Grey Peak and the Kechika Formation adjacent to the study area in the northern Cordillera. Nowlan (unpublished reports, 1983, 1995) lists conodonts identified from one Road River sample from Grey Peak and from three samples from the Kechika from Mount Sheffield. Tipnis (unpublished report, 1979) reported ages of Grey Peak conodonts to range from Cambrian to Early Ordovician, however, these collections have been misplaced. Orchard (unpublished report, 1995) reported on sparse faunas from 11 samples from the Kechika River (94L) map area, four of which were from Ordovician Kechika and Road River. Jackson (unpublished report, 1980) reported on graptolites from the Grey Peak section, near the top of the Kechika, ranging from late Tremadocian to early Arenigian. Data from these reports are included in the discussion of each lithostratigraphic unit. Information from the existing graptolite zonation for the region are integrated with the conodont data where possible, based on Jackson's (1980) report, the unpublished thesis by Davies (1966), and Lower to Middle Ordovician graptolite zonation by Lenz and Jackson (1986).

Because of the remote location of the study area (Fig. 1), few previous detailed stratigraphic studies have been carried out and the conodont biostratigraphy for the Lower Paleozoic has not been established. The main objectives of this study are to measure, describe in detail, and establish the Upper Cambrian to Lower Silurian stratigraphic framework (Kechika Formation, Road River Group, and Skoki Formation) across a platform to basin transect of the northern Canadian Cordilleran miogeocline; document the conodont fauna through detailed taxonomic identifications and descriptions of select taxa; interpret the conodont evolutionary relationships where data permit; establish, using the taxonomic database from the continuous, thick stratigraphic succession, the Lower Paleozoic conodont biostratigraphy across the platform to basin transect; refine and adjust the standard Midcontinent and Atlantic Realm zonations for warm, shallow and cooler, deeper water facies, respectively, and temporally constrain sequence boundaries and units in the stratigraphic framework.

Lithostratigraphy

The general stratigraphic framework has been established from previous regional mapping projects. The stratigraphic units in the study area range from Upper Cambrian (Trempealeauan) to Lower Silurian (Llandovery) and include, in stratigraphic order, the Kechika Formation and Skoki Formation and equivalent Road River Group (Fig. 3). The stratigraphy is revised in detail (Pyle and Barnes 2000) to divide the Kechika Formation (late Cambrian to early Arenigian in age) into five formal members: Lloyd George, Quentin, Grey Peak, Haworth, and Mount Sheffield. The Skoki Formation (early to late Arenigian in age) comprises three new formal members defined as Sikanni Chief, Keily, and Redfern. The Road River Group is divided into three new formations: Ospika (early Arenigian to Llanvirn in age), Pesika (Lower Silurian in age), and Kwadacha (Lower Silurian to Middle Devonian in age, formerly the Silurian Siltstone). The Ospika Formation is further subdivided into five formal members: Cloudmaker, Finlay Limestone, Chesterfield, Finbow Shale, and Ware (Fig. 3).

Fig. 3. Correlation of lithostratigraphic units. Time scale after Webby (1998) and Cooper and Nowlan (1999), regional correlations for the Kechika Trough and Macdonald Platform modified from Norford (1991).

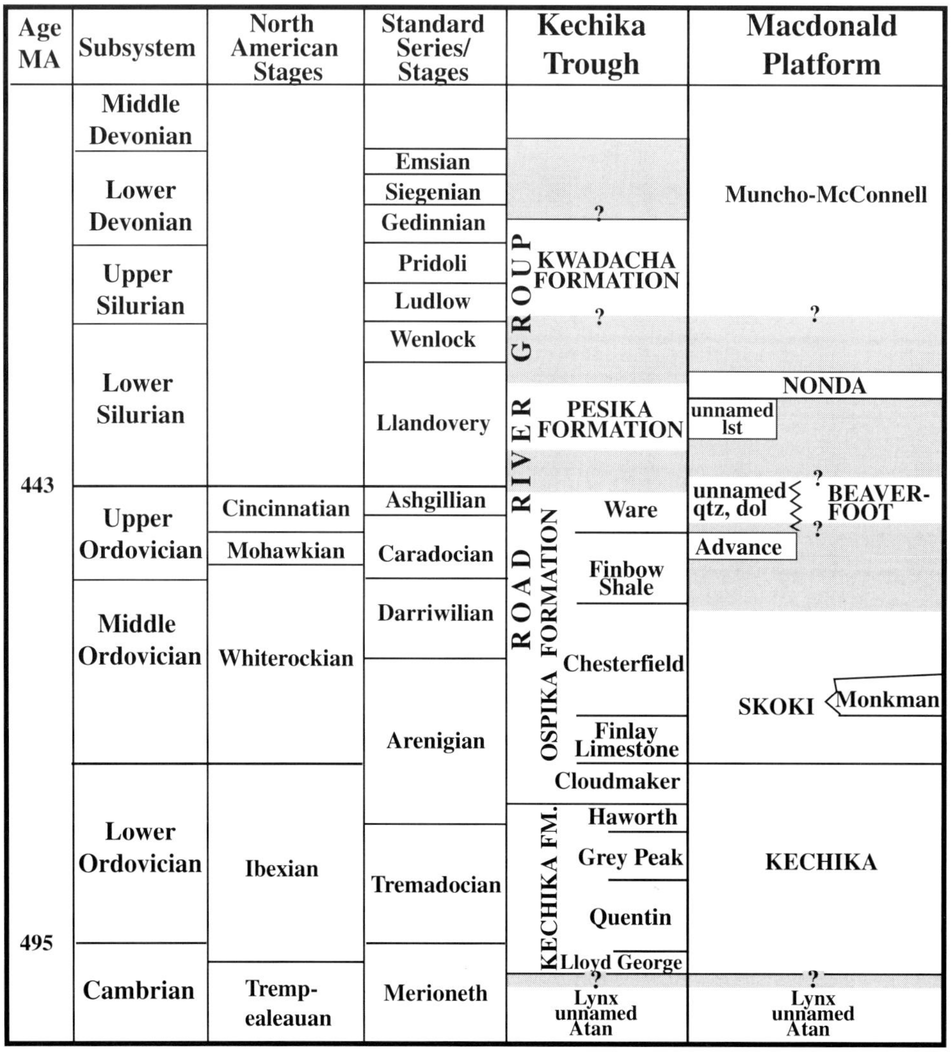

Kechika Formation

Gabrielse (1963) was the first to describe the Kechika (as a Group) at its type locality in the McDame map area and applied the name to the Ware map area (Gabrielse et al. 1977). In the study area, the Kechika typically comprises a succession of grey and light grey, cleaved, phyllitic calcareous shale and thin-bedded argillaceous limestone to grainstone. The Kechika is laterally extensive within the study area and ranges in thickness from 200 m in the east, thickening westward up to 1400 m. It represents a sedimentary package which accumulated on a broad, gentle ramp, as indicated by an east-to-west transition from shallow to deeper water facies (Cecile and Norford 1979). The unit is truncated to the east by the regional sub-Devonian unconformity.

Skoki Formation

The Skoki Formation has its type section at Skoki Mountain and reference section at Mount Wilson, southwest Alberta (Norford 1969). The Skoki Formation was named by Walcott (1928) and is widespread in the southern Rockies, comprising dolostone and rare limestone, commonly with oncolitic facies, beds with quartz silt and sand and argillaceous layers (Norford 1969). The Skoki ranges from 62 to 186 m thick and is present in several map areas in the southern Rockies (Walcott in Glass 1990). A much thicker succession, over 1000 m, has been measured in the study area. An anomalous thickness of 1260 m has been recorded by Thompson (1989) in the Halfway River map area.

In the study area, the Skoki is a sequence of thick- to massive-bedded, light-grey-weathering dolostone, limestone, and shale and nodular to thin-bedded limestone within massive-bedded dolomitic shale. It represents subtidal platformal facies in the east and thickens westward to the platform margin. As Thompson (1989) and Norford (1991) described, this margin shifted through time, extending farthest west during early Middle Ordovician, with a narrow shelfbreak facies developed between the platform margin and equivalent basinal facies (see Fig. 1 for position of shale-out).

Road River Group

The Road River was named as a formation by Jackson and Lenz (1962) with a type section in the Richardson Mountains (Richardson Trough), northern Yukon. It is a thick succession (over 910 m) of dark graptolitic shale, argillaceous limestone, and minor chert, dolostone, siltstone, and sandstone. It varies in thickness and has been described from the Yukon, east-central Alaska, and northeastern British Columbia (Jackson and Lenz in Hills 1981). The Road River is exposed extensively in the Northern Cordillera from the Kechika Trough to the Selwyn Basin and Richardson Trough. It is a thick succession of several unnamed basinal equivalents of Lower Ordovician to Devonian shallow-water units and contains a variety of facies. Hiatuses may occur within the Road River Group in some areas, but are poorly defined.

In the study area, Cecile and Norford (1979) reported a thickness over 1320 m for the Road River Group, which includes mappable units. Section 4 (57°26′N, 124°35′W), north of the Akie and Ospika rivers, is the type section for the three new formations of the Road River Group, the Ospika, Pesika, and Kwadacha formations that exceed 700 m in thickness. In this study, conodonts were recovered from the Ospika Formation, which is 455 m thick at its type section and is defined as a succession of largely black, organic-rich shale and brown- to orange-weathering shale, minor thin-bedded limestone, dolostone and siltstone and quartzite in the upper member. Its lower contact is unconformable with the Kechika at the type section, but abrupt and conformable with the Kechika at other sections within the Kechika Trough. The upper contact with the Pesika Formation is conformable at the type section, but regionally unconformable (Cecile and Norford 1979). The Ospika Formation is a recessive package that is widely distributed in the Ware, Halfway River, Kechika, and McDame map areas.

The Pesika Formation is 230 m thick at its type section and comprises mostly thin-bedded, grey, dolomitic limestone and dark grey, graptolitic shale with some thin beds of medium grey calcareous lime mudstone to grainstone. It ranges from 20 to 100 m thick and also includes slope breccia, thin-bedded limestone, minor shale, and dolostone as observed by Cecile and Norford (1979).

The Kwadacha Formation is extensive in the study area and has previously been referred to informally as the Silurian Siltstone of the upper Road River Group (Ferri et al. 1995–1997; MacIntyre 1992). The unit is exposed in the Ware and Kechika map areas where thicknesses vary considerably from 75 to 200 m (Gabrielse 1981; McClay et al. 1988), 250 m to more than 500 m thick (Cecile and Norford 1979; Gabrielse 1981), and up to 1000 m in the northern Kechika map area (Ferri et al. 1996). The Kwadacha Formation has been mapped in several thrust panels in the Gataga map area where it thins westward (Ferri et al. 1995, 1996).

At its type section, the Kwadacha Formation is estimated to be 300 m thick. It is a uniform, resistant, cliff-forming succession which weathers orange to brown. It is predominantly thin-bedded, well-laminated, medium- to coarse-grained, medium grey, argillaceous siltstone and light grey dolomitic siltstone. The platy talus contains abundant, large horizontal feeding traces in the basal part of the formation. No conodonts were recovered from this unit.

Biostratigraphy

Conodont biostratigraphy and provinciality

The correlation of shallow- and deep-water facies is complicated by the provincialism of conodonts in the Ordovician as first noted by Sweet et al. (1959) and Sweet and Bergström (1962). The two faunal realms, the (North) Atlantic and (North American) Midcontinent realms, necessitate two separate biozonations, and a biozonation is proposed for each (Figs. 4–10). Provincialism within each Realm is controlled by ecologic factors in which conodont faunas are adapted to warmer, more saline waters (Midcontinent) or cooler waters of normal salinity (Atlantic) (Pohler and Barnes 1990).

The zonation for the Atlantic Realm has been well established from the Baltic area of Europe (Bergström 1971; Lindström 1971; van Wamel 1974; Löfgren 1978, 1985, 1994–1996) and from Newfoundland (Bagnoli et al. 1987; Pohler et al. 1987; Barnes 1988; Stouge and Bagnoli 1988; Johnston and Barnes 1999). The zonation for the Lower Ordovican Midcontinent Realm was informally established by Ethington and Clark (1971, 1982) and Repetski (1982) and for the Cambro–Ordovican boundary by Miller (1980, 1984, 1988) in the southwestern United States. A significant contribution was made by Ji and Barnes (1994) who established eight shallow-water assemblage zones and six deeper water lineage zones for the St. George Group, western Newfoundland. Correlation of the 10 lineage and assemblage zones for the Survey Peak Formation established by Ji and Barnes (1996) is possible with the Lower Ordovician zones proposed herein. Ross et al. (1997) proposed a revised conodont zonation for the Ibexian strata of Utah, which is used as the standard in Fig. 9.

Preservation of conodont fauna

The majority of the conodonts recovered are well preserved. Those from the Kechika Formation are the best preserved and numerous, especially within the Mt. Sheffield Section (Section #13). Those from the Skoki Formation and Road River Group, particularly from dolomitic facies, are less pristine, bearing crystal overgrowths and commonly broken, but still identifiable. The conodont faunas are most diverse and abundant within the platform and shelf facies of the Kechika and Skoki formations and less diverse and abundant within basinal Road River Group equivalents. Samples from basinal facies of the Kechika and Road River from three localities in the Kechika Trough were sparse to barren. Yields from productive samples ranged from less than 10 to over 2000 elements per sample.

The colour alteration index (CAI) of conodonts was quantified by Epstein et al. (1977) by experimental calibration and a scale was proposed, ranging from values for pale yellow elements (CAI 1) through light and dark brown to black (CAI 5). Rejebian et al. (1987) extended the calibration to CAI 6–8 for black to grey (CAI 6), white (CAI 7), and clear (8). The CAI values of the specimens in the study area range from 3 to 5, indicating burial temperatures of 110° to over 300°C.

Conodont zonation

Based on the conodont data from the most productive sections, that is, Sections 1, 4, 5, 13, and Grey Peak, the stratigraphic ranges of conodonts have been plotted for each section to determine the proposed zonation (Figs. 4–8).

In this study, 9 zones, some with subzones, are defined based on the Midcontinent Realm faunas, and 11 zones are defined based on the Atlantic Realm faunas. Two sets of zones, those for shallow, warm water Midcontinent and those for colder water Atlantic Realm faunas are proposed, allowing for precise zonal tie points where mixing of the two realms occurs. New zones are named in accordance with the North American Stratigraphic Code (Hedberg 1976). The zones for the Cambrian–Ordovician boundary, that is, the *Cordylodus* zones and *Iapetognathus* Zone, are useful in both realms

Fig. 4A. Stratigraphic ranges of conodonts from the Kechika Formation, Section 13.

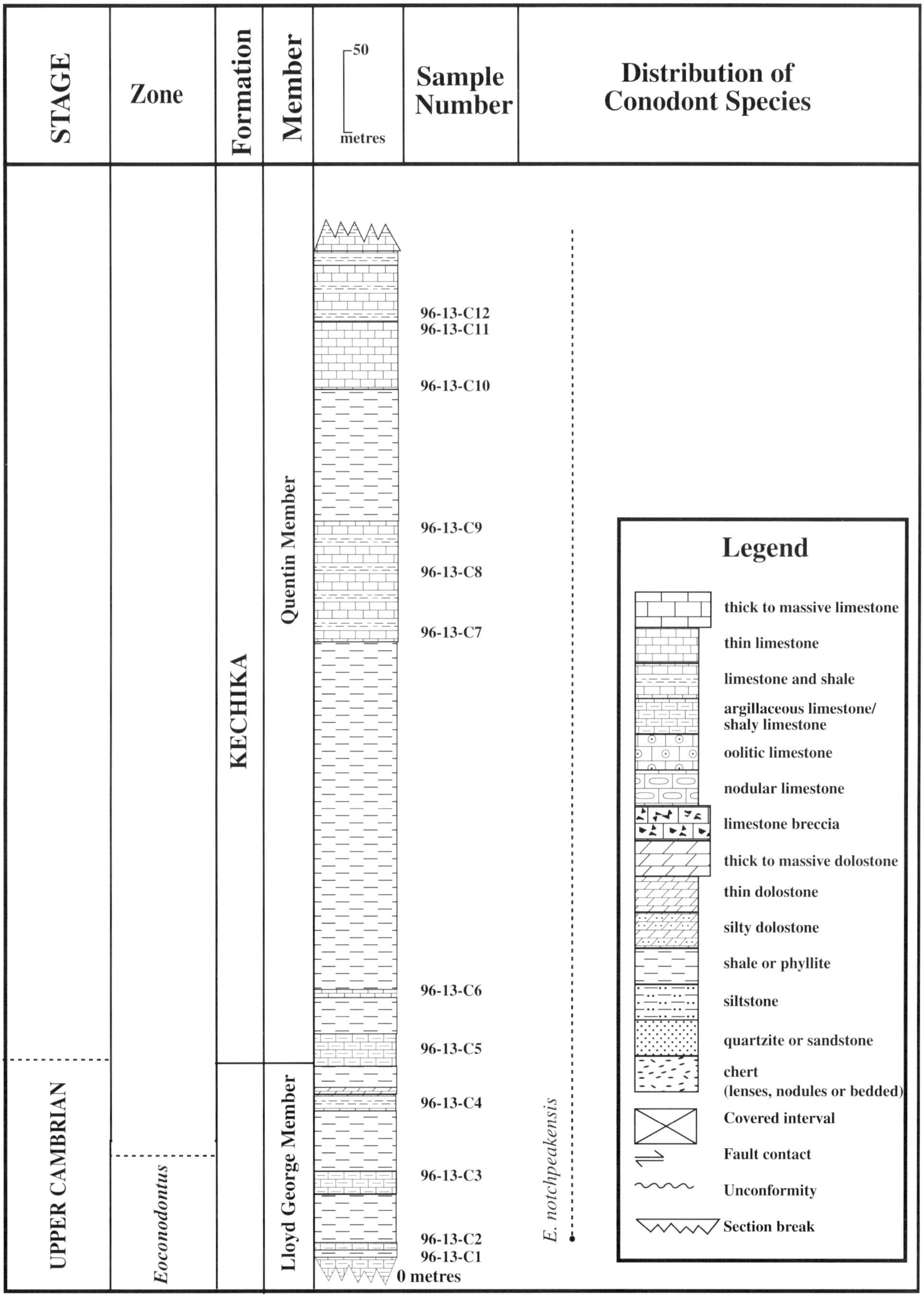

Fig. 4B. Stratigraphic ranges of conodonts from the Kechika Formation, Section 13.

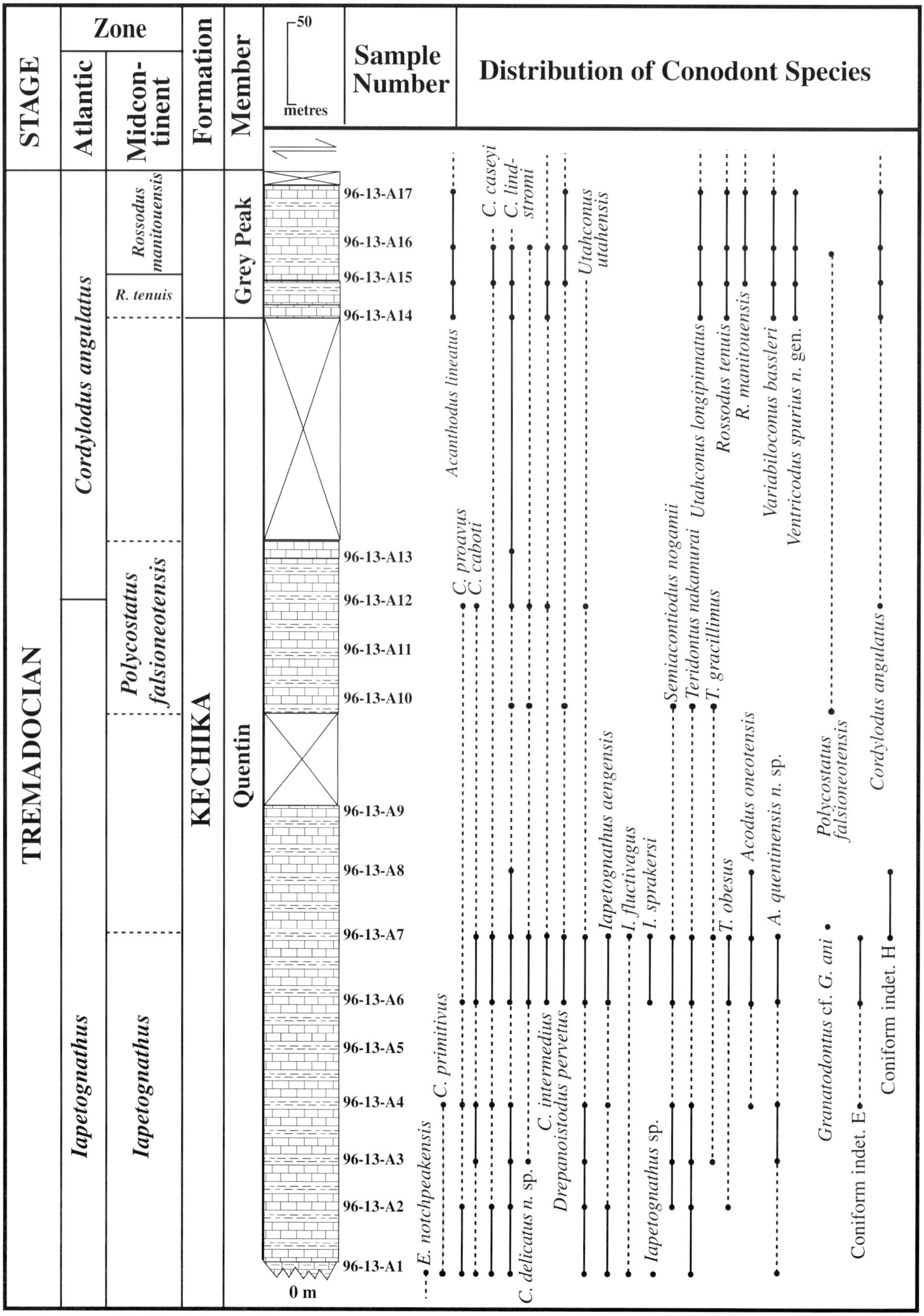

Fig. 4C. Stratigraphic ranges of conodonts from the Kechika Formation, Section 13.

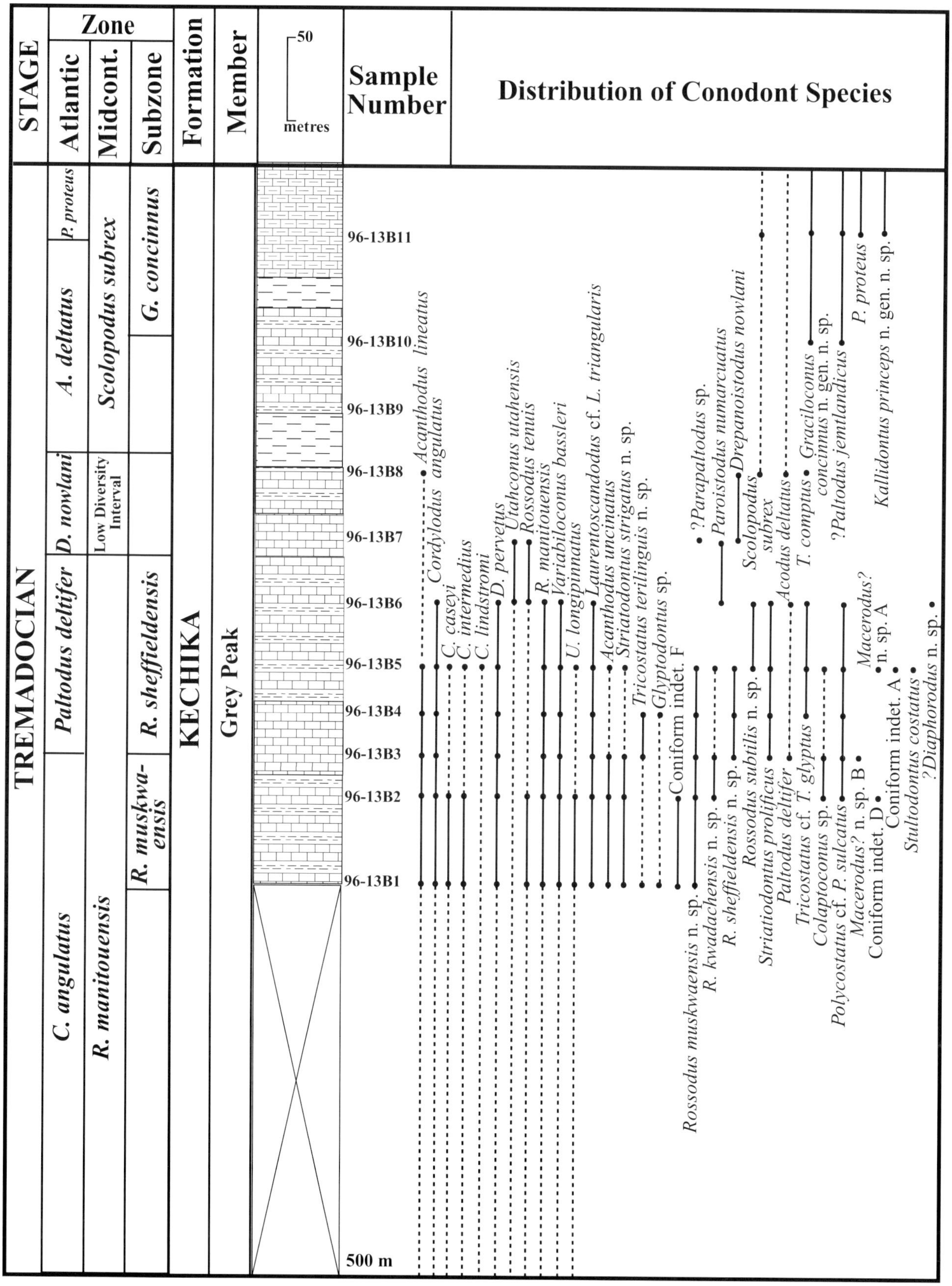

Fig. 4D. Stratigraphic ranges of conodonts from the Kechika Formation, Section 13.

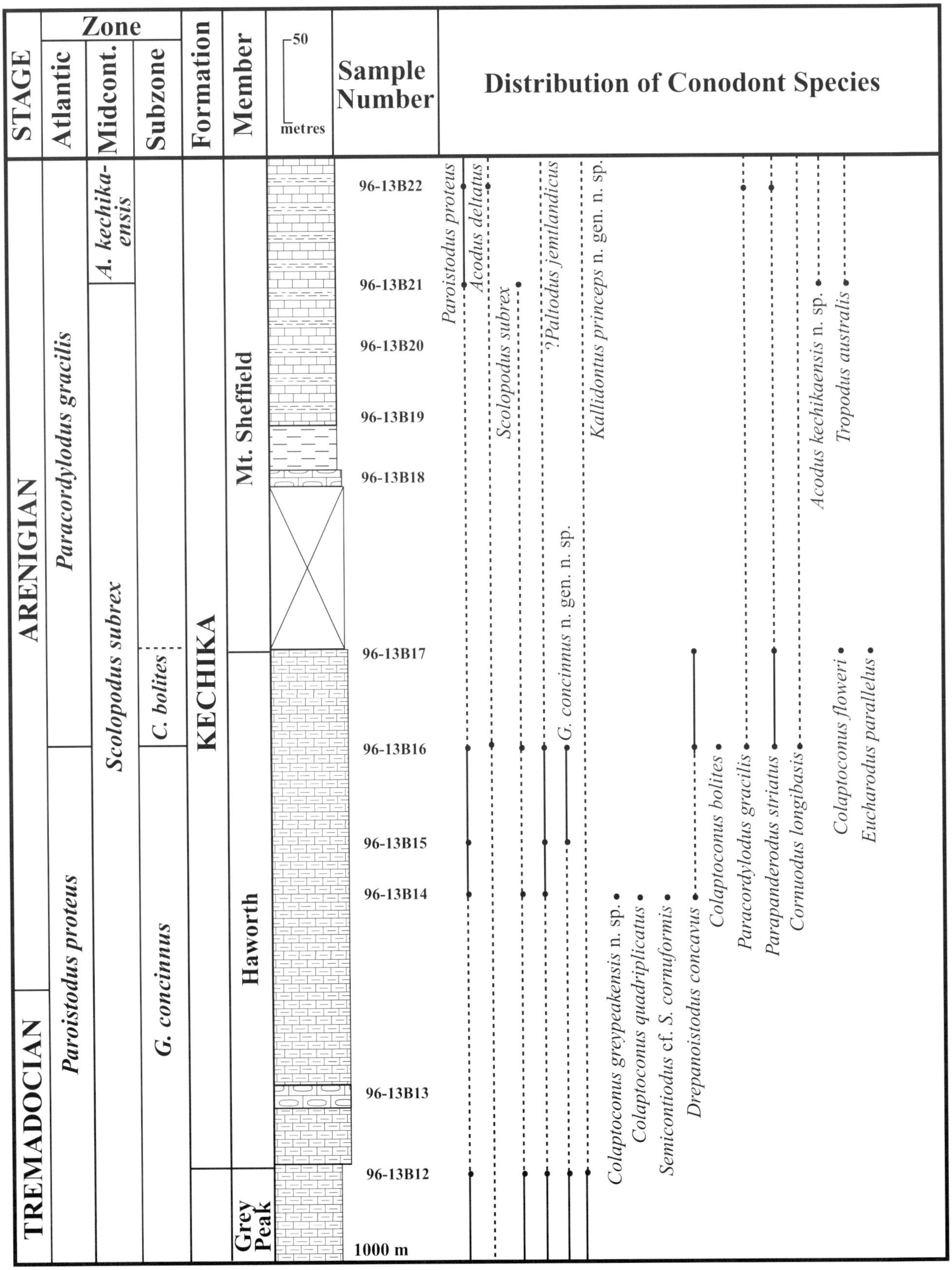

Fig. 4E. Stratigraphic ranges of conodonts from the Kechika and Skoki formations, Section 13.

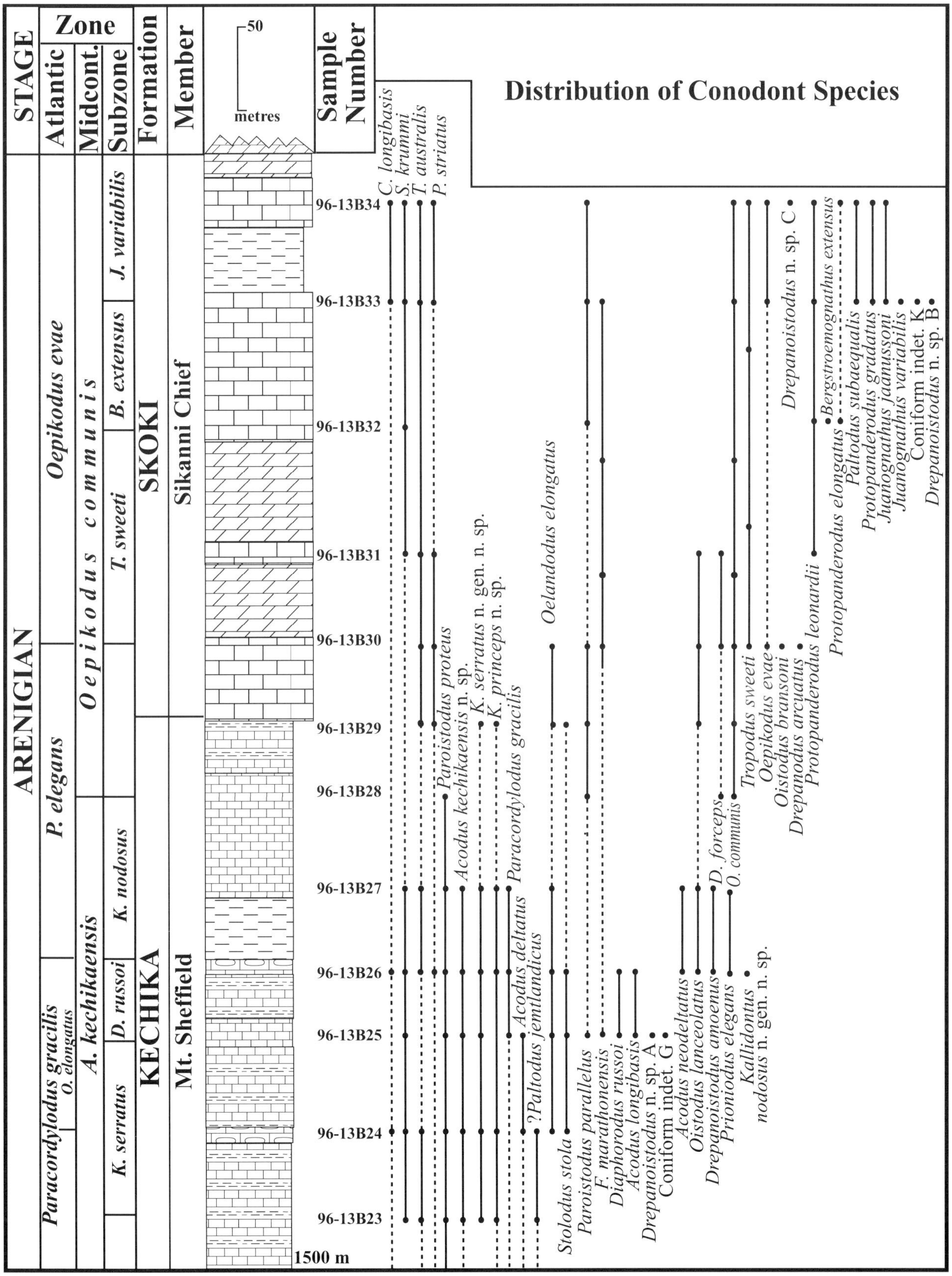

Fig. 5. Stratigraphic ranges of conodonts from the Kechika and Skoki formations, Section 1.

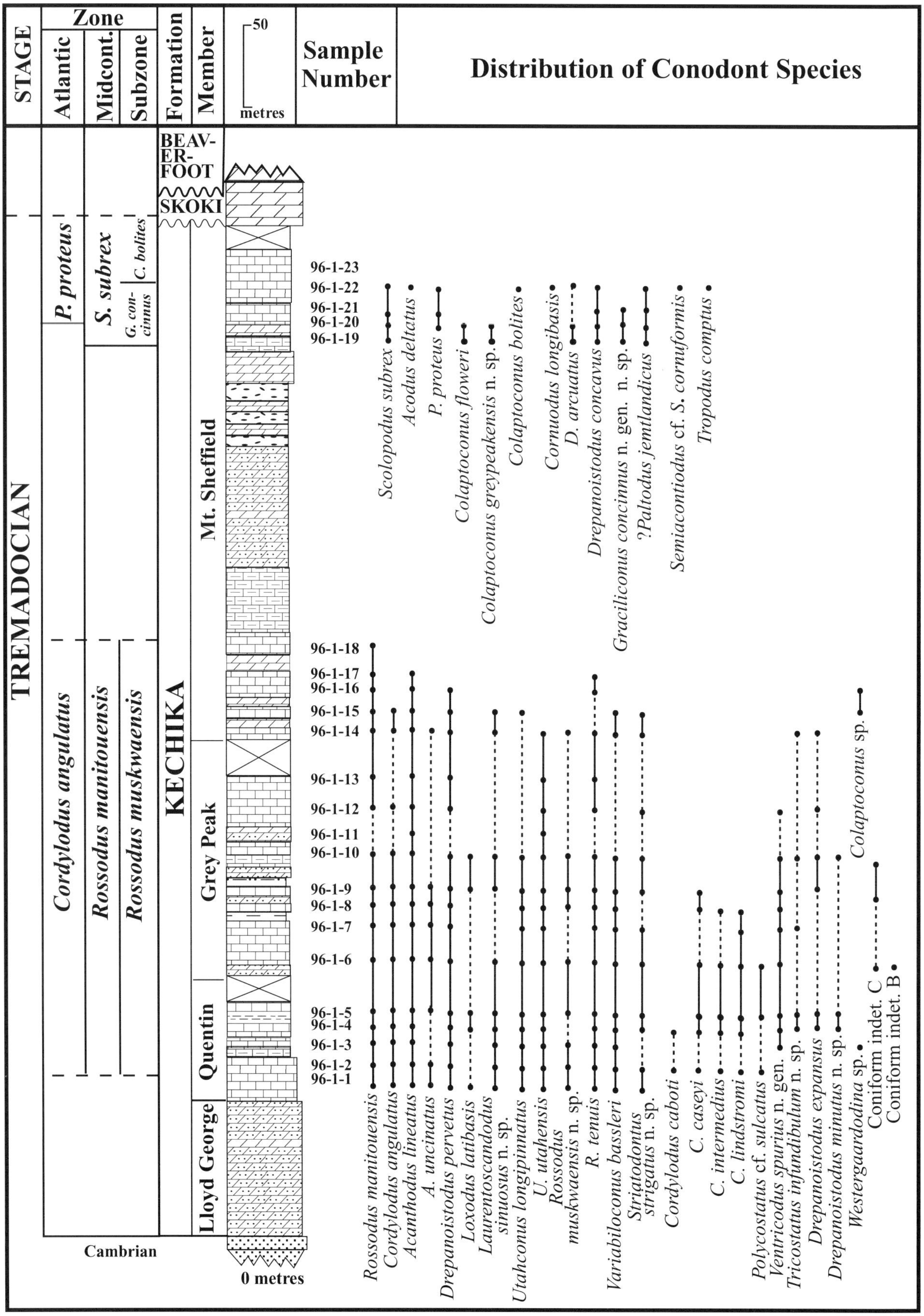

Fig. 6A. Stratigraphic ranges of conodonts from the upper Kechika and Skoki formations, Section 5.

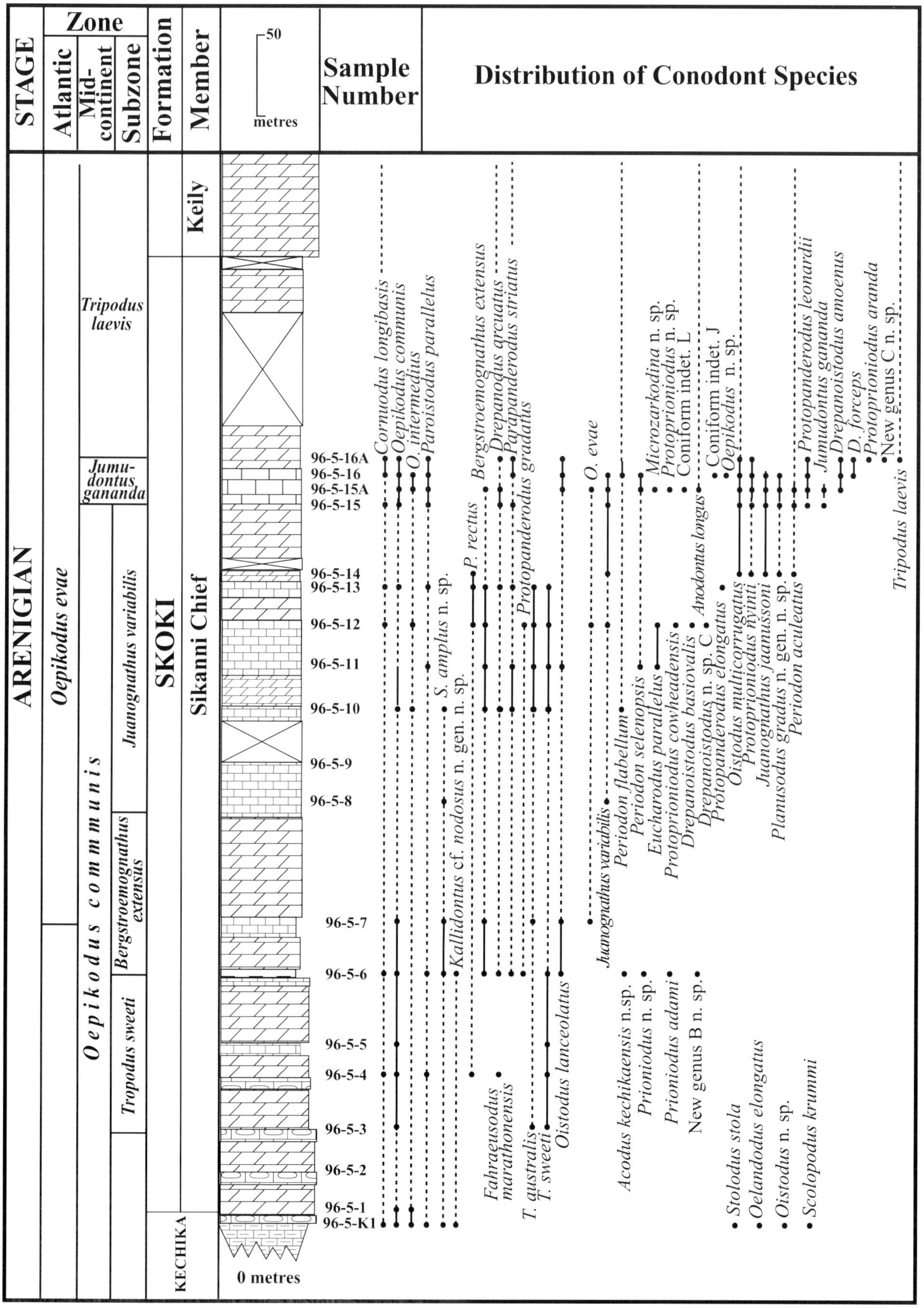

Fig. 6B. Stratigraphic ranges of conodonts from the Skoki Formation, Section 5.

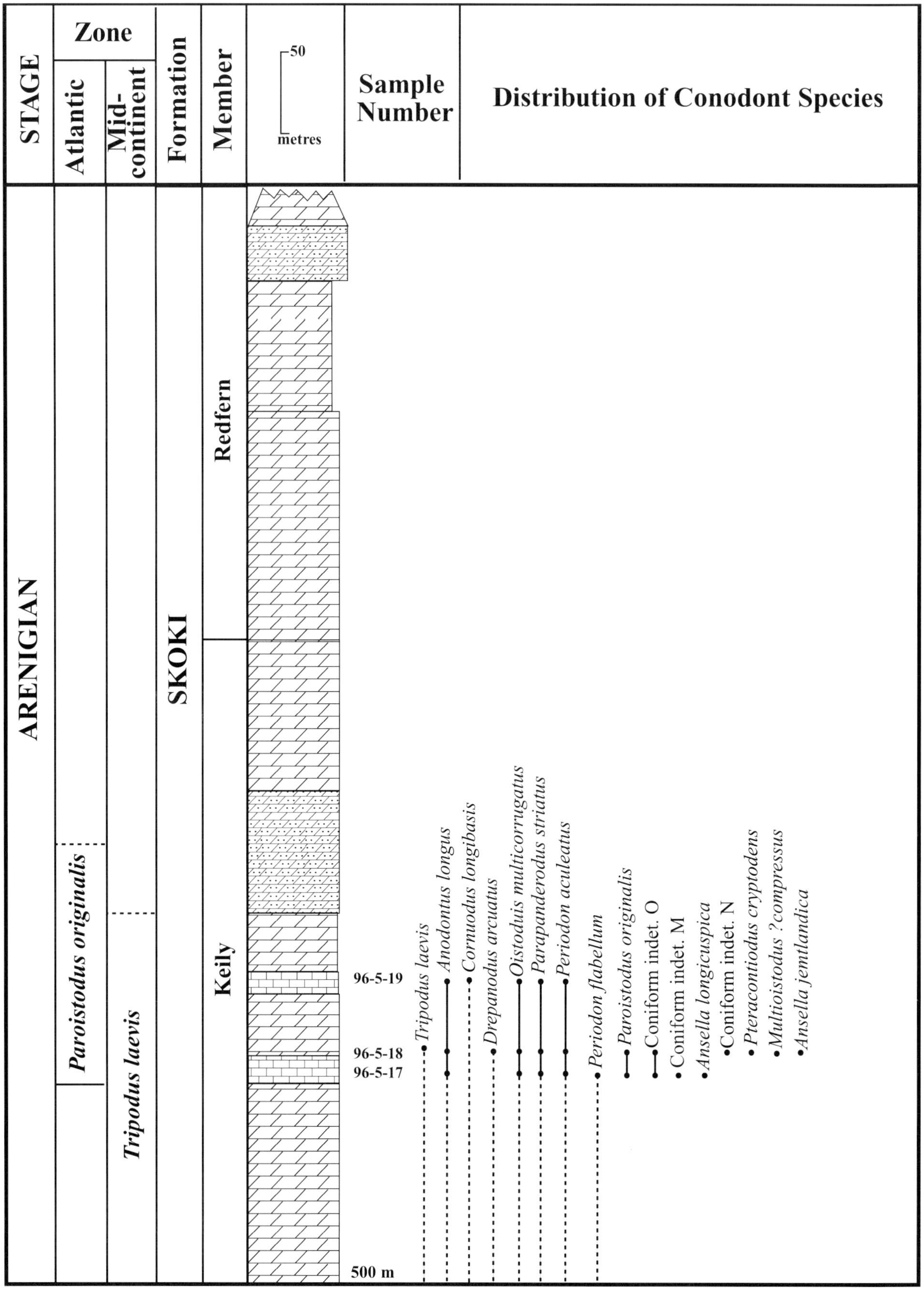

Fig. 7A. Stratigraphic ranges of conodonts from the Kechika Formation, Grey Peak Section.

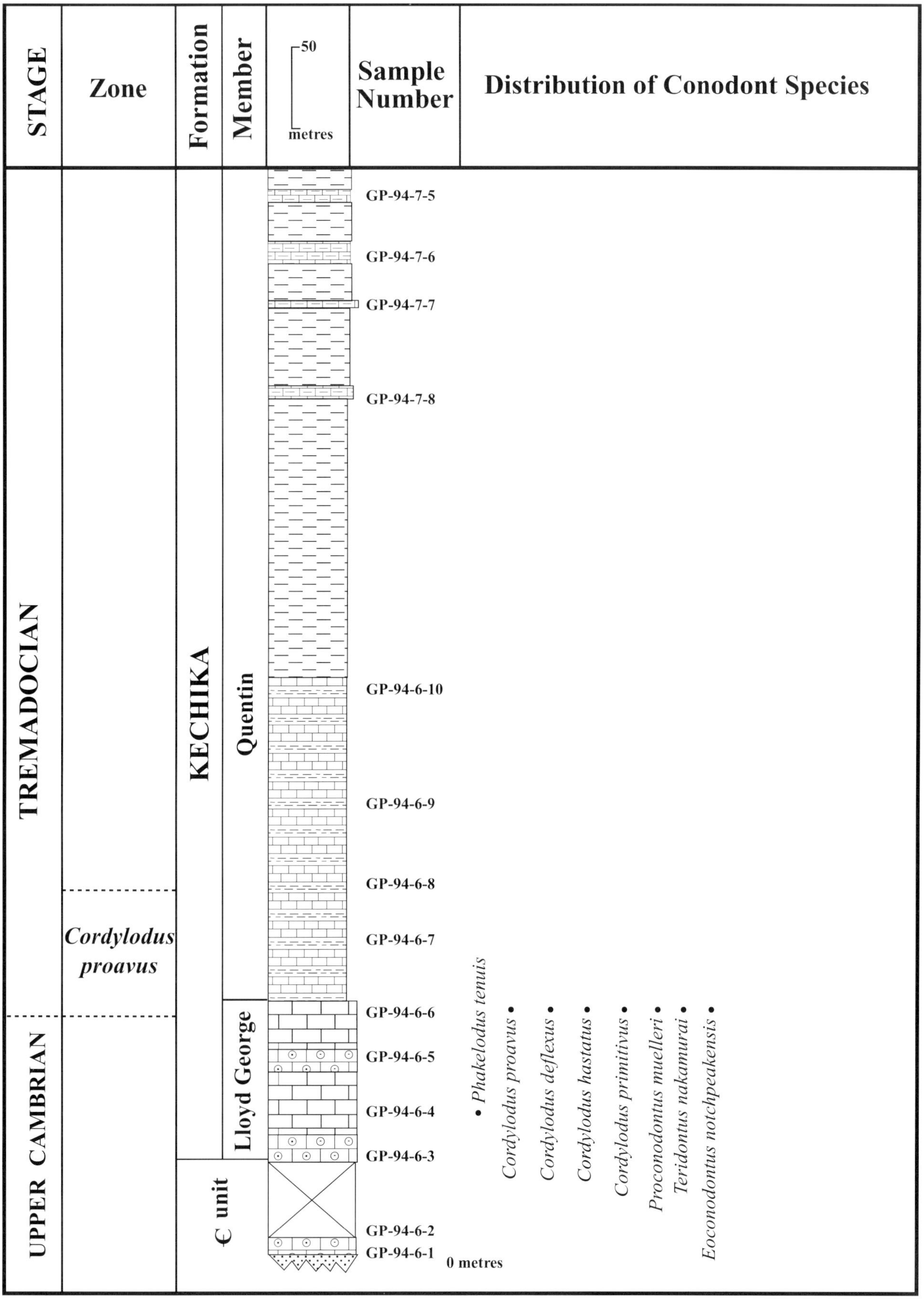

Fig. 7B. Stratigraphic ranges of conodonts from the Kechika Formation, Grey Peak Section.

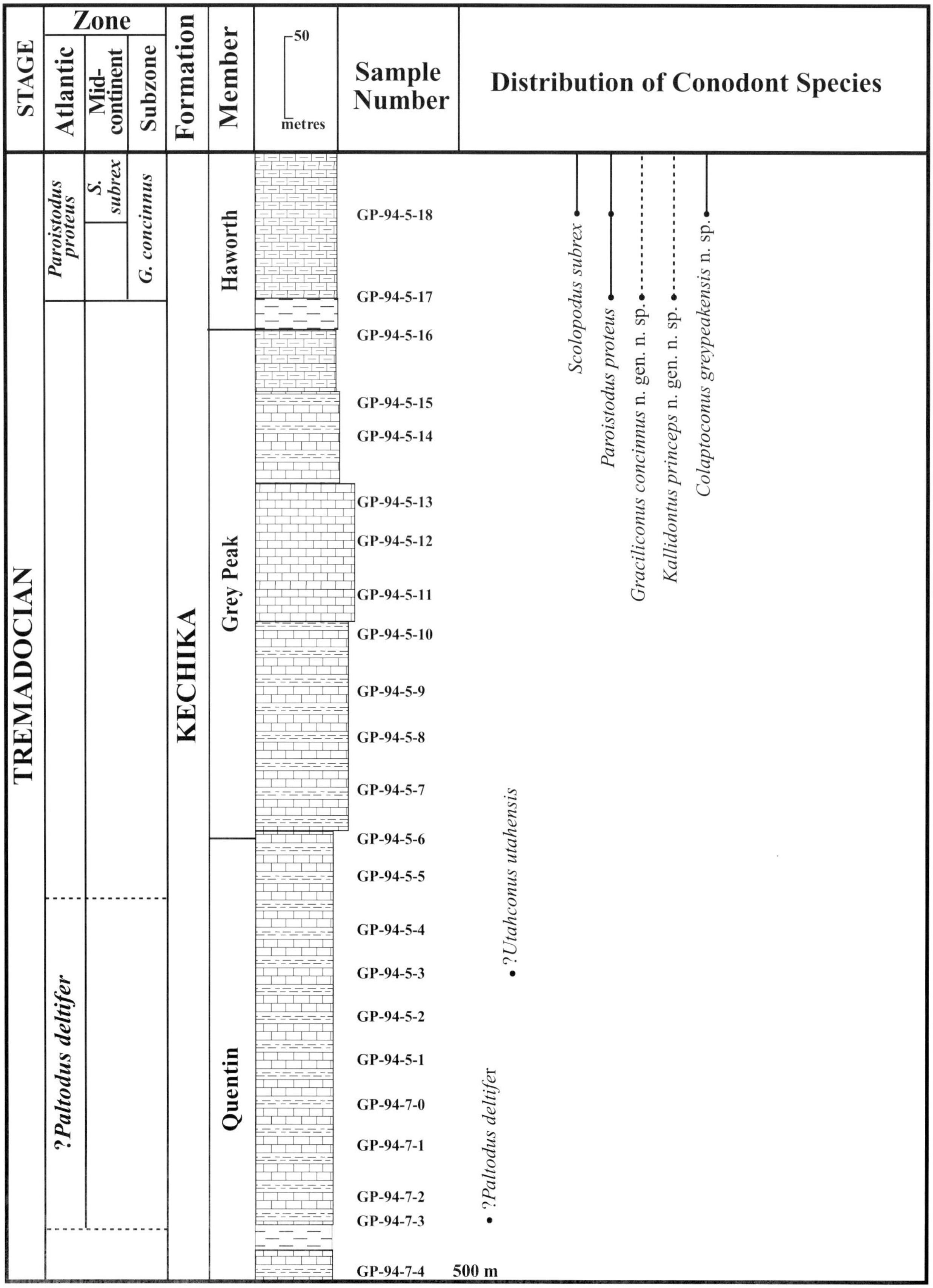

Fig. 7C. Stratigraphic ranges of conodonts from the Kechika and Ospika formations, Grey Peak Section.

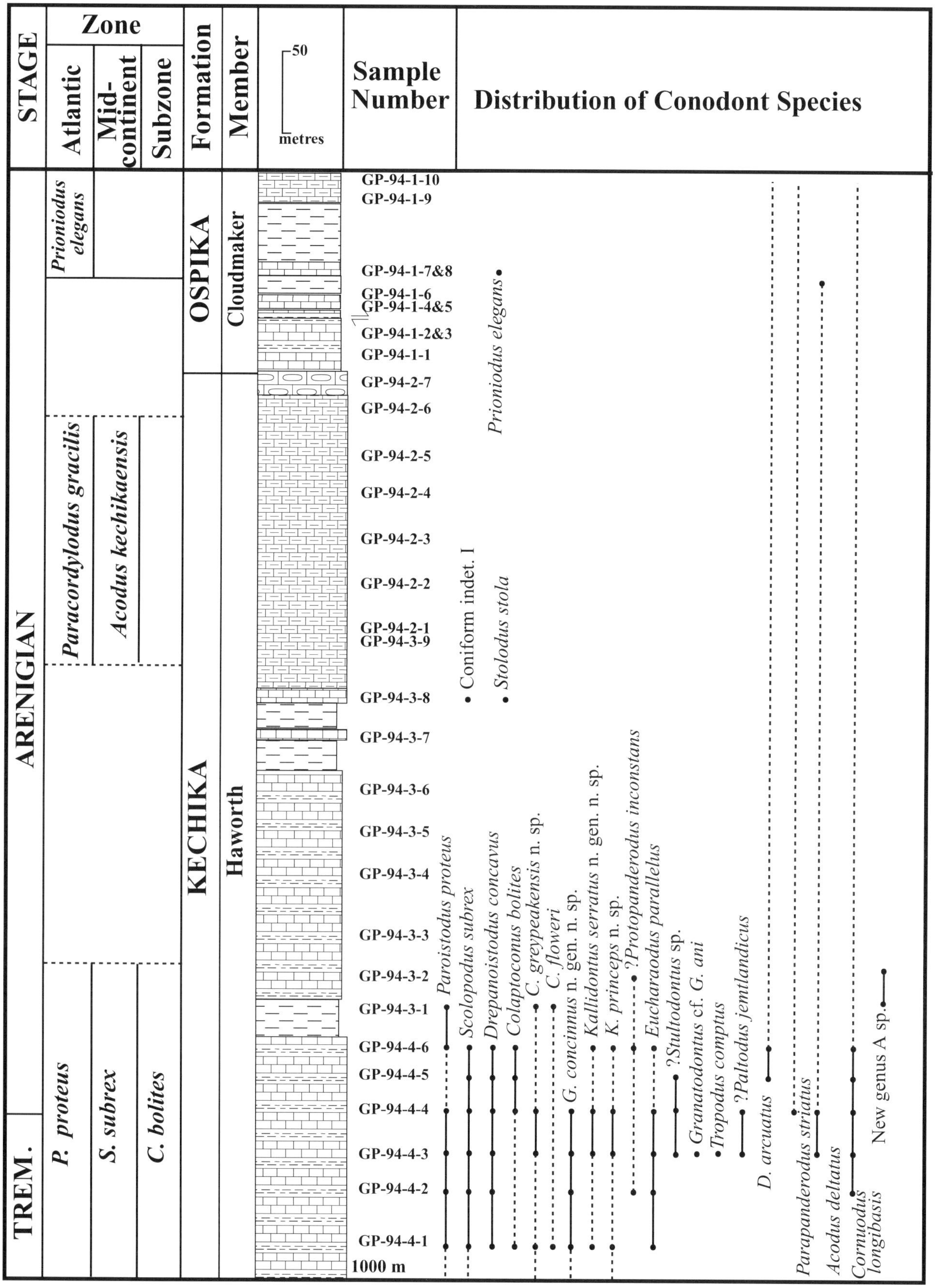

Fig. 7D. Stratigraphic ranges of conodonts from the Ospika Formation, Grey Peak Section.

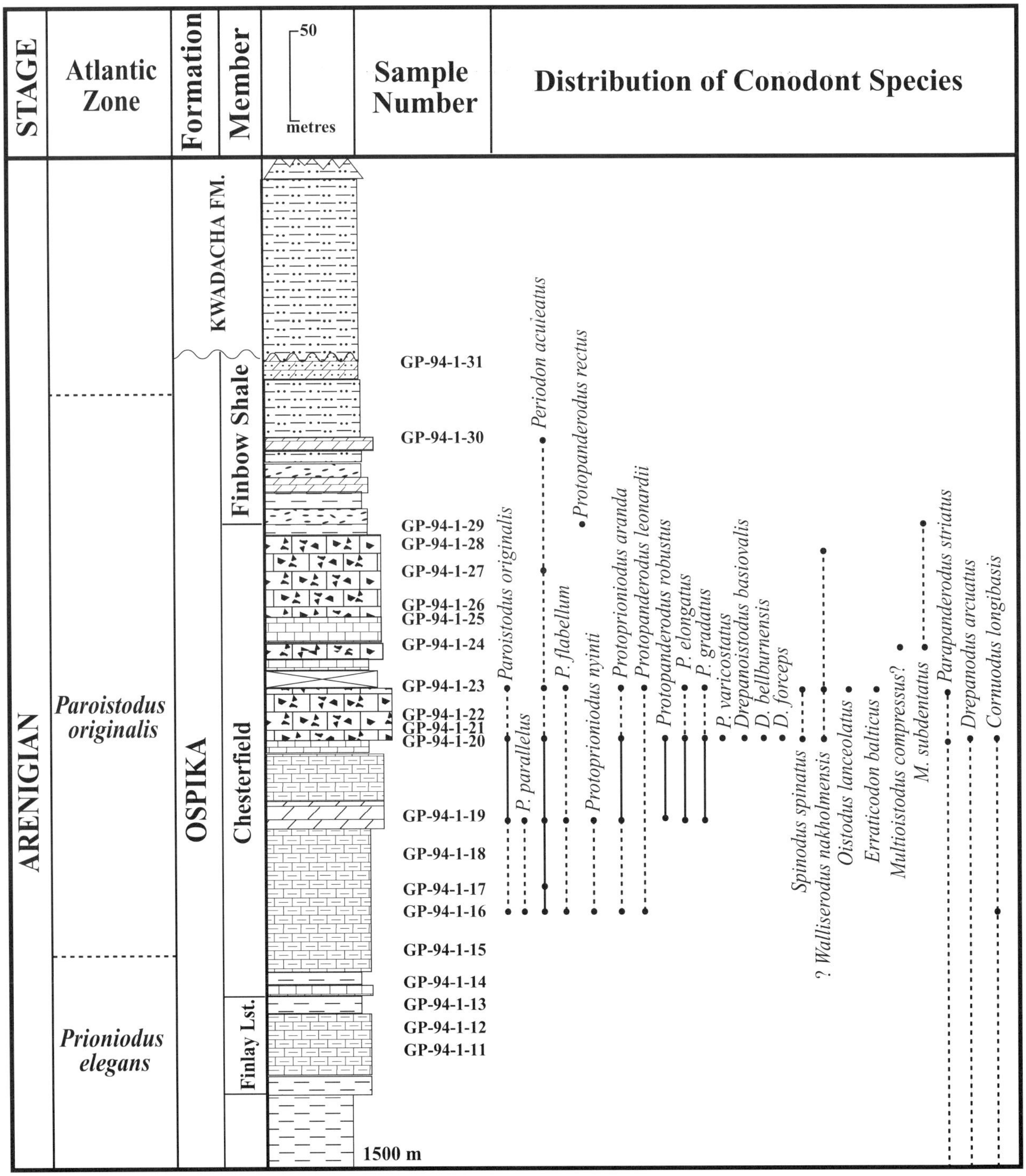

Fig. 8A. Stratigraphic ranges of conodonts from the upper Kechika Formation and Ospika Formation, Section 4.

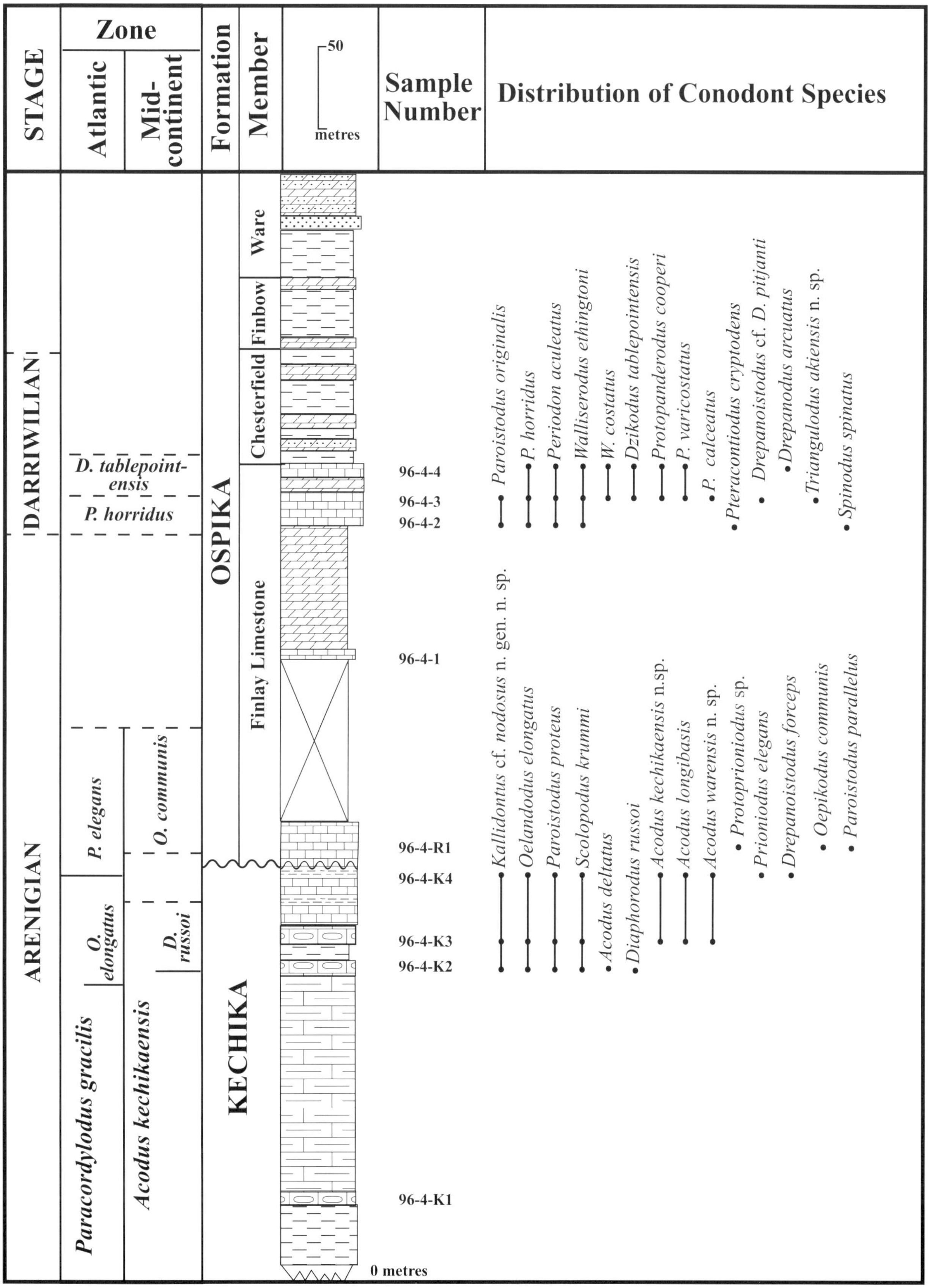

Fig. 8B. Stratigraphic ranges of conodonts from the Pesika Formation, Section 4.

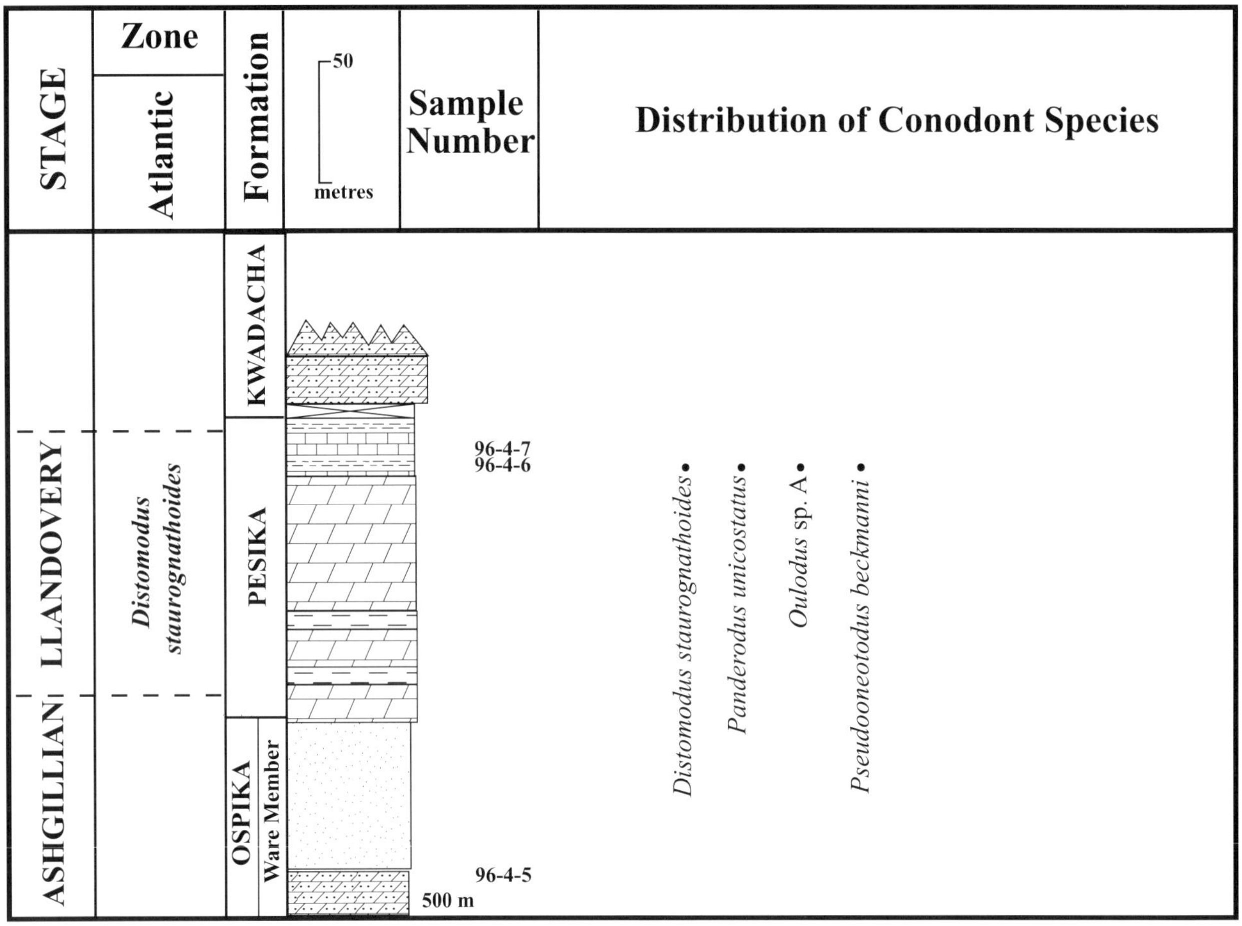

and are discussed first. Some zones do not have a defined base because of breaks in some sections, and their base is therefore indicated by a dashed line in Figs. 9 and 10. The use of subzones is proposed to further refine zones based on the expanded stratigraphic sequences preserved in the study area. The zones are described in ascending stratigraphic order and shown in Figs. 9 and 10. Those zones and subzones which are new are indicated, and those which correspond to the standards are defined based on reference sections from the study area. In the following section, a comparison of the zones recognized and proposed in this study to that of the standard is discussed.

Cosmopolitan zones

The following zones are based on taxa which are cosmopolitan in their distribution.

Eoconodontus Zone

Definition — The base of the zone is defined as the first appearance of *Eoconodontus notchpeakensis* (Miller) and was originally proposed as a subzone by Miller (1980) and as a zone by Miller (1988).

Reference section — Section 13 (57°42´N, 124°35´W), 4 km south of Mount Sheffield, Ware map area. The base of the zone is taken at 10.5 m above the point where measurement of the Lloyd George Member began (sample 96-13-C2, Fig. 4A). However, the contact with underlying Cambrian strata was not observed, and as data comes from one sample only, the top of the zone and therefore thickness of the zone has not been determined.

Fig. 9. Biostratigraphic correlation chart for the Ordovician System. Conodont and graptolite zones after Fortey et al. (1995), Harris et al., (1995*a*), Webby (1995), and time scale after Webby (1998) and Cooper and Nowlan (1999). See Fig. 10 for full names of abbreviations used for subzones from this study.

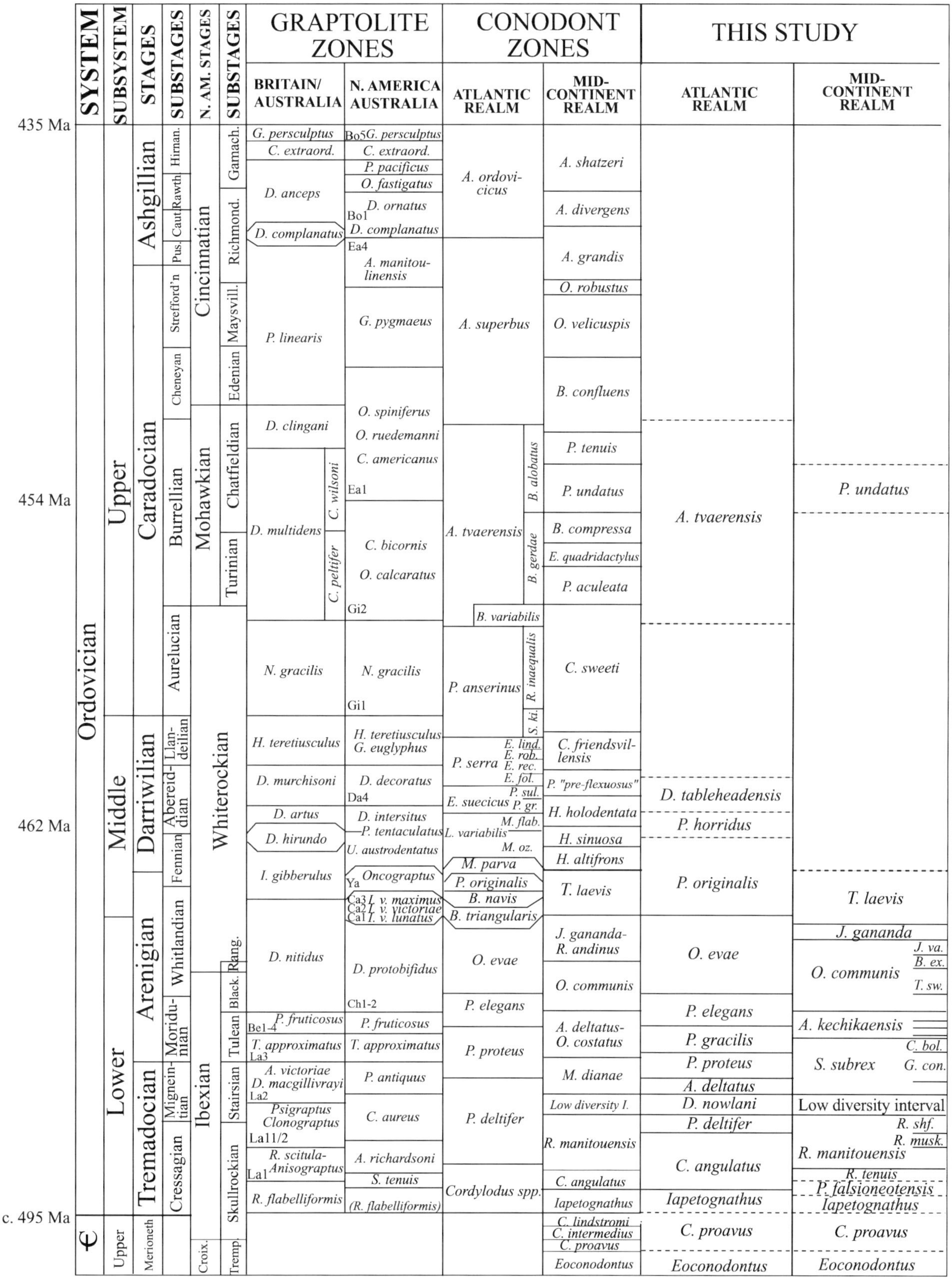

Fig. 10. Conodont biostratigraphy of the Kechika, Skoki, and Ospika formations compared to the standard Atlantic and Midcontinent Realm Conodont zones from Fig. 9.

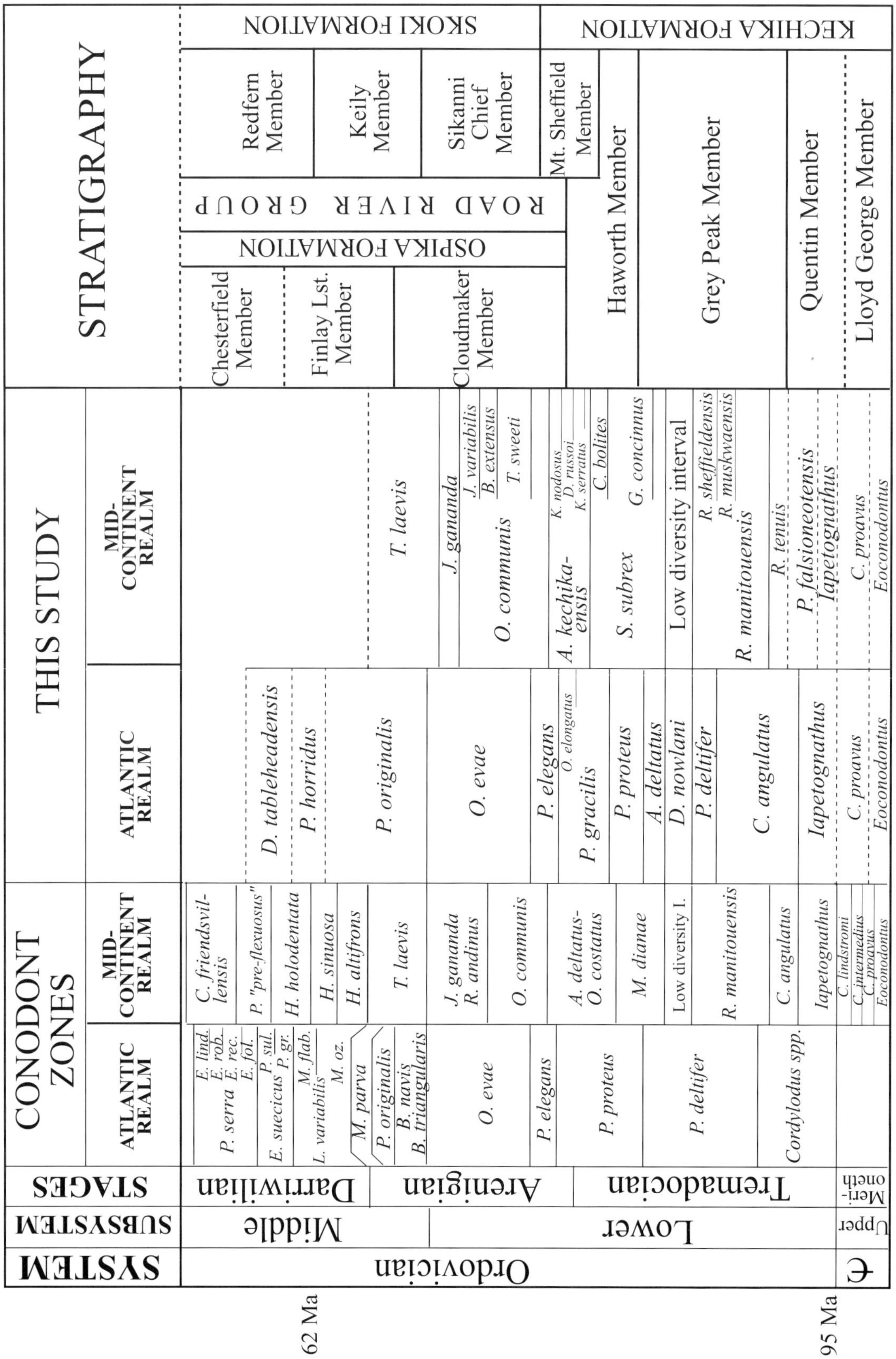

Type of zone — The zone is an interval zone which ranges from the level of the first occurrence of *E. notchpeakensis* to the level of appearance of *Cordylodus proavus* Müller.

Discussion — Miller (1988) proposed the *Eoconodontus* Zone in Utah as comprising a lower *E. notchpeakensis* Subzone and *Cambrooistodus minutus* Subzone. The latter of these subzones is not recognized, and as conodont data come from one sample, the occurrence within the Lloyd George Member of the Kechika Formation is assigned to the *Eoconodontus* Zone rather than to a subzone.

Cordylodus proavus Zone

Definition — The base of the zone is defined as the first appearance of *Cordylodus proavus* and was emended to include subzones by Miller (1988).

Reference section — Grey Peak (57°48´N, 125°13´W), Ware map area. The zone includes strata 112 m above the base of the measured section and 66 m above the base of the Lloyd George Member of the Kechika Formation (sample GP-94-6-6, Fig. 7A). As data comes from one sample only, the base and top of the zone and therefore thickness of the zone have not been determined.

Discussion — The zone marks the appearance of *C. proavus* and has been divided into subzones for strata in the type area of the Ibex (Miller 1988; Ross et al. 1997). Although the total thickness of the zone has not been determined in this study, a number of important species occur with *C. proavus* at its first occurrence in the Kechika Formation. These include the continuation of *E. notchpeakensis* and *Protoconodontus muelleri* Miller from strata below, *Cordylodus primitivus* Bagnoli, Barnes, and Stevens, *C. hastatus* Barnes, *C. deflexus* s. f. Bagnoli, Barnes, and Stevens, and *Teridontus nakamurai* (Nogami).

Iapetognathus Zone

Definition — The base of the zone is defined as the first appearance of *Iapetognathus aengensis* (Lindström) and *I. fluctivagus* Nicoll et al. and was proposed by Nicoll et al. (1999).

Reference section — Section 13 (57°42´N, 124°35´W), 4 km south of Mount Sheffield, Ware map area. The base of the zone is not constrained but taken at 0 m within limestone of the Quentin Member of the Kechika Formation (sample 96-13-A1, Fig. 4B), after a section break where measurement for the Kechika begins. The zone is at least 154.5 m thick and occurs in the lower part of the Quentin Member. The top of the zone is not constrained because an interval of 110 m within the Quentin Member comprises a covered interval and a sampled interval which yields long-ranging species from the *Iapetognathus* Zone, but no species of *Iapetognathus*.

Type of zone — The zone is an interval zone based on the first appearance of *I. fluctivagus*.

Discussion — Nicoll et al. (1999) discuss the importance of the *Iapetognathus* Zone in defining the base of the Ordovician and the utility of this cosmopolitan genus for global correlation. The zone is characterized by species of *I. aengensis* and *I. fluctivagus*, and *I. sprakersi* Landing appears in younger strata, as observed in the Ibex area, Utah. *Cordylodus lindstromi* Druce and Jones appears at the same level as *I. fluctivagus*, so there is no evidence for a sequential nature of these zones. Nicoll et al. (1999) indicate the two species are likely coincident but maintained a *C. lindstromi sensu lato* zone because of the different interpretations of *C. lindstromi*.

Other important taxa occur in the zone such as *Cordylodus caboti* Bagnoli, Barnes, and Stevens, *C. caseyi* Druce and Jones, and *C. delicatus* n. sp., *Utahconus utahensis* Miller, and *Semiacontiodus nogamii* (Miller).

Midcontinent Realm zones

Ten Midcontinent Realm zones are recognized in the Kechika and Skoki formations of which five are new.

Polycostatus falsioneotensis Zone, new

Definition — The base of the zone is defined as the first appearance of *Polycostatus falsioneotensis* Ji and Barnes and is a new zone.

Reference section — Section 13 (57°42´N, 124°35´W), 4 km south of Mount Sheffield, Ware map area. The base of the zone occurs within the upper part of the Quentin Member at 264.5 m above the base of the section (sample 96-13-A10, Fig. 4B). The top of the zone coincides with a covered interval below the base of the Grey Peak Member, at 325 m, so that the zone is 60.5 m thick.

Type of zone — The zone is an interval zone, ranging between the level of first occurrence of *P. falsioneotensis* and the first occurrence of *Rossodus tenuis* (Miller).

Discussion — *P. falsioneotensis* has been reported mainly from the Midcontinent Realm of North America (Landing and Barnes 1981; Nowlan 1985; Ji and Barnes 1994) and also from China (Zhang in An et al. 1983). Ji and Barnes (1994) defined a *P. falsioneotensis* – *R. tenuis* Zone, however, *R. tenuis* occurs in younger strata at the reference section and its first occurrence marks the level of the overlying *R. tenuis* Zone.

There is lower species diversity through the *P. falsioneotensis* Zone, and no new species appear. Of the species ranging from strata below, *Teridontus nakamurai* (Nogamii) and *T. gracillimus* Nowlan occur at the level of first appearance of *P. falsioneotensis*. *Utahconus utahensis* and several long-ranging species of *Cordylodus* range through the zone. The zone is partly coeval with the *C. angulatus* Zone.

Rossodus tenuis Zone, new

Definition — The base of the zone is defined as the first appearance of *Rossodus tenuis* and is a new zone.

Reference section — Section 13 (57°42´N, 124°35´W). The base of the zone coincides with the base of the Grey Peak Member, 430 m above the base of the section (sample 96-13-A14, Fig. 4B). The zone is a 21 m thick interval.

Type of zone — The zone is an interval zone extending from the first appearance of *R. tenuis* to the level of the appearance of its descendant, *R. manitouensis.*

Discussion — *R. tenuis* is a Midcontinent Realm species reported from western Newfoundland (Ji and Barnes 1994), Western United States (Miller 1980; Taylor and Landing 1982), and China (Wang 1986). The zone is characterized by the appearance of its nominate species and the first appearance of *Ventricodus spurius* n. gen. (Ethington and Clark). Characteristic species include *Variabiloconus bassleri* (Furnish) and *Utahconus longipinnatus* Ji and Barnes. It is coeval with the middle part of the *C. angulatus* Zone. The zone contains long-ranging species such as *C. caseyi, C. lindstromi, C. delicatus, C. intermedius* Furnish, and *Drepanoistodus pervetus* Nowlan. *R. tenuis* is considered an immediate predecessor of *R. manitouensis* and is therefore useful to distinguish strata underlying the first appearance of *R. manitouensis.*

Rossodus manitouensis Zone

Rossodus muskwaensis Subzone, new

Rossodus sheffieldensis Subzone, new

Definition — The base of the zone is defined as the first appearance of *Rossodus manitouensis*, which was designated by Landing et al. (1986). The base of the *Rossodus muskwaensis* Subzone is defined by the first appearance of *R. muskwaensis* n. sp. and is a new subzone. The base of the *R. sheffieldensis* Subzone is defined by the first appearance of the nominate species and is a new subzone.

Reference section — Section 13 (57°42´N, 124°35´W), 4 km south of Mount Sheffield, Ware map area. The base of the zone lies within the Kechika Formation, Grey Peak Member, 451 m above the base of the section (sample 96-13-A15, Fig. 4B). The base of the *R. muskwaensis* Subzone is at 679 m (sample 96-13-B1, Fig. 4C), and the subzone is 57 m thick, but overlies a covered interval. The base of the *R. sheffieldensis* Subzone occurs at a level 736 m above the base of the section (sample 96-13-B3, Fig. 4C) and is 97 m thick. The total thickness of the *R. manitouensis* Zone is 382 m.

Type of zone — The zone is an interval zone extending from the first appearance of *R. manitouensis* to that of the level of the Low Diversity Interval, which is marked by the disappearance of *R. manitouensis*. The *R. muskwaensis* Subzone is an interval zone extending from the first appearance of *R. muskwaensis* to the level of the first appearance of *R. sheffieldensis*, which marks the next subzone, and its top is marked by the base of the Low Diversity Interval.

Discussion — The *R. manitouensis* Zone was established by Landing et al. (1986) as a North American biostratigraphic unit for marginal and open shelf sequences. It is equivalent to Fauna C of Ethington and Clark (1971) and has mainly been reported from North America, however, Löfgren et al. (1998) reported *R. manitouensis* in low abundance from the *P. deltifer* Zone in Baltoscandia.

The division of the *R. manitouensis* Zone into two subzones provides refinement of the upper part of the zone. The lower part of the zone is characterized by the appearance of *R. manitouensis*, which occurs with *R. tenuis* and *Ve. spurius*, *Va. bassleri*, and *U. longipinnatus*. Species of *Cordylodus* also range through the Zone, including *C. angulatus*, *C. intermedius*, *C. lindstromi*, *C. caseyi*, and also *Drepanoistodus pervetus* and *Utahconus utahensis*. The top of the Zone is characterized by the disappearance of all of these species but the latter two and *Acanthodus lineatus* Furnish, which range into the overlying Low Diversity Interval.

The termination of several lineages at the top of the Zone has been noted by Ethington and Clark (1982) and Ji and Barnes (1994) and is not only evident among the long-ranging species, but among new species which are introduced in both of the subzones. The *R. muskwaensis* Subzone is also characterized by the first appearance of *Laurentoscandodus* cf. *L. triangularis* (Furnish), *Tricostatus terilinguis* n. sp., and *Striatodontus strigatus* n. sp.

New appearances within the *R. sheffieldensis* Subzone include *Striatodontus prolificus* Ji and Barnes, *Polycostatus* cf. *P. sulcatus* Ji and Barnes, *Macerodus*? n. sp. A and B, *Rossodus subtilis* n. sp., and *Paltodus deltifer* Lindström. The Subzone is coeval with the *Paltodus deltifer* Zone.

Low Diversity Interval

Definition — The interval is an informal zone designated by Ethington and Clark (1982) for an interval of less diverse and abundant conodonts.

Reference section — Section 13 (57°42´N, 124°35´W), 4 km south of Mount Sheffield, Ware map area. The base of the zone is within the Grey Peak Member at 833 m above the base of the section (sample 96-13-B7, Fig. 4C). The zone is 30 m thick.

Type of zone — The zone is an informal interval zone which coincides with the disappearance of *Rossodus manitouensis* at its base and ranges to the first appearance of *Scolopodus subrex*.

Discussion — The interval is characterized by the low number and diversity of taxa into which range only a few long-ranging species such as *Utahconus utahensis* and *Rossodus tenuis* from the underlying *R. manitouensis* Zone. *Paroistodus numarcuatus* (Lindström) occurs within this zone, and the first appearance of *Drepanoistodus nowlani* Ji and Barnes at the base of the zone is useful for correlation with the deeper water *D. nowlani* Zone. In Utah, Ethington and Clark (1982) did not define a zone for this interval because of the appearance of long-ranging taxa only, and Ross et al. (1997) followed this designation, suggesting the zone is equivalent to the lower part of Fauna D (Ethington and Clark 1971).

Scolopodus subrex Zone, new

Graciloconus concinnus Subzone, new

Colaptoconus bolites Subzone, new

Definition — The base of the zone is defined as the level of the first appearance of *Scolopodus subrex* Ji and Barnes. The base of the *Graciloconus concinnus* Subzone is defined as the level of the first appearance of *Graciloconus concinnus* n. gen. n. sp., and the overlying *Colaptoconus bolites* Subzone is marked by the first appearance of *Colaptoconus bolites* (Repetski).

Reference section — Section 13 (57°42′N, 124°35′W), 4 km south of Mount Sheffield, Ware map area. The base of the zone is within the Grey Peak Member at 863 m above the base of the section (sample 96-13-B8, Fig. 4C). The base of the *G. concinnus* Subzone is at 923 m above the base of the section (sample 96-13-B10, Fig. 4C) and is 367 m thick. It spans the upper part of the Grey Peak Member and into the Haworth Member. The base of the *C. bolites* Subzone is at 1230 m above the base of the section (sample 96-13-B16, Fig. 4D) within the uppermost Haworth Member. The upper boundary of the subzone is not determined because of an overlying covered interval and an interval of shale, both at the base of the Mt. Sheffield Member.

Type of zone — The zone is an interval zone extending from the first appearance of *Scolopodus subrex* to the first appearance of *Acodus kechikaensis* n. sp. The *G. concinnus* Subzone is an interval zone extending from the first appearance of *G. concinnus* to the first appearance of *C. bolites*. The *C. bolites* Subzone is an interval zone extending from the first appearance of *C. bolites* to the first appearance of *Acodus kechikaensis* n. sp.

Discussion — As discussed by Ji and Barnes (1994), *S. subrex* is the Midcontinent Realm form which differs from the younger Atlantic realm *S. rex* (see Systematic Paleontology). *S. subrex* occurs in North America (e.g., Ji and Barnes 1994, 1996) and also has been reported from Korea (Lee 1970, 1975). As a new genus and species, *G. concinnus* is reported as a Midcontinent Realm species which occurs in platformal and marginal shelf facies. *C. bolites* has been reported by Repetski (1982) and Ji and Barnes (1994).

The zone is characterized by an assemblage including abundant *S. subrex*, ?*Paltodus jemtlandicus* Löfgren, and *Paroistodus proteus* Lindström throughout the zone. The *S. subrex* – *G. concinnus* Subzone is characterized by the appearance of *Kallidontus princeps* n. gen. n. sp., *Colaptoconus greypeakensis* n. sp., and *C. quadraplicatus* (Branson and Mehl) and *Semiacontiodus* cf. *S. cornuformis* (Sergeeva). The *S. subrex* – *C. bolites* Subzone is coeval with the *Paracordylodus gracilis* Zone and is characterized by the appearance of *C. floweri* (Repetski), and long-ranging taxa such as *Parapanderodus striatus* (Graves and Ellison), *Cornuodus longibasis* (Lindström), and *Eucharodus parallelus* (Branson and Mehl). The *Scolopodus subrex* Zone is equivalent to the deeper water *Acodus deltatus* and *Paroistodus proteus* zones.

Acodus kechikaensis Zone, new
Kallidontus serratus Subzone, new
Diaphorous russoi Subzone, new
Kallidontus nodosus Subzone, new

Definition — The base of the zone is defined as the level of the first appearance of *Acodus kechikaensis* n. sp. The base of the *Kallidontus serratus* Subzone is the level of the first appearance of *Kallidontus serratus* n. gen. n. sp. The base of the *Diaphorodus russoi* Subzone is the level of the first appearance of the nominate species. The base of the *Kallidontus nodosus* Subzone is the level of the first appearance of the nominate species.

Reference section — Section 13 (57°42´N, 124°35´W), 4 km south of Mount Sheffield, Ware map area. The base of the zone is within the Mt. Sheffield Member at 1444.5 m above the base of the section (sample 96-13-B21, Fig. 4D). The zone is 262 m thick. The base of the *Kallidontus serratus* Subzone is at 1518 m above the base of the section (sample 96-13-B23, Fig. 4E). The subzone is 75 m thick. The base of the *Diaphorodus russoi* Subzone is at 1593 m above the base of the section (sample 96-13-B25, Fig. 4E), and the subzone is 30 m thick. The base of the *Kallidontus nodosus* Subzone is at 1623 m above the base of the section (sample 96-13-B26, Fig. 4E), and the subzone is 83.5 m thick.

Type of zone — The zone is an interval zone extending from the first appearance of *A. kechikaensis* n. sp. to the first appearance of *O. communis*. Each of the three subzones is an interval subzone based on the first appearance of the nominate species.

Discussion — As discussed within Systematic Paleontology, *A. kechikaensis* n. sp. is broadly similar to Australian forms reported by McTavish (1973) but is here considered to be a new North American form, reported also by Ethington and Clark (1982). It extends into deeper water facies but is dominant in shallow-water facies. The species which designate the subzones are from a new genus, *Kallidontus*, which is most abundant in shelf facies but also extends into platformal and off-shelf facies. *D. russoi* (Serpagli) has been reported from North America by Repetski (1982), from the Argentine Precordillera by Serpagli (1974), Lehnert (1995), and Albanesi (1998), and from Greenland by Smith (1991).

The zone is characterized by an assemblage including the first appearances of *Tropodus australis* (Serpagli), *Paroistodus parallelus* (Pander), *Fahraeusodus marathonensis* (Bradshaw), and *Kallidontus serratus* n. gen. n. sp. The first appearances of *D. russoi* coincides with that of *Acodus longibasis* McTavish and with *Oelandodus elongatus* van Wamel and *Stolodus stola* (Lindström). The *D. russoi* Subzone is coeval with *Oelandodus elongatus* Subzone of the *P. gracilis* Zone. The *K. nododus* Subzone includes the first appearances of *Acodus neodeltatus* n. sp., *Oistodus lanceolatus* (Pander), *Drepanoistodus amoenus* (Lindström), and *Prioniodus elegans* Pander and is coeval with the lower *Prioniodus elegans* Zone. Long-ranging taxa within the zone include *Cornuodus longibasis* (Lindström), *Scolopodus krummi* (Serpagli), and *Paracordylodus striatus* (Graves and Ellison).

Oepikodus communis Zone
Tropodus sweeti Subzone, new
Bergstroemognathus extensus Subzone, new
Juanognathus variabilis Subzone, new

Definition — The base of the zone is defined as the level of the first appearance of *Oepikodus communis*, established by Repetski and Ethington (1983). Each of the subzones are new and based on the first appearance of the nominate species.

Reference section — Section 5 (57°30′N, 124°26–W), Ware map area. The base of the zone is not observed but sample 96-5-K1 (0 m) is within the zone, which is at least 339 m thick. At Section 13 (57°42′N, 124°35′W), the base of the zone is at 1706.5 m above the base of the section (sample 96-13-B28, Fig. 4E) and lies within the platy limestone of the upper Mt. Sheffield Member of the Kechika Formation and ranges into the Skoki Formation. The zone is at least 264 m thick at Section 13, but the top of the zone is not observed.

At Section 5, the base of the *Tropodus sweeti* Subzone occurs at 44 m above the base of the Skoki Formation (sample 96-5-3, Fig. 6A). The subzone is 70 m thick. The base of the *Bergstroemognathus extensus* Subzone occurs at 114 m above the base of the Skoki Formation (sample 96-5-6, Fig. 6A), and the subzone is 80 m thick. The base of the *Juanognathus variabilis* Subzone occurs at 194 m above the base of the Skoki Formation (sample 96-5-8, Fig. 6A), and the subzone is 145 m thick. All of these subzones lie within the Sikanni Chief Member of the Skoki Formation.

Type of zone — The zone is an interval zone extending from the level of the first appearance of *Oepikodus communis* to the level of the first occurrence of *Jumudontus gananda* Cooper. Each of the subzones is based on the first appearances of *Tropodus sweeti* (Serpagli), *Bergstroemognathus extensus* Serpagli, and *Juanognathus variabilis* Serpagli, respectively.

Discussion — The *O. communis* Zone is widely recognized in North America (Ross et al. 1997), established by Repetski and Ethington (1983), and is equivalent to the lower part of Fauna E of Ethington and Clark (1971).

The zone includes characteristic species in each subzone as well as some long-ranging species from the *A. kechikaensis* Zone below. The *T. sweeti* Subzone includes species which range from below such as *Fahraeusodus marathonensis*, *Tropodus australis*, and *Oepikodus intermedius* (Serpagli). The *B. extensus* Subzone includes species which range from below and the first appearance of *Prioniodus adami* Stouge and Bagnoli. Species characteristic of the *J. variabilis* Subzone include *Periodon selenopsis* (Serpagli), *P. flabellum* (Lindström), *P. aculeatus* Hadding, *Oistodus multicorrugatus* Harris, *Protoprioniodus nyinti* Cooper, *P. cowheadensis* Stouge and Bagnoli, *Juanognathus jaanussoni* Serpagli, and *Planusodus gradus* n. gen. n. sp.

Jumudontus gananda Zone, new

Definition — The base of the zone is defined as the level of the first appearance of *Jumudontus gananda* Cooper and is new.

Reference section — Section 5 (57°30′N, 124°26′W), Ware map area. The lower boundary of the zone occurs at 339 m above the base of the Skoki Formation (sample 96-5-15, Fig. 6A), and the top is at 343.5 m (sample 96-5-15A). The zone is 4.5 m thick and lies within the uppermost Sikanni Chief Member.

Type of zone — The zone is an interval zone extending from the level of the first appearance of *Jumudontus gananda* to the level of the first appearance of *Tripodus laevis* Bradshaw.

Discussion — This zone is used instead of the *Reutterodus andinus* Zone proposed by Ross et al. (1997), as *J. gananda* is an associated species in that zone. They based the *R. andinus* Zone on the occurrence of coniform elements assigned to that species, which have herein been assigned to *Juanognathus variabilis* Serpagli. *J. gananda* is not usually an abundant species, but is a common taxon in the early Arenigian in several areas of the world. In Australia, it has been reported from the Horn Valley Siltstone, Amadeus Basin (Cooper 1981), and Canning Basin (Nicoll 1992). Other occurrences include those within the Precordilleran Argentina (e.g., Serpagli 1974; Lehnert 1995; Albanesi 1998), Scotland (Ethington and Austin 1991), Baltoscandia (Bergström 1988), and southwest United States (Ethington and Clark 1982; Repetski 1982).

Characteristic species within the zone include the first appearances of *Protoprioniodus* n. sp., *Microzarkodina* n. sp., and *Drepanoistodus forceps* (Lindström) as well as long-ranging species from the underlying *O. communis* Zone.

Tripodus laevis Zone

Definition — The base of the zone is defined as the level of the first appearance of *Tripodus laevis* and was defined by Ross et al. (1997).

Reference section — Section 5 (57°30´N, 124°26´W), Ware map area. The lower boundary of the zone occurs at 353 m above the base of the Skoki Formation (sample 96-5-16A, Figs. 6A, 6B) and is at least 249.5 m thick, however, the top of the zone has not been defined in this study. The zone ranges from the upper part of the Sikanni Chief Member and into the Keily Member.

Type of zone — The zone is an interval zone extending from the level of the first appearance of *Tripodus laevis*.

Discussion — The base of this zone has been used to define the base of the Whiterockian stage and the Middle Ordovician Series in the Ibex area, Utah (Webby 1998). Characteristic species in the zone include *Protoprioniodus aranda* Cooper, *Pteracontiodus cryptodens* (Mound), and *Ansella jemtlandica* (Löfgren).

Atlantic Realm zones

Eleven cooler, and commonly deeper water, Atlantic Realm zones are recognized and some redefined for the Kechika Formation and Road River Group.

Cordylodus angulatus Zone

Definition — The base of the zone is defined as the first appearance of *Cordylodus angulatus* as proposed by Miller (1988).

Reference section — Section 13 (57°42´N, 124°35´W), 4 km south of Mount Sheffield, Ware map area. The base of the zone occurs at 305.5 m above the base of the section (sample 96-13-A12, Fig. 4B). The zone is 430.5 m thick and ranges through the upper part of the Quentin Member into the Grey Peak Member.

Type of zone — The zone is an interval zone which ranges from the level of the first occurrence of *C. angulatus* to the level of the first occurrence of *P. deltifer* (Lindström).

Discussion — This zone was proposed as a Midcontinent Realm zone which underlies the *Rossodus manitouensis* Zone and is equivalent to the upper part of Fauna B (of Ethington and Clark 1971) by Ross et al. (1997). In this study, it is defined as a deeper water zone which is equivalent to the upper part of the *P. falsioneotensis* Zone, *R. tenuis* Zone, and all but the upper part of the *R. manitouensis* Zone. *C. angulatus* is a cosmopolitan species which occurs in abundance in both platformal and off-shelf facies, but is proposed to define a deeper water zone and provide tie points to the Midcontinent Realm zones. The species which range in those zones occur with *C. angulatus* and are not listed again here.

Paltodus deltifer Zone

Definition — The base of the zone is defined as the first appearance of *Paltodus deltifer* as proposed by Lindström (1971).

Reference section — Section 13 (57°42´N, 124°35´W), 4 km south of Mount Sheffield, Ware map area. The base of the zone occurs at 736 m above the base of the section (sample 96-13-B3, Fig. 4C). The zone is 97 m thick and lies within the middle of the Grey Peak Member of the Kechika Formation.

Type of zone — The zone is an interval zone which

ranges from the level of the first occurrence of *Paltodus deltifer* to the level of the first occurrence of *Drepanoistodus nowlani*.

Discussion — Although *P. deltifer* occurs in low abundance, it provides correlation to the upper part of the Midcontinent *R. manitouensis* Zone. The *P. deltifer* Zone is well established in Sweden (e.g., Lindström 1955; Löfgren 1985; Bagnoli et al. 1988), and the zonal index species occurs in low abundance from the Argentine Precordillera as reported by Albanesi (1998). The *Paltodus deltifer* Zone is coeval with the *R. sheffieldensis* Subzone of the *R. manitouensis* Zone.

Drepanoistodus nowlani Zone, new

Definition — The base of the zone is defined as the first appearance of *Drepanoistodus nowlani* and is a new zone.

Reference section — Section 13 (57°42´N, 124°35´W), 4 km south of Mount Sheffield, Ware map area. The base of the zone occurs at 833 m above the base of the section (sample 96-13-B7, Fig. 4C). The zone is 30 m thick and occurs within the middle of the Grey Peak Member of the Kechika Formation.

Type of zone — The zone is an interval zone which ranges from the level of the first occurrence of *Drepanoistodus nowlani* to the level of the first appearance of *Acodus deltatus*.

Discussion — Ji and Barnes (1994) recognized a *D. nowlani – M. dianae* Assemblage Zone which overlies their deeper water *C. angulatus* Zone in the Boat Harbour Formation. *D. nowlani* has also been reported by Repetski (1982). Although the species is low in abundance in this study, the recognition of this zone is significant because it is coeval with the Low Diversity Interval, where there are too few diagnostic species of the shallow-water realm to define a zone.

Acodus deltatus Zone, emended

Definition — The base of the zone is defined as the level of the first appearance of *Acodus deltatus* and is emended as an Atlantic Realm Zone.

Reference section — Section 13 (57°42´N, 124°35´W), 4 km south of Mount Sheffield, Ware map area. The base of the zone occurs at 863 m above the base of the section (sample 96-13-B8, Fig. 4C). The zone is 108 m thick and occurs within the upper part of the Grey Peak Member of the Kechika Formation.

Type of zone — The zone is an interval zone which ranges from the level of the first occurrence of *Acodus deltatus* to the level of the first occurrence of *Paroistodus proteus*.

Discussion — Ross et al. (1997) proposed an *Acodus deltatus/Oneotodus costatus* Zone, which is equivalent to the upper part of Fauna D and overlies the *Macerodus dianae* Zone and Lower Diversity Interval. *A. deltatus* is a widely reported species, which Ross et al. (1997) note is not widely distributed in the continental interior in the western United States. Therefore, in this study, the *A. deltatus* Zone is used as a cooler water zone to improve the correlation of strata, as this zone is coeval with the lower part of the *Scolopodus subrex* Zone.

Paroistodus proteus Zone

Definition — The base of the zone is defined as the level of the first appearance of *Paroistodus proteus* as was originally proposed by Lindström (1971).

Reference section — Section 13 (57°42´N, 124°35´W), 4 km south of Mount Sheffield, Ware map area. The base of the zone occurs at 923 m above the base of the section (sample 96-13-B10, Figs. 4C, D). The zone is 307 m thick and ranges from the upper part of the Grey Peak Member to the upper part of the Haworth Member of the Kechika Formation.

Type of zone — In this study, the zone is an interval zone which ranges from the level of the first occurrence of *Paroistodus proteus* to the level of the first occurrence of *Paracordylodus gracilis*.

Discussion — The *P. proteus* Zone is widely recognized in Sweden, and Löfgren (1994) divided the *P. proteus* Zone into four subzones. Characteristic species used by Löfgren to subdivide the lower part of the zone are not observed in this study, however, two species used for her upper two subzones, *Paracordylodus gracilis* and *Oelanodus elongatus*, are here proposed to comprise a separate zone and subzone, respectively. The species which occur within the *P. proteus* Zone in this study include *Scolopodus subrex*, *Acodus deltatus*, *Graciloconus concinnus* n. gen. n. sp., ?*Paltodus jemtlandicus*, *Kallidontus princeps* n. gen. n. sp., and the first appearances of *Colaptoconus greypeakensis* n. sp., *C. quadriplicatus*, *Semiacontiodus* cf. *S. cornuformis*, and *Drepanoistodus concavus*. In this study, the zone is coeval with the *S. subrex* Zone.

Paracordylodus gracilis Zone

Oelandodus elongatus Subzone, new

Definition — The base of the zone is defined as the level of the first appearance of *Paracordylodus gracilis* as proposed by Johnston and Barnes (1999). The base of the *Oelandodus elongatus* Subzone is defined by the first appearance of *Oelandodus elongatus* (Lindström) and is a new subzone.

Reference section — Section 13 (57°42´N, 124°35´W), 4 km south of Mount Sheffield, Ware map area. The base of the zone occurs at 1230 m above the base of the section (sample 96-13-B16, Fig. 4D). The zone is 393 m thick and ranges from the uppermost Haworth Member to the upper Mt. Sheffield Member of the Kechika Formation.

Type of zone — The zone is an interval zone which ranges from the level of the first occurrence of *Paracordylodus gracilis* to the level of the first occurrence of *Prioniodus elegans*. The *Oelandodus elongatus* Subzone is an interval zone which ranges from the level of the first occurrence of *O. elongatus* to the first occurrence of *P. elegans*.

Discussion — The zone contains characteristic species, including those which first appear within the *C. bolites* Subzone of the *Scolopodus subrex* Zone and within the *Acodus kechikaensis* Zone. The *Oelandodus elongatus* Subzone marks the first appearance of *Stolodus stola*, *Paroistodus parallelus*, and *Fahraeusodus marathonensis*. The *Paracordylodus gracilis* Zone has also been proposed by Johnston and Barnes (1999) as underlying the *P. elegans* Zone in the Cow Head Group, western Newfoundland.

Prioniodus elegans Zone

Definition — The base of the zone is defined as the level of the first appearance of *Prioniodus elegans* and was originally proposed by Lindström (1971).

Reference section — Section 13 (57°42´N, 124°35´W), 4 km south of Mount Sheffield, Ware map area. The base of the zone occurs at 1623 m above the base of the section (sample 96-13-B26, Fig. 4E). The zone is 151 m thick. It ranges through the upper part of the platy limestone of the Mt. Sheffield Member of the Kechika Formation and into the lower part of the Skoki Formation.

Type of zone — The zone is an interval zone which ranges from the level of the first appearance of *Prioniodus elegans* and ranges to the level of the first appearance of *Oepikodus evae* (Lindström).

Discussion — The occurrence of *P. elegans*, although in low abundance, provides for correlation to the zone recognized in Sweden, where it is often 0.1 m thick (Lindström 1971; Löfgren 1993). It has been reported from North America by Stouge and Bagnoli (1988) and Pohler and Orchard (1990) and from the Argentine Precordillera by Lehnert (1995) and Albanesi (1998).

Species which first occur in this zone include the nominate species of that subzone and *Oistodus lanceolatus* and *Drepanoistodus amoenus*. Species that range from the underlying zone include *Paroistodus proteus*, which is replaced by increasing abundance of *P. parallelus* (Pander). *Paracordylodus gracilis* and *Oelandodus elongatus* also occur in this zone. The lower half of the zone is coeval with the *K. nodosus* Subzone of the *A. kechikaensis* Zone. The upper half of the zone is coeval with the lowermost part of the *Oepikodus communis* Zone.

Oepikodus evae Zone

Definition — The base of the zone is defined as the level of the first appearance of *Oepikodus evae* as proposed by Lindström (1971).

Reference section — Section 13 (57°42´N, 124°35´W), 4 km south of Mount Sheffield, Ware map area. The base of the zone occurs at 1774 m above the base of the section (sample 96-13-B30, Fig. 4E). The base occurs at 33 m above the base of the Skoki Formation. The zone is at least 196.5 m thick at Section 13. At Section 5 (57°30 N, 124°26 W), Ware map area, the base of the zone occurs at 144 m above the base of the Skoki (sample 96-5-7, Fig. 6A). However, this section is dominated by Midcontinent Realm species. The upper boundary of the zone was observed at 590.5 m above the base of the Skoki Formation.

Type of zone — The zone is an interval zone which ranges from the level of the first appearance of *Oepikodus evae* to the level of the first appearance of *Paroistodus originalis* (Sergeeva).

Discussion — The zone is well established in Baltoscandia (e.g., Lindström 1955, 1971; Löfgren 1978, 1993), and the species has been reported in North America by Pohler and Orchard (1990) and Stouge and Bagnoli (1988) and in the Argentine Precordillera by Serpagli (1974), Lehnert (1995), and Albanesi (1998). The zone is largely coeval with the *O. communis* Zone, although its base is younger than that of the *O. communis* Zone. It is also coeval with at least the lower half of the *Tripodus laevis* Zone.

Paroistodus originalis Zone

Definition — The base of the zone is defined as the level of the first appearance of *Paroistodus originalis* as defined by Lindström (1971).

Reference section — Section 5 (57°30´N, 124°26´W), Ware map area. The base of the zone occurs at 590.5 m above the base of the Skoki, within the Keily Member (sample 96-5-17, Fig. 6B). The top of the zone is not observed at Section 5. At the Grey Peak Section (57°48´N, 125°13´W), Ware map area, the base of the zone is taken at 184.5 m above the base of the Ospika Formation, Road River Group, within the Chesterfield Member (Fig. 7D).

Type of zone — The zone is an interval zone which ranges from the level of the first appearance of *Paroistodus originalis*.

Discussion — Löfgren (1994) commented that the base of the *P. originalis* Zone should be drawn at the boundary of increasing frequencies of *P. originalis* and *D. basiovalis* over *P. parallelus* and *D. forceps*, rather than at the first appearances of these taxa. In this study, as data from underlying strata at Section 5 and Grey Peak is sparse, the boundary of the zone is tentatively suggested to be at the first occurrence of *P. originalis*, as comparision of abundances is not available.

The zone is partially coeval with the *Tripodus laevis* Zone. At Section 5, species which first occur in the zone include *Ansella longicuspica* Zhang and *Ansella jemtlandica* (Löfgren), *Pteracontiodus cryptodens* (Mound) and *Multioistodus compressus*? Cullison. At the Grey Peak Section, long-ranging taxa from underlying zones occur such as *Paroistodus parallelus*, *Periodon flabellum*, *P. aculeatus*, *Protopanderodus leonardi* Serpagli, and *P. gradatus* Serpagli. New taxa which appear include *Protopanderodus robustus* (Hadding) and *P. varicostatus* (Sweet and Bergström), *Drepanoistodus basiovalis* (Sergeeva) and *D. bellburnensis* Stouge, *Spinodus spinatus* (Hadding) and *Erraticodon balticus* Dzik.

Paroistodus horridus Zone, new

Definition — The base of the zone is defined as the level of the first appearance of *Paroistodus horridus* (Barnes and Poplawski) and is a new zone.

Reference section — Section 4 (57°26´N, 124°35´W), north of the Akie and Ospika rivers. The base of the zone occurs at 135 m above the base of the Ospika Formation, Road River Group, within the upper part of the Finlay Limestone Member (sample 96-4-2, Fig. 8A). The zone is 14.5 m thick.

Type of zone — The zone is an interval zone which ranges from the level of the first appearance of *Paroistodus horridus* and ranges to the level of the first appearance of *Dzikodus tablepointensis* (Stouge).

Discussion — The zone has been used as a subzone by Albanesi (1998) within the *Lenodus variabilis* Zone, which overlies the *Paroistodus originalis* and *Microzarkodina parva* zones proposed for the Atlantic Realm. The *P. horridus* Zone is therefore proposed as a deep-water zone which lies above the *P. originalis* Zone.

Characteristic species which occur within this zone at Section 4 include *Paroistodus originalis* and *Periodon aculeatus* from underlying strata, *Spinodus spinatus* and *Pteracontiodus cryptodens*, and the appearance of *Walliserodus ethingtoni* (Fåhræus).

Dzikodus tablepointensis Zone, emended

Definition — The base of the zone is defined as the level of the first appearance of *Dzikodus tablepointensis* as proposed by Zhang (1998).

Reference section — Section 4 (57°26´N, 124°35´W), north of the Akie and Ospika rivers. The base of the zone occurs at 148.5 m above the base of the Ospika Formation, Road River Group, within the upper part of the Finlay Limestone Member (sample 96-4-3, Fig. 8A). The zone is at least 13.5 m thick. Its upper boundary is tentatively drawn at the base of the Chesterfield Member, which is a shale and dolostone unit at Section 4 and contains graptolites of the *Paraglossograptus tentaculatus* Zone.

Type of zone — The zone is an interval zone which ranges from the level of the first appearance of *Dzikodus tablepointensis*.

Discussion — Zhang (1998) proposed the *Dzikodus tablepointensis* Zone as overlying the *Lenodus variabilis* and *Yangtzeplacognathus crassus* zones. The taxon is distinct and useful for zonation of deeper water strata. Characteristic species which appear with *D. tablepointensis* at Section 4 include *Walliserodus costatus* Dzik, *Protopanderodus cooperi* (Sweet and Bergström), *P. calceatus* Bagnoli and Stouge, and *Triangulodus akiensis* n. sp.

Distomodus staurognathoides Zone

Definition — The base of the zone is defined as the level of the first appearance of *Distomodus staurognathoides* (Walliser) and was originally proposed by Walliser (1964).

Reference section — Section 4 (57°26´N, 124°35´W), north of the Akie and Ospika rivers. The base of the zone occurs at 620 m above the base of the Ospika Formation, Road River Group, and 192 m above the base of the Pesika Formation, Road River Group (sample 96-4-6, fig. 8B). The top of the zone is not constrained.

Type of zone — The zone is an interval zone which

ranges from the level of the first appearance of *Distomodus staurognathoides*. On Anticosti Island, it is overlain by the *Pterospathodus celloni* Zone (Uyeno and Barnes 1983).

Discussion — The *Distomodus staurognathoides* Zone has been recognized as a global zone for the middle Llandovery (Aldridge and Schönlaub 1989) and has been reported in Europe (Viira and Männik 1997), Australia (Simpson 1995), and Anticosti Island (Uyeno and Barnes 1983). Associated taxa observed in this study include *Panderodus unicostatus* (Branson and Mehl), *Oulodus* sp. A, and *Pseudooneotodus beckmanni* (Bischoff and Sannemann).

Comparison with standard zonations

The following discussion compares the conodont biozonation established by this study to those established as standards from well-studied regions elsewhere in the world. Adjustments are made to the standard by the inclusion of new zones and by introduction of new subzones. Rather than include a list of correlations with sections studied in Ordovician successions around the world, the discussion focuses on those correlations to which key tie points of zonal levels can be made (Fig. 9). Correlations with other key sections in the world are shown in Fig. 11.

Comparison with zonation across the Cambro–Ordovician boundary

In the Kechika, only the *Eoconodontus*, *C. proavus*, and *Iapetognathus* zones, established for both realms, are recognized. The *C. intermedius* Zone, established by Miller (1988) between the *C. proavus* and *Iapetognathus* zones, is not recognized because of a section break below the *Iapetognathus* Zone. The Cambrian–Ordovician boundary has been defined at the base of the *Iapetognathus fluctivagus* Zone (Nicoll et al. 1999). The base of the Tremadocian therefore lies within the Quentin Member of the Kechika Formation, where measurement began above the section break, so the first appearance of *Iapetognathus* may not have been observed.

The *C. angulatus* Zone was established as a shallow-water zone by Ross et al. (1997) as underlying the *R. manitouensis* Zone. In this study, *C. angulatus* is used as a deep-water zone, which, in agreement with the standard, occurs in older strata than *R. manitouensis* in which it correlates with the upper one-third of the *P. falsioneotensis* Zone, all of the *R. tenuis* Zone, but also with the lower two-thirds of the *R. manitouensis*. The range of the zone has therefore been adjusted to extend into younger strata, thus shortening the overlying *P. deltifer* Zone of the standard Atlantic zonation.

Comparison with standard Midcontinent Realm zonation

The main comparison of the Midcontinent Realm zonation proposed in this study will be made to that of the composite stratotype (836 m) described by Ross et al. (1997) in the type Ibex area, Utah, and to that proposed in western Newfoundland by Ji and Barnes (1994).

The long, continuous Section 13, the reference section for all of the new zones of the Kechika Formation, together with Section 5, the reference section for the zones of the Skoki Formation, exceed 2500 m. Several adjustments are made to the Midcontinent Realm zonation proposed in this study compared to the zonations by Ross et al. (1997) and Ji and Barnes (1994).

The first adjustment is the addition of two new zones between the *Iapetognathus* Zone and the *R. manitouensis* Zone, the *Polycostatus falsioneotensis* and *R. tenuis* zones. Ji and Barnes (1994) proposed a *P. falsioneotensis – R. tenuis* Zone, however, *R. tenuis* occurs in younger strata in the reference section of this study and therefore defines a separate zone. The recognition of *P. falsioneotensis* in the Cordillera provides for correlation to strata in western Newfoundland.

The *R. manitouensis* Zone is refined by the definition of two new subzones, the *R. muskwaensis* and *R. sheffieldensis* subzones. The overlying Low Diversity Interval could not be revised and, as noted by Ethington and Clark (1982) and Ross et al. (1997), will remain as an informal zone because of sparse fauna in the interval. However, the equivalency of this interval to the new *D. nowlani* Zone proposed for the Atlantic Realm provides a tie point for this interval.

The *S. subrex* Zone is new and replaces the *Macerodus dianae* Zone of the standard Midcontinent Realm zonation as proposed by Ross et al. (1997). *M. dianae* was first described from slope deposits (Fåhræus and Nowlan 1978) and may be restricted to deeper waters and was not recognized in this study. *S. subrex*, however, is an abundant and distinct species,

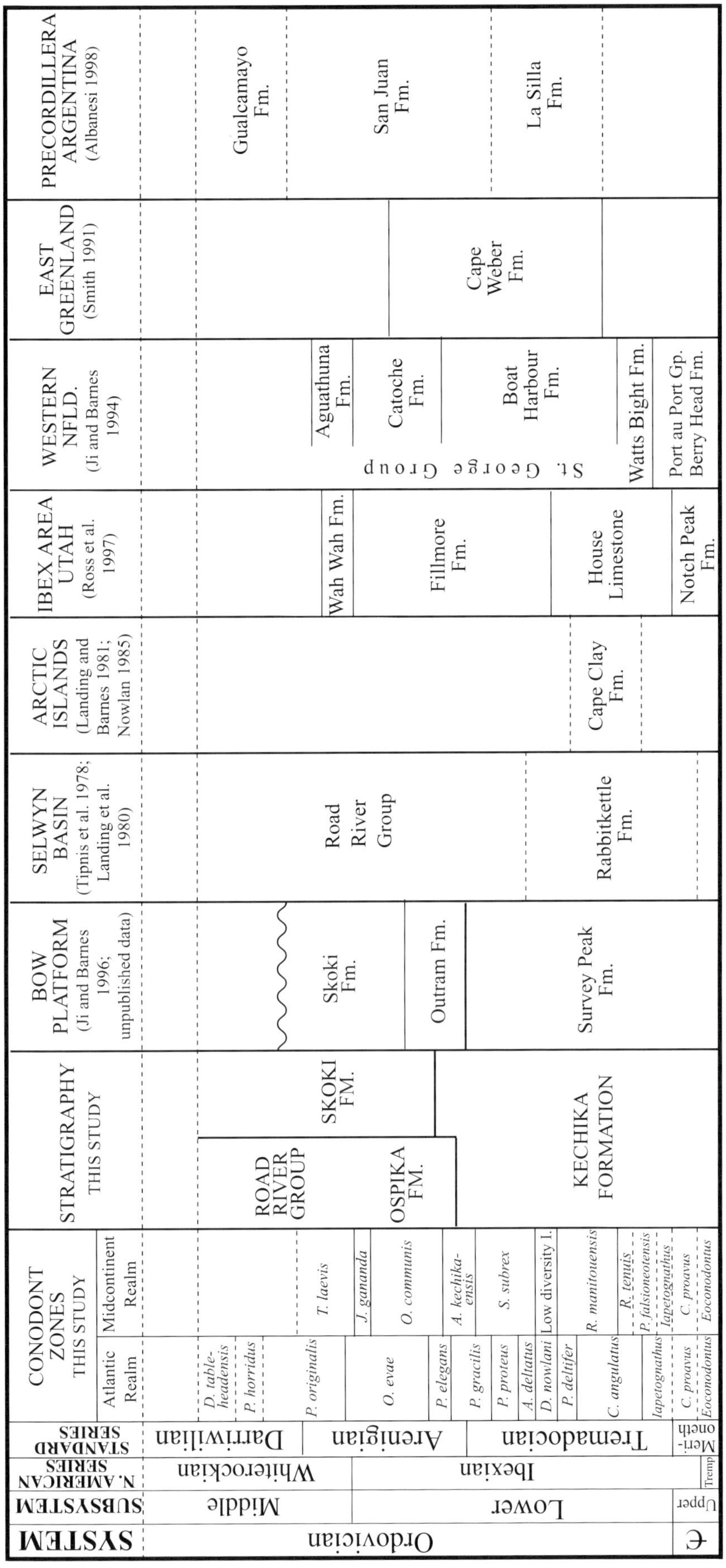

Fig. 11. Correlation chart showing conodont-based correlations of the Kechika and Skoki formations and Road River Group in northern British Columbia, with stratigraphic units of other selected regions referred to in the text. The source of each is given at the top of each column. Details of subzones and stratigraphic members for the study area are shown in Fig. 10.

and its zone has been subdivided into the *Graciloconus concinnus* and *Colaptoconus bolites* subzones. The *S. subrex* Zone extends into the lower half of the *A. deltatus – O. costatus* Zone of the standard Midcontinent Realm zonation. Ji and Barnes (1994) recognized a shallow-water *Protopanderodus inconstans – Scolopodus subrex* Zone, which is in part correlative with the *S. subrex* Zone of this study. The upper half of the *A. deltatus – O. costatus* Zone is replaced by the new *A. kechikaensis* Zone, which is subdivided into three new subzones.

The *O. communis* Zone is recognized at the same level as in the standard, but is further refined by division into three subzones, the *Tropodus sweeti*, *Bergstroemognathus extensus*, and *Juanognathus variabilis* subzones. Ji and Barnes (1994) recognized a *Protoprioniodus simplicissimus – Oepikodus communis* Zone for this interval.

The *J. gananda – R. andinus* Zone is revised as the *J. gananda* Zone and has a slightly higher occurrence based on the first appearance of the nominate species. As discussed within Systematic Paleontology, the coniform element of *R. andinus* that was used as a basis for the lowest occurrence of the standard zone by Ross et al. (1997) has been reassigned to the *J. variabilis* apparatus so that the base of the *J. variabilis* Subzone of the *O. communis* Zone correlates to the base of *R. andinus* Zone of Ross et al. (1997).

The *T. laevis* Zone is the same as that proposed for the standard. Based on the zonation proposed in this study, a number of local and regional correlations can be drawn.

Comparison with standard Atlantic Realm zonation

Conodont faunas provide critical tie points with the standard Atlantic Realm zonation. In the Kechika Formation, the *P. deltifer*, *P. proteus*, and *P. elegans* zones established within the Atlantic Realm (Lindström 1971; Löfgren 1978, 1985, 1993–1996) are recognized. This indicates that the Kechika spans the Tremadocian–Arenigian boundary.

Adjustments to this standard include a younger first appearance of *P. deltifer* and expansion of the *C. angulatus* Zone, which occupies the lower half of the standard *P. deltifer* Zone. *C. angulatus* is a much more abundant and distinct species than *P. deltifer* within the middle Tremadocian. The *P. deltifer* Zone as used in this study is also much shorter compared to the standard because two new zones which are equivalent to the upper half of the standard *P. deltifer* Zone are proposed. These are the *D. nowlani* and *A. deltatus* zones. The *D. nowlani* Zone correlates to the Low Diversity Interval of the standard Midcontinent Realm and is thus an important tie point.

The *P. proteus* Zone has also been adjusted compared to the standard because a new overlying zone, the *Paracordylodus gracilis* Zone of Johnston and Barnes (1999), has been adopted. Löfgren (1994) proposed the *P. gracilis* and *O. elongatus – A. deltatus* subzones within the upper *P. proteus* Zone. The *P. gracilis* Zone in this present study contains an uppermost *O. elongatus* Subzone. *A. deltatus* has a lower first appearance in the study area than in Baltoscandia. Johnston and Barnes (1999) regard the *P. gracilis* Zone as uppermost Tremadocian and suggest that *P. gracilis* has a lower first appearance in the Cow Head Group, western Newfoundland, than in Baltoscandia, and this is supported by the material from the Kechika.

The base of the Arenigian is taken at the base of the *Tetragraptus approximatus* Zone, and although the global stratotype has yet to be formally decided (Williams et al. 1999), the base is just below the base of the *P. elegans* Zone recognized in Cow Head strata (Johnston and Barnes 1999). The recognition of the *P. elegans* Zone within the upper Kechika allows correlation to that in the Cow Head Group and in Baltoscandia.

The *O. evae* Zone is recognized in the Skoki Formation and can be correlated to that recognized within the Cow Head Group. The zones overlying this interval are different from those proposed in Baltoscandia. The *Baltoniodus triangularis*, *B. navis*, and *Microzarkodina parva* zones (Löfgren 1978, 1993–1996; Lindström 1971; Stouge and Bagnoli 1990) are not recognized. Instead, the *Paroistodus originalis* Zone is proposed as overlying the *O. evae* Zone. The absence of *Baltoniodus* may be due to the endemic nature of the genus in Baltoscandia, however, it has been recognized in China (Zhang 1998) and Argentina (Albanesi 1998). The *P. originalis* Zone is, therefore, much thicker than that in Baltoscandia.

The *P. horridus* Zone is proposed and was used as a subzone of the *Lenodus variabilis* Zone by Albanesi (1998). In this present study, *Lenodus variabilis* was not present. The other emended zone is the *Dzikodus tablepointensis* Zone, which was proposed by Zhang (1998) from the Hubei Province, south-central China. Again, as some species that are used in the standard Atlantic Realm zonation are likely endemic, the use of these latter two zones can provide tie points within the early Darriwilian (Fig. 9).

Discussion of key evolutionary lineages

The species diversity of conodonts was highest in the early Ordovician (Sweet 1988), and the diversification of several lineages is documented in this study. The lineages discussed in this section are those that are important to the biostratigraphic zonation. The faunas from this study can be compared to the interpretations by Sweet (1988) who suggested phylogenetic relationships among the major conodont groups and to those of Ji and Barnes (1994) who more specifically depicted several phylogenies of Early Ordovician conodont taxa. The following discussion provides an overview of some of the major lineages, and further details and justification are discussed in Systematic Paleontology.

The most dramatic diversification within an early Tremadocian lineage is that within *Cordylodus*. The transition from *Proconodontus muelleri* to *Eoconodontus notchpeakensis* to *Cordylodus primitivus* in the late Cambrian is evident (Systematic Paleontology). By the early Tremadocian, seven species of *Cordylodus* occured and the last descendant, *C. angulatus*, appeared in the middle Tremadocian (Fig. 12). Another early denticulate genus, *Iapetognathus*, also diversified rapidly in the earliest Tremadocian into three species recognized in this study. Its origins are discussed by Nicoll et al. (1999). Neither *Cordylodus* nor *Iapetognathus* were ancestral to later conodonts, but are key in the zonation of the Cambrian–Ordovician boundary.

Sweet (1988) interpreted the *Proconodontus* stock to have also given rise to the Family Fryxellodontidae within which an important genus for the zonation of lower Arenigian strata is *Kallidontus* n. gen. The first representative of the *Kallidontus* lineage is *K. princeps* n. sp., which appeared in the *S. subrex* Zone and speciated first into *K. serratus* n. sp., which in turn gave rise to *K. nodosus* n. sp. (Fig. 13). The latter two species are used to refine the *A. kechikaensis* Zone. The lineage (Fig. 13) may represent a separate lineage within the Fryxellodontidae.

Ji and Barnes (1994) described the *Teridontus* complex, which includes the *Teridontus*, *Semiacontiodus*, *Polycostatus*, *Variabiloconus*, *Striatodontus*, and *Colaptoconus* lineages. The material from this study supports the proposal of *Teridontus nakamurai* as an ancestor for *T. gracillimus*, *T. obesus*, and *Semiacontiodus nogamii*, the latter of which gave rise to the costate species of members of the Family Protopanderodontidae such as *Polycostatus*, *Variabiloconus*, and *Striatodontus* (Fig. 12). The *Colaptoconus* lineage appears in younger strata, within a second diversification event of the middle Tremadocian. Ji and Barnes (1994) suggested a close evolutionary relationship of the *Colaptoconus* lineage to the *Striatodontus* lineage based on their comparable morphology and stratigraphic range. The apparatus structure of the two genera is similar, but the stratigraphic ranges noted in this study are not as close as those in western Newfoundland (Ji and Barnes 1994), which suggests that *Striatodontus* may have an earlier appearance along the Cordilleran margin in comparison to the Appalachian margin.

The families Drepanoistodontidae and Acanthodontidae also have origins among oneotodontid ancestors such as *Teridontus*, but the origins of *Drepanoistodus* and *Acanthodus* are not as clear (Fig. 12). *Drepanoistodus pervetus* is the earliest representative of that lineage and within the Midcontinent Realm gave rise to *D. minutus* n. sp., which in turn possibly gave rise to *D. expansus*. *D. nowlani* appears in the late Tremadocian, possibly from the long-ranging *D. pervetus*. Sweet (1988) suggested *Paltodus* as the rootstock of the Drepanoistodontidae, however, from this study, species of *Drepanoistodus* appear earlier than the first occurrence of *Paltodus*.

Acanthodus lineatus is an abundant species in the early Tremadocian (Table 15) and is the earliest representative of the Acanthodontidae. The genus has a simple apparatus of a and e elements. *A. lineatus* ranges across the Tremadocian–Arenigian boundary, and its late form develops strong lineations on the cusp. These latter forms with strong lineations are similar to early forms of *Scolopodus subrex*, which bear few, weakly developed costae on the cusp. *S. subrex* also has a simple apparatus of a, b, and e elements and becomes more strongly costate in later forms. This interpretation differs from that of Ji and Barnes (1994, text-fig. 16) in which the *Scolopodus* lineage originates from the *Acodus* lineage.

The next major group of importance for biostratigraphy is the Prioniodontida. Sweet (1988) suggested that the oldest genus within the order was *Rossodus*. This present study suggests that the origin of the prioniodinids extends back to the appearance of *Utahconus*

Fig. 12. Interpreted phylogeny of *Cordylodus, Teridontus, Semiacontiodus, Polycostatus, Striatodontus, Drepanoistodus,* and *Acanthodus* lineages represented in the Lloyd George, Quentin, and Grey Peak members of the Kechika Formation. Exact stratigraphic measurements not shown, as lineages span several zones. See Fig. 4 for stratigraphic ranges.

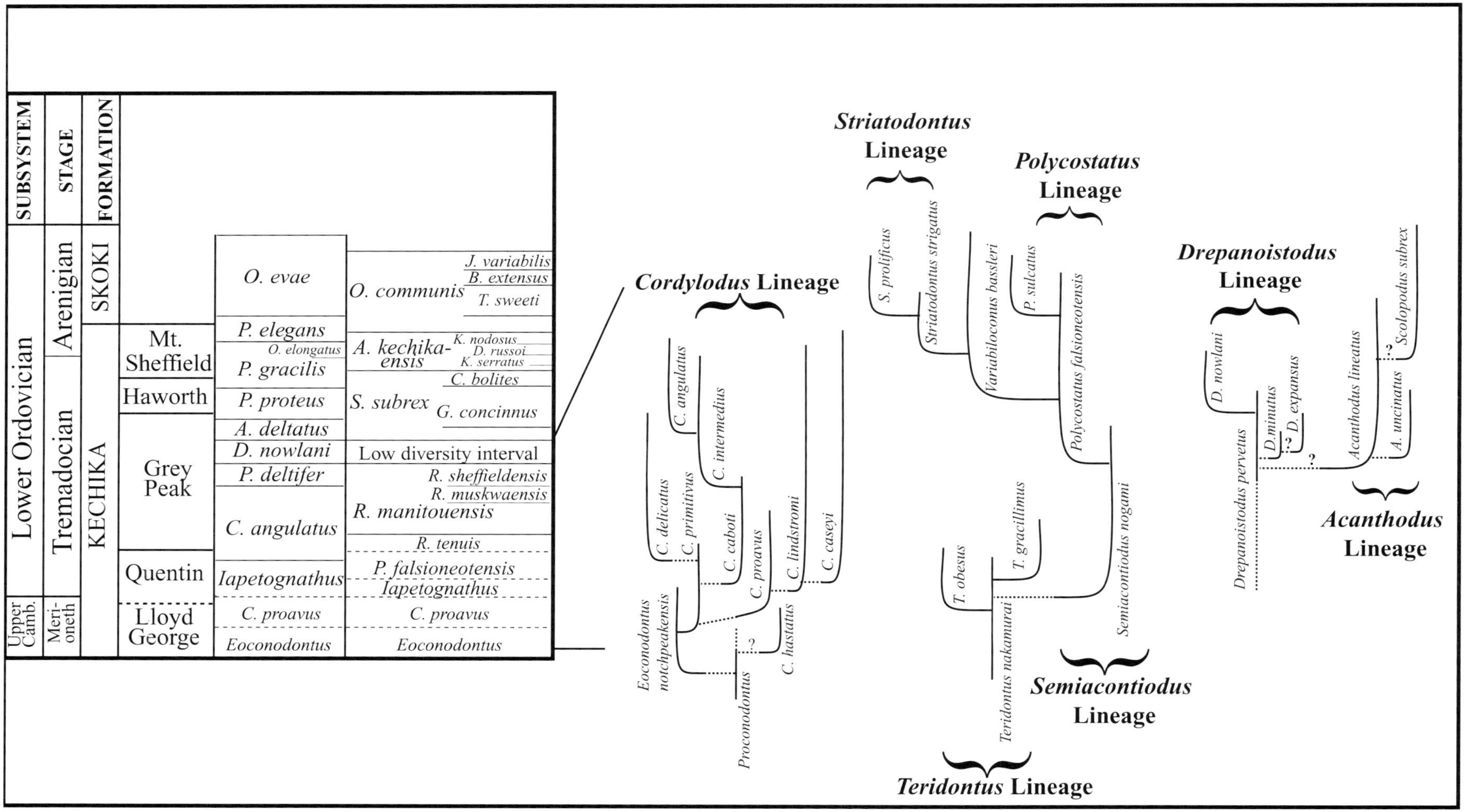

Fig. 13. Interpreted phylogeny of the *Kallidontus* lineage represented in the Grey Peak to Mt. Sheffield members, Kechika Formation. Stratigraphic measurements not to scale.

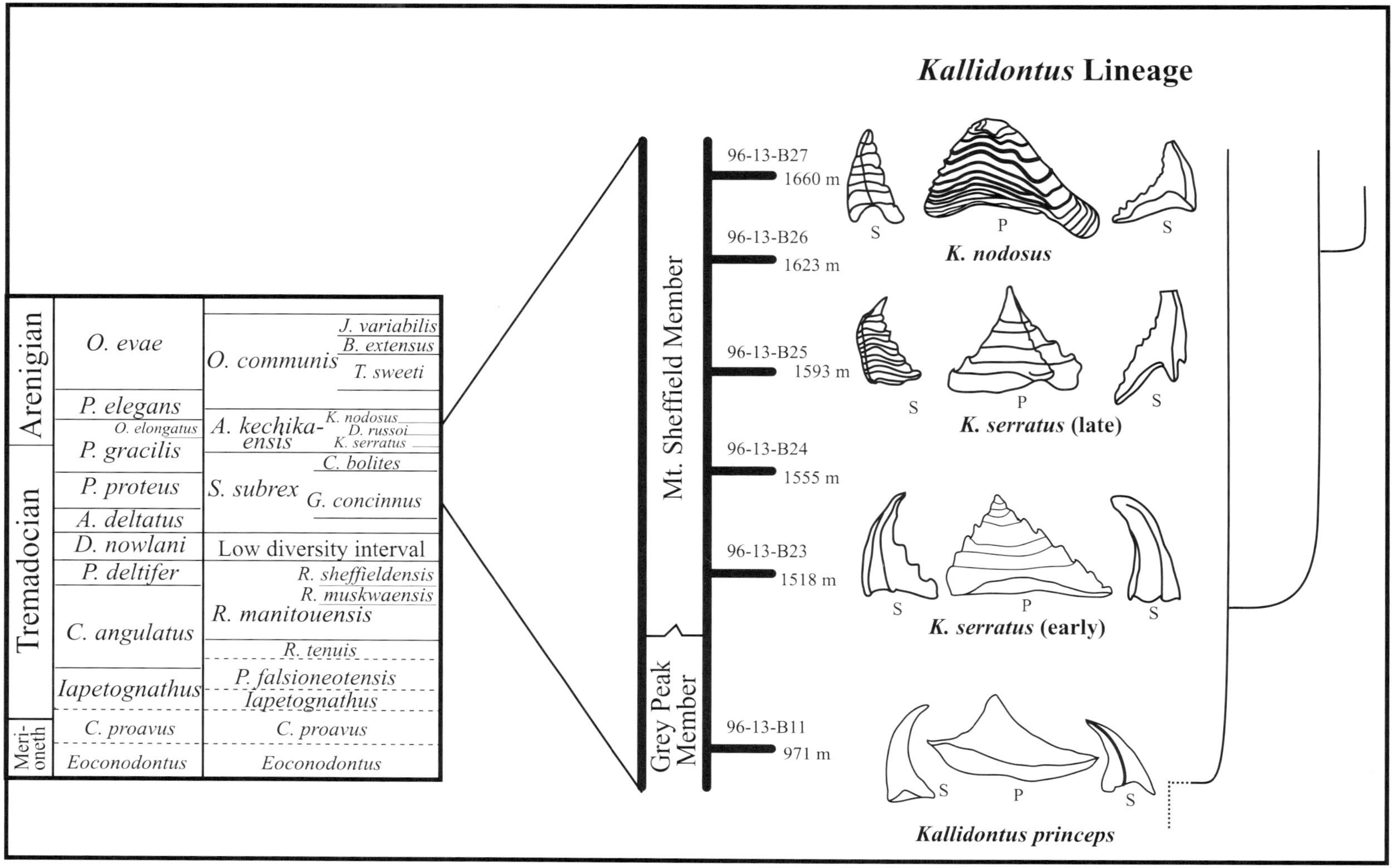

utahensis, *Acodus quentinensis*, and *A. oneotensis* in the early Tremadocian. These three early prioniodontids may have been derived from a *Teridontus* lineage ancestor (Figs. 14, 15). *Acodus* is, therefore, a much more common genus in the earliest Tremadocian and is considered to belong within a separate family, the Acodontidae (Dzik 1994) rather than among the Oistodontidae, which contains *Rossodus*. *Acodus* is not a *nomen dubium* as suggested by Sweet (1988).

Utahconus utahensis clearly gave rise to *Rossodus tenuis*, the direct ancestor of *Rossodus manitouensis* (see Systematic Paleontology). In the late Tremadocian, the *Rossodus* lineage underwent a speciation event to produce four new species, two of which have herein been used to define subzones (*R. muskwaenesis* and *R. sheffieldensis*) within the *R. manitouensis* Zone (Fig. 14). *Tricostatus*, which is likely another descendent of *U. utahensis*, also diversified during this interval into three species (Fig. 14). Ji and Barnes (1994) suggested that *T. glyptus* is closely related to *R. tenuis*, and the similarity of the apparatus of *Tricostatus* to *Rossodus* suggests that the two genera may have had a common ancestor, *U. utahensis*.

The Acodontidae diversified rapidly within the *Acodus kechikaensis* Zone (Fig. 15). The *Acodus* lineage diversified from *Acodus deltatus*, which likely decended from *A. quentinensis* n. sp. *A. kechikaensis* n. sp. diverged from *A. deltatus* in the latest Tremadocian and itself likely gave rise to two species, *A. longibasis* and *A. warensis* n. sp. *A. deltatus* is a long-ranging species and possibly gave rise to *A. neodeltatus* n. sp. in the earliest Arenigian. The abundant and diverse conodont fauna from the *A. kechikaensis* Zone indicates that *A. neodeltatus* n. sp. was the direct ancestor of *Oepikodus communis* through denticulation of the keeled cusp margins (Fig. 15).

Tropodus australis and *Diaphorodus russoi* are also abundant in this interval. Although there has been much confusion in the past about the assignment of the genera *Acodus*, *Tropodus*, and *Diaphorodus*, the faunas from this study support the apparatus reconstructions proposed by Albanesi (1998) and favour the assignment of the genera to the Acodontidae rather than to the Oistodontidae as proposed by Sweet (1988). Thus, the early forms of *Acodus* are better candidates as the root stock of the Acodontidae, rather than *Rossodus* as proposed by Sweet (1988, fig. 5.11). However, the Oistodontidae (including *Rossodus*) are an important root stock for lineages within the Prioniodontida, particularly the Oistodontidae. The Acodontidae likely gave rise to lineages such as the Prioniodontidae based on the interpreted relationship of *A. neodeltatus* and *O. communis* from this study.

In summary, this overview illustrates the significance of the northern Cordilleran faunas in modifying the understanding of early conodont evolutionary lineages. There is clearly high potential for studying lineages preserved within the expanded stratigraphic successions of the northern Cordilleran miogeocline. Speciation events are well documented, such as those within the *Rossodus* and *Acodus* lineages (Figs. 14 and 15), and provide refinement of the conodont biostratigraphy (Figs. 10 and 11). The initial phases of diversification of latest Cambrian to Early Ordovician conodonts are critical events in the 300 million year history of the group, yet for which few studies have been devoted.

The initial adaptive radiation of conodonts in the latest Cambrian and Early Ordovician is documented in this study by the number of species and their total abundance within the *Iapetognathus* Zone (Table 1). In the Early Tremadocian, the genus *Cordylodus* comprises 16.44% of the total fauna (Table 15). The highest peak in species diversity occurs within the *Rossodus manitouensis* Zone in which genera such as *Acanthodus*, *Rossodus*, *Utahconus*, and *Variabiloconus* together comprise over 62% of the total fauna. These initial phases of speciation contribute to the first diversity peak illustrated by Sweet (1988, fig. 6.1), which is highest in the middle Ibexian (mid-Tremadocian).

The total species abundance decreased through the middle to late Tremadocian (Low Diversity Interval), but increases through the latest Tremadocian in which over 40% of the fauna is assigned to *Acodus* (Table 16). In the early Arenigian, a second peak is reached with the initiation of the *Oepikodus* lineage, a genus that comprises over 28% of the total fauna (Table 17). This diversity diminishes at the base of the *Tripodus laevis* Zone, or base of the Whiterockian. The faunas within the Skoki are high in diversity but not as abundant as compared to those of the Kechika, however, future studies may provide more material for the study of several lineages of the early and middle Arenigian.

Fig. 14. Interpreted phylogeny of *Rossodus* and *Tricostatus* lineages represented in the Quentin and Grey Peak members, Kechika Formation. Stratigraphic measurements not to scale.

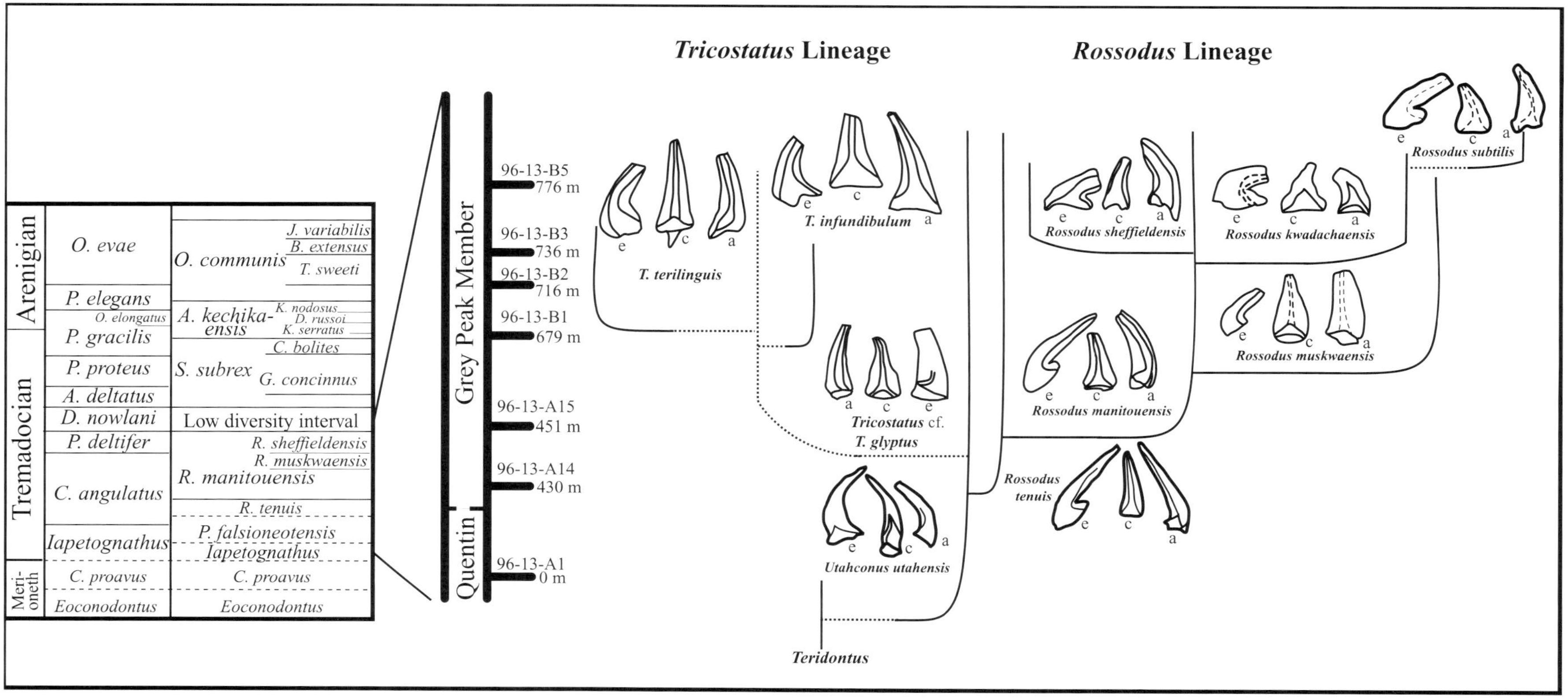

Fig. 15. Interpreted phylogeny of *Acodus* lineage represented from the Quentin through Mt. Sheffield members, Kechika Formation. Stratigraphic measurements not to scale.

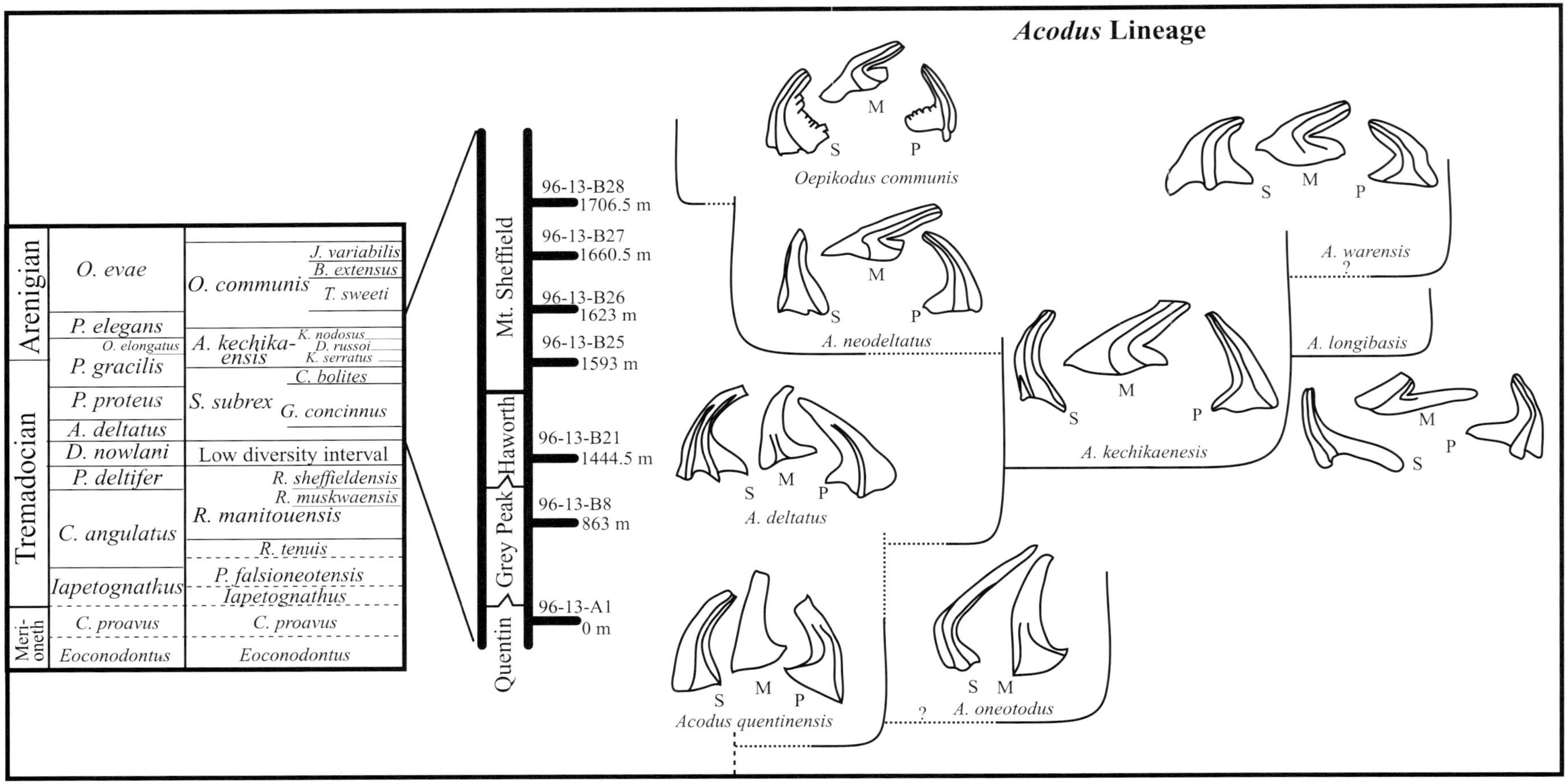

Systematic paleontology

Introduction

The suprageneric classification of orders of conodonts proposed in Sweet (1988) and modified by Aldridge and Smith (1993) is followed herein. Additional suprageneric ranks of Craniata Linnaeus, 1758, Vertebrata Linnaeus, 1758, and Gnathostomata Cope, 1889, have been determined through cladistic analysis by Donoghue et al. (2000).

Within the "Systematic Paleontology," taxa are listed in alphabetical order and assigned to the appropriate family and order. Species for which there are few comments or that require no revision are simply listed with occurrence, material, and repository information. The synonymy lists are kept to a minimum for described species by referring the reader to the most complete synonymy in a recent publication for which all the references listed are in agreement with the taxonomic assignment.

In most identifications, a multielement apparatus has been determined. However, in some situations, the suffix s. f. (*sensu formo*) (Barnes and Poplawski 1973) designates a form taxon which has not yet been synonymized in a multielement scheme (e.g., *Cordylodus deflexus* s. f.). In situations where full identification is not possible, genera are reported in open nomenclature. The uncertainty of assignments is indicated by a question mark. A provisional identification is indicated by cf. (e.g., *Agenus* cf. *aspecies*). A new, undescribed taxon is related to a known taxon using aff. (e.g., *Agenus* aff. *aspecies*). When specific identification is not attempted or not possible, the abbreviation sp. is used. Recognition of a new genus is indicated by n. gen. and new species by n. sp. (Bengtson 1988).

The terminology used for multielement taxonomy follows the letter designations after Sweet and Schönlaub (1975) for the location of elements within well-known septimembrate apparatuses (S, M, P notation). For coniform apparatuses, the letter designations proposed by Barnes et al. (1979) and adopted by Ji and Barnes (1994) are followed (a, b, c, e, f). The study favours nomenclaturally neutral terms, and the naming of element types after form genera or species which they resemble (e.g., oistodiform) is avoided as much as possible.

All conodonts illustrated were photographed by a Hitachi S-3500N scanning electron microscope. All figured specimens are deposited in the National Type Collections of the Geological Survey of Canada (GSC).

Phylum CHORDATA Bateson, 1886

Class CONODONTA Pander, 1856

Order PROCONODONTIDA Sweet, 1988

Family CORDYLODONTIDAE Lindström, 1970

Genus *Cordylodus* Pander, 1856

Type species — *Cordylodus angulatus* Pander, 1856

Discussion — *Cordylodus* Pander is a common and important faunal component in uppermost Cambrian and Lower Tremadocian strata. Its apparatus was described by Miller (1980) and Bagnoli et al. (1987) as bimembrate. Nicoll (1990) has interpreted the apparatus as septimembrate, however, the M elements he illustrated are not recognized in this collection, and the P elements he illustrated are here interpreted to be compressed e elements. Therefore, the apparatus of all species is described as having three morphotypes as defined by Ji and Barnes (1994) and Andres (1988). These are a (subrounded), with two variants, and c (suberect) elements within the first transition series and e (compressed) with two variants having slight to strong compression, making up the second transition series. However, in early forms, not all of these morphotypes are well defined.

Cordylodus angulatus Pander, 1856

Plate 4, figs. 1–3

a element — *Cordylodus angulatus* Pander, 1856, p. 33, pl. 2, fig. 31, pl. 3, fig. 10; Nicoll, 1990, pp. 536–543, figs. 3, 4b (S element), figs. 6, 7, 11.4 (S elements) only; Ji and Barnes, 1994, pp. 31–32, pl. 5, figs. 1–4 (contains synonymy through 1994); Harris et al., 1995*b*, p. 24, pl. I, fig. AB.

c element — *Cordylodus angulatus* Pander, 1856, p. 33, pl. 2, fig. 30; Nicoll, 1990, pp. 536–543, figs. 5.3, 5.4, figs. 10.3, 10.4 (Sa element) only; Ji and Barnes, 1994, pp. 31–32, pl. 5, fig. 5; Löfgren, 1997, p. 260, fig. 4E.

e element — *Cordylodus angulatus* Pander, 1856, p. 33, pl. 2, figs. 27–29; Nicoll, 1990, pp. 536–543, fig. 3 (4c), fig. 8 (Pb element), fig. 9 (Pa element), fig. 12 (Pa and Pb elements); Ji and Barnes, 1994, pp. 31–32, pl. 5, figs. 6–9 (contains synonymy to 1994); Löfgren, 1997, p. 260, fig. 4D.

not *Cordylodus angulatus* Pander. Nicoll, 1990, fig. 3 (4a), figs. 5.1, 5.2, figs. 10.1, 10.2.

Discussion — The apparatus of *C. angulatus* contains three morphotypes as discussed in Ji and Barnes (1994): a (subround, with symmetry variants), c (suberect), and e (compressed, with two variants). The distinguishing characteristic of *C. angulatus* is the shape of the shallow basal cavity in lateral view in which the anterior margin of the cavity is deeply concave, and the posterior margin originates under the posterior process and runs parallel to the process, turning down to form the cavity tip below or on line with the process. In robust elements, where the cavity cannot be observed, the a elements vary in the degree of lateral compression and have a notch or wide angle between the cusp and first denticle. They have an extremely high base, the margin of which is straight. The c elements are characterized by a suberect element morphology and often have an arched basal margin and process. The e elements are characterized by one variant in which a prominent basal flare is developed on the inner lateral margin or the basal margin (previously described as *C. rotundatus*). The other variant of the e element is compressed along the cusp and basal margin and does not develop the basal flare. Both, however, are characterized by the notch between the cusp and first denticle. The septimembrate apparatus plan proposed by Nicoll (1990) has been partially synonymized here, however, the P elements he figures are the variants of the compressed e elements and the M element is not observed in this material.

Occurrence — *Cordylodus angulatus* Zone, Kechika Formation.

Material — 1000 a elements, 76 c elements, 303 e elements.

Repository — GSC 119078, 119079, 119080.

Cordylodus caboti Bagnoli, Barnes, and Stevens, 1987

Plate 3, figs. 17–20

a element — *Cordylodus caboti* Bagnoli, Barnes, and Stevens, 1987, p. 152, pl. 1, figs. 10–14 (contains synonymy to 1987).

e element — *Cordylodus oklahomensis* Müller. Landing, 1983, p. 1161, text-fig. 7f; Fortey et al., 1982, p. 119, text-figs. 6v, x, 8g, o.

C. intermedius? Furnish. Miller, 1980, pp. 17–18, pl. fig. 17, text-fig. 4M.

not *Cordylodus caboti* Bagnoli et al., 1987, p. 152, pl. 1, fig. 10.

Discussion — As described by Bagnoli et al. (1987), the apparatus comprises rounded elements (p) which are the a and c elements and compressed (q) elements which are the e elements, often referred to the species *C. prion*. The specimens of *C. oklahomensis* listed in the synonymy are those which have straight-edged basal cavities. However, the compressed elements, referred to as *C. oklahomensis* possibly represent the compressed elements of other species, differentiated only by the shape of the basal cavity. The basal cavity of the elements is a characteristic feature in that it is not as deep as in *C. proavus*, extending to a point slightly higher than the posterior process. The basal cavity has straight margins to convex near the tip, and its apex is centrally located. Both features distinguish it from *C. intermedius*.

Occurrence — *Iapetognathus* Zone to *Rossodus manitouensis* Zone, Kechika Formation.

Material — 52 a elements, 9 c elements, 353 e elements.

Repository — GSC 119068, 119069, 119070, 119071.

Cordylodus caseyi Druce and Jones, 1971

Plate 4, figs. 4–7

Cordylodus caseyi Druce and Jones, 1971, pp. 67–68, pl. 2, figs. 9–12, text-figs. 23d, e; Nicoll, 1990, pp. 543–545, figs. 13 (4, 5), 14 (1, 2) only; Ji and Barnes, 1996, p. 881, figs. 12.17–12.23.

Description — Apparatus of three morphotypes characterized by robust elements with lateral costae. Basal cavity has similar shape to that of *C. intermedius* because anterior basal cavity margin is concave and tip of cavity is located anteriorly. Basal cavity is broad, extending under posterior process. Basal margin is drawn out anteriorly to form anticusp.

The a elements are subsymmetrical, subround, and comprise two variants within symmetry transition series. Cusp is long and slender, cusp margins are rounded, cusp cross section is circular. Cusp has inner postero-lateral keel and inner lateral compression. Cusp is twisted laterally, and denticulated posterior process is twisted toward inner side, denticles are slightly compressed. Basal margin is arched and drawn out antero-laterally into anticusp. Other variant has antero-lateral buttress drawn out onto anticusp and is more stout. Outer surface of cusp is smooth, and inner lateral surface is compressed postero-laterally. Base is compressed.

The symmetrical c element has keel which originates basally parallel to anterior cusp margin and turns posterior and downward. Posterior process is arched. Element is slightly laterally compressed; cusp cross section is oval with rounded margins.

The compressed e elements are of two variants in which degree of arching of the basal margin differs. In both variants, anticusp is developed, more strongly in more arched specimens. Denticles are discrete to slightly confluent along their bases.

Discussion — *C. caseyi* is characterized by rounded, robust elements with long slender cusps and lateral or antero-lateral buttresses. The basal cavity shape is similar to that of *C. intermedius*, and the two species are likely closely related. The original description by Druce and Jones (1971) was only of the a elements. In this study, compressed e elements and symmetrical c elements are included in the apparatus reconstruction. The e elements are similar to those of coeval *Cordylodus* species and are distinguished by the presence of the anticusp.

Occurrence — *Iapetognathus* Zone to *Rossodus manitouensis* Zone, Kechika Formation.

Material — 71 a elements, 7 c elements, 19 e elements.

Repository — GSC 119074, 119075, 119076, 119077.

Cordylodus deflexus s. f. Bagnoli, Barnes, and Stevens, 1987

Plate 3, fig. 21

Cordylodus deflexus Bagnoli, Barnes, and Stevens, 1987, p. 153, figs. 1–4 (contains synonymy to 1987).

?*Cordylodus tortus* Barnes, 1988, p. 412, figs. 13p–r, 14b.

Discussion — As noted in the synonymy within Bagnoli et al. (1987), it is difficult to distinguish rounded and compressed elements from previous studies. They describe p elements which are rounded with an anticusp. The q elements are compressed and the process is laterally deflected. In both element types, the cusp is twisted laterally and the process is twisted. The specimens in this study are in agreement with their observations, and are likely part of another *Cordylodus* apparatus, but sufficient material is lacking in this collection to confirm this.

Occurrence — *Cordylodus proavus* Zone, Kechika Formation.

Material — 2 elements.

Repository — GSC 119043.

Cordylodus delicatus n. sp.

Plate 4, figs. 8–11

Cordylodus andresi Barnes, 1988, pp. 410–411, figs. 13e, 14a only (c element).

Cordylodus proavus Müller. Miller, 1969, p. 432, pl. 65, figs. 38, 39 only; Druce and Jones, 1971, p. 120, pl. 1, figs. 2a, b only; Müller, 1973, p. 35, pl. 9, fig. 8 only; Miller, 1980, pp. 19–20, pl. 1, figs. 14, 15 only; Landing, 1983, p. 1168, fig. 9C only; Chen and Gong, 1986, pp. 130–133, pl. 35, fig. 8, pl. 36, figs. 2, 11, 12, 15, text-fig. 40 (1, 2) only.

C. sp. cf. *C. proavus* Müller. Nowlan, 1985, p. 111, fig. 4 (4) only.

Cordylodus n. sp. Pyle and Barnes (2001), Pl. 1, fig. 5.

Diagnosis — A small, delicate species of *Cordylodus* with extremely wide angle between slender, elongate cusp and first denticle of posterior denticulated process. Basal cavity deep, with anterior margin concave and located close to anterior margin of cusp.

Description — Apparatus of three elemental morphotypes, all with long, slender cusp and denticulated posterior process. Characteristic feature of elements is broad, sweeping angle between first discrete denticle of posterior process and slender cusp.

The subrounded a elements are laterally compressed, some slightly and others strongly, and comprise first symmetry transition series. Cusp is slender, reclined to recurved, cusp margins are rounded. Basal cavity is deep and broadly based, extending under long posterior cusp margin but not into discrete denticles. Basal cavity narrows at the point of curvature of cusp, but extends deep into cusp. Cusp tip and denticles are filled with white matter. Anterior margin of basal cavity is concave from antero-basal margin and is close to anterior cusp margin. Posterior basal cavity margin is directly under posterior cusp margin and curves anteriorly to join anterior basal margin, forming central apex at or above point of curvature and flexure of cusp. Shape of basal margin and degree of lateral compression of cusp is variable. Most have horizontal basal margin and posterior process and oval basal cross section. Antero-basal corner is smoothly rounded and laterally compressed. Denticles are laterally compressed, reclined posteriorly, and oval in cross section.

The c element shares similar characteristics with a elements but is symmetrical and not as compressed as a element. Cusp is erect. Basal margin is slightly arched and antero-basal corner is sharp, not rounded.

The compressed e elements are not as abundant. They have strongly compressed cusp and basal flare, usually between first denticle and cusp. Degree of compression along inner anterior lateral margin varies from slight to strong. Some elements have basal margin with inner flare and downwardly directed posterior cusp margin.

Etymology — Named for its delicate nature.

Discussion — *C. delicatus* is similar to *C. primitivus* in its primitive characters such as a deep basal cavity, and simple morphology within an undifferentiated apparatus. As Barnes (1988) described, variation in the flexure of the posterior process is a distinguishing feature, however, in this study the elements vary in the degree of compression and curvature of the cusp. The p2 and p3 elements of Barnes (1988) have wider basal cavities and a narrow angle between the cusp and process and are not observed in this collection. Possibly they are other morphotypes in the apparatus.

C. andresi was originally proposed for elements with large, deep basal cavities, extending nearly to the tip (Viira et al. 1987). The elements of *C. andresi* are variable but are denticulated at a point high on the posterior margin of the cusp. As suggested by Müller and Hinz (1991), *C. andresi* as described by Barnes (1988) is neither *C. primitivus* nor *C. andresi* sensu Viira et al. (1987). Hence, this new species, *C. delicatus*, is erected.

Occurrence — *Iapetognathus* Zone to *Rossodus manitouensis* Zone, Kechika Formation.

Material — 28 a elements, 6 c elements, 13 e elements.

Repository — GSC 119050, 119051, 119052.

Cordylodus hastatus Barnes, 1988

Plate 4, figs. 12–14

Cordylodus hastatus Barnes, 1988, pp. 411–412, Figs. 13s–x, Fig. 14d (contains synonymy to 1988); Fåhræus and Roy, 1993, p. 18, pl. 1, figs. 15, 17 only.

Description — Basal cavity extends beneath denticulated posterior process in all elements. White matter is present in tip of cusp and within denticles. The a elements comprise subround, subsymmetrical elements. Cusp suberect to reclined. Cusp margins are rounded to keeled and vary in degree of lateral compression. Posterior process and anterior margin are slightly downwardly directed. In some, angle between first denticle and posterior margin is wide, in others more acute. Denticles are laterally compressed and confluent.

The c element is suberect and symmetrical. Cusp is broad and keeled, process and anterior margin are downwardly directed. Element retains symmetry, as process is not deflected nor cusp curved laterally. Denticles are inclined posteriorly. Inner cusp margin bears faint carina.

The e element is distinctive, broad, and robust with laterally compressed cusp with strongly keeled margins and well-developed inner carina on some and slight lateral curvature of cusp. Posterior process contains laterally compressed, broad denticles, confluent and inclined posteriorly. Process slightly upwardly deflected in some specimens. Angle between first denticle and posterior margin of cusp is sharp and acute. Base is nearly flat in only one specimen, in all others anterior margin is directed downward.

Discussion — *C. hastatus* contains three morphotypes, which Barnes (1988) distinguished as p1, p2, p3, and q. The p1 and p2 are the two variants of the a element, the p3 is equivalent to e elements, and c elements are equivalent to q elements. The specimens are similar to those described by Barnes (1988). If the a elements are broken, they can resemble rounded *C. primitivus*, or if the basal cavity is not observed, elements can resemble *C. proavus*. The basal cavity differs from that of *C. proavus* in having a straight, not curved, anterior margin. The a elements of *C. hastatus* are distinct from *C. proavus* in having confluent bases of the denticles.

Occurrence — *Cordylodus proavus* Zone, Kechika Formation.

Material — 2 a elements, 2 c elements, 9 e elements.

Repository — GSC 119053, 119054, 119055.

Cordylodus intermedius Furnish, 1938

Plate 4, figs. 15–17

a element — *Cordylodus intermedius* Furnish, 1938, p. 338, pl. 42, fig. 31, text-fig. 2c; Ji and Barnes, 1994, p. 32, pl. 5, figs. 10–13 (contains synonymy to 1994); Harris et al., 1995*b*, p. 27, pl. II, fig. G.

c element — *Cordylodus lenzi* Müller, 1973, p. 31, text-fig. 2F, pl. 10, fig. 7 (only); Ji and Barnes, 1996, pl. 5, figs. 14, 16.

e element — *Cordylodus oklahomensis* Druce and Jones, 1971, p. 69, pl. 5, figs. 7b, c only, text-fig. 23j.

Cordylodus intermedius Furnish. Ji and Barnes, 1996, pl. 5 (17, 18).

Discussion — As described by Barnes (1988), *C. intermedius* is an evolutionary intermediate between *C. proavus* and *C. angulatus*. Of the three morphotypes described by Ji and Barnes (1994), the a elements are most distinct in having a shallow basal cavity extending to the level of the process, its anterior margin is concave and the basal cavity tip is located anteriorly. The compressed elements are more difficult to distinguish among coeval species, but can be identified by the anterior concavity of the basal cavity margin in *C. intermedius*.

Occurrence — *Iapetognathus* Zone to *Rossodus manitouensis* Zone, Kechika Formation.

Material — 86 a elements, 3 c elements, 15 e elements.

Repository — GSC 119063, 119064, 119065.

Cordylodus lindstromi Druce and Jones, 1971

Plate 4, figs. 18–21

a element — *Cordylodus lindstromi* Druce and Jones, 1971, pp. 68–69, pl. 1, figs. 7a–8b (only), pl. 2, figs. 8a–c; Jones, 1971, p. 47, pl. 2, figs. 4a–c; Müller, 1973, p. 32, pl. 9, fig. 10 only, text-figs. 2D, 6A, B; Miller, 1980, pp. 18–19, pl. 1, fig. 18 only; Viira et al., 1987, pl. 2, fig. 7, pl. 4, figs. 7, 8 only; Andres, 1988, pp. 134–136, Abb. 36 (middle fig.), Abb. 37 (in part); Barnes, 1988, pp. 408–409, figs. 13 k, l only (p element); Harris et al., 1995*b*, p. 27, pl. II, fig. H. Ji and Barnes, 1996, p. 881, figs. 12.6–12.8, 12.11–12.12.

not Ji and Barnes, 1994, p. 32, pl. 5, figs. 19, 20 (e element variant)

c element — *Cordylodus lindstromi* Druce and Jones. Müller, 1973, p. 32, pl. 9, fig. 11; Andres, 1988, pp. 134–136, Abb. 36 (top fig.), Abb. 37 (in part); Ji and Barnes, 1996, p. 881, figs. 12.9, 12.10.

e element — *Cordylodus lindstromi* Druce and Jones, 1971, pp. 68–69, pl. 1, figs. 9a, b only, text-fig. 23h; Miller, 1980, pp. 18–19, pl. 1, fig. 19 only; Andres, 1988, pp. 134–136, Abb. 36 (lower fig.), Abb. 37 (in part); Barnes, 1988, pp. 408–409, fig. 13j; Ji and Barnes, 1994, p. 32, pl. 5, figs. 19–22; Ji and Barnes, 1996, p. 881, fig. 12.13.

Discussion — The apparatus was initially illustrated by Druce and Jones (1971) and described by Miller (1980) as having two elemental morphotypes. Andres (1988) figured three morphotypes, which are recognized in this study. The characteristic feature of *C. lindstromi* is the biapical basal cavity, however, this character must be used with caution, as it may be exhibited by other species. The shape of the basal cavity is a distinguishing feature in that it is narrow and its anterior margin follows the margin of the cusp and its apex is situated anteriorly. Another characterisitic is the proximity to the cusp of the first denticle of the posterior process. The elements of *C. lindstromi* can be distinguished from *C. proavus* because the cavity of *C. proavus* is much broader and deeper, and the majority of the elements observed in this material are robust compared to the slender, more delicate cusps of *C. lindstromi.* As Ji and Barnes (1994) described, the a elements are much more common than the e elements, and both the a and e elements show variations in curvature. The c elements are distinguished by being symmetrical, the cusp is suberect and basal margin is arched.

Occurrence — *Iapetognathus* Zone to *Rossodus manitouensis* Zone, Kechika Formation.

Material — 1221 a elements, 27 c elements, 615 e elements.

Repository — GSC 119059, 119060, 119061.

Cordylodus primitivus Bagnoli, Barnes, and Stevens, 1987

Plate 5, figs. 1–3

Cordylodus primitivus Bagnoli, Barnes, and Stevens, 1987, p. 154, pl. 1, figs. 1–6 (contains synonymy to 1987); Lehnert, 1995, p. 80, pl. 14, fig. 2; Mens et al., 1996, pl. 1, figs. 7, 8.

not *Cordylodus primitivus* Bagnoli, Barnes, and Stevens sensu Müller and Hinz, 1991, p. 54, pl. 43, figs. 4, 5, 8–15, 21A–F (*C. andresi* sensu Viira, Sergeyeva, and Popov, 1987).

Description — Three elemental morphotypes are recognized. Subrounded (a) elements have symmetry variants of nearly symmetrical and rounded, asymmetrical and compressed on one side, and fully laterally compressed. Cusps are slender to broader based. Basal cavity is broad and deep, with its margins close to anterior and posterior margins of cusp, expands under posterior margin below denticles. Cusp is long, reclined to recurved, smooth in rounded forms and keeled anteriorly and (or) posteriorly in variants. Basal margin is flat to extended anteriorly and posteriorly. Denticulated extension of posterior margin bears laterally compressed denticles with variation in point of origin along margin; some arise just below point of curvature of the cusp, others are along flat basal margin. No distinct process developed. Denticles are discrete, vary in number from one to four along margin, at right angles to posterior margin or reclined posteriorly.

Symmetrical (c) element is rare and has suberect cusp which is smoothly rounded and contains white matter in cusp tip. The basal margin is slightly arched. Denticles are reclined, process is directed downward.

Compressed asymmetrical (e) elements have compression along inner postero-lateral margin, some have minor inner carina and basal flare, others are smooth along basal margin. Cusp is twisted laterally and bent inward. White matter is present in cusp tip, above basal cavity, and within denticles.

Discussion — *C. primitivus* is a transitional species from *E. notchpeakensis* to *C. proavus*, and although it lacks a well-developed posterior process, its posterior margin is denticulate, and when broken, elements are indisguishable from *E. notchpeakensis*. The apparatus contains elements of variable morphology, which Bagnoli et al. (1987) did not differentiate into three morphotypes but described as approximate to Miller's (1980) rounded and compressed elements. We agree with Mens et al. (1996) that *C. primitivus* is a separate species from *C. andresi* Viira and Sergeyeva, based on the distribution of white matter, however, Szaniawski and Bengtson (1998) caution using white matter as a diagnostic feature especially depending on the state of preservation. *C. primitivus* is different from *C. andresi* Viira and Sergeyeva in the overall shape of the cusp and the nature of denticulation in which the denticles arise much higher on the cusp margin in *C. andresi* than in *C. primitivus*. Szaniawski and Bengtson (1998) suggest that *C. andresi* is derived from *Proconodontus* stock. *C. primitivus* is clearly a descendant of *E. notchpeakensis*.

Occurrence — *Cordylodus proavus* Zone to *Iapetognathus* Zone, Kechika Formation.

Material — 47 a elements, 6 c elements, 9 e elements.

Repository — GSC 119047, 119048, 119049.

Cordylodus proavus Müller, 1959

Plate 5, figs. 4–6

a element — *Cordylodus proavus* Müller, 1959, pp. 448–449, pl. 15, figs. 11, 12, 18, text-fig. 3B; Miller, 1980, pp. 19–20, pl. 1, fig. 14, text-fig. 4G; Bagnoli et al., 1987, p. 154, pl. 1, fig. 8 (see synonymy within); Barnes, 1988, figs. 13A, B; Harris et al., 1995*b*, p. 25, pl. I, fig. AE; Ji and Barnes, 1996, p. 881, fig. 12 (1, 2, 4, 5); Nicoll, 1990, p. 550, figs. 20, 21 only; Parsons and Clark, 1999, p. 6, figs. 5.24, 5.25.

c element — *Cordylodus proavus* Müller. Nowlan, 1985, p. 111, fig. 5 (12, 13, 18, 19); Nicoll, 1990, p. 550, fig. 19 (3, 4) only; Ji and Barnes, 1996, fig. 12 (3).

e element — *Cordylodus oklahomensis* Müller, 1959, p. 447, pl. 15, figs. 15–16, text-fig. 3A.

Cordylodus proavus Müller. Miller, 1980, pp. 19–20, pl. 1, fig. 15, text-fig. 4H; Bagnoli et al., 1987, p. 154, Pl. 1, fig. 9 only (see synonymy within); Buggisch and Repetski, 1987, p. 154, pl. 4, figs. 1, 2, 5, 6, pl. 5., fig.1, pl. 8, fig. 7; Barnes, 1988, fig. 13C; Nicoll, 1990, p. 550, fig. 22 (2, 4, 5) only.

Description — Apparatus of three morphotypes in which elements have distinct posterior denticulated process and rounded to keeled cusp. Basal cavity is deep and its margins are parallel to the cusp margins, extending into cusp tip beyond point of curvature of cusp and under process. Cusp tip and denticles are filled with white matter. Basal cavity is biapical in some elements.

The a elements are subround and comprise variants with a symmetry transition series. Asymmetry is attributed to lateral flexure of the posterior process. Basal outline is flat to slightly arched. Denticles are discrete and rounded.

The c element is characterized by symmetry and arching of basal margin. Cusp margins are rounded and anterior cusp margin is extended in some specimens. Posterior denticulated process contains up to five laterally compressed, discrete denticles.

The e elements are strongly laterally compressed. Posterior cusp margin is keeled more strongly than anterior, and cusp is laterally compressed along length of inner lateral posterior margin and extends along basal margin. Denticles are compressed and confluent along their bases. Inner basal margin is flared.

Discussion — Bagnoli et al. (1987) describe *C. proavus* as having a more differentiated apparatus in which elements develop a distinct posterior process than the denticulated posterior margin of *C. primitivus*. The basal cavity is a distinguishing characteristic in *C. proavus* elements because it is deep, extends parallel to the cusp margins, and its anterior outline curves convexly. The e elements are distinguished from *C. hastatus* by the development of a faint carina along the inner compressed lateral margin and flare to the basal margin, whereas the cusp of *C. hastatus* is smooth and evenly compressed along the length of the cusp. Elements within a septimembrate plan by Nicoll (1990) have been partially synonymized. No M elements are observed as he illustrated, and the P elements are synonymized as laterally compressed e elements, the Sa element is symmetrical and arched and synonymized as the c element, and the Sc, Sb, and Sd elements described by Nicoll (1990) are variants of the a element.

Occurrence — *Cordylodus proavus* Zone to *Iapetognathus* Zone, Kechika Formation.

Material — 52 a elements, 3 c elements, 6 e elements.

Repository — GSC 119056, 119057, 119058.

Genus *Eoconodontus* Miller, 1980

Type species — *Proconodontus notchpeakensis* Miller, 1969

Eoconodontus notchpeakensis (Miller, 1980)

Plate 7, figs. 1–3

Proconodontus notchpeakensis Miller, 1969, p. 438, pl. 66, figs. 21–29, text-fig. 5G

Eoconodontus notchpeakensis (Miller). Miller, 1980, pp. 22–23, pl. 1, figs. 10–12; text-figs. 3D, E (includes synonymy to 1980); Bagnoli, Barnes, and Stevens, 1987, pp. 155–156, pl. 2, figs. 5–7 (includes synonymy to 1987); Buggisch and Repetski, 1987, p. 155, pl. 6, figs. 8a, pl. 8, fig. 12 only; Viira et al., 1987, p. 148, pl. 1, figs. 9–11; Lehnert, 1995, p. 85, pl. 1, figs. 1, 2; Mens et al., 1996, pl. 2, figs. 4, 5; Harris et al., 1995*b*, p. 24, pl. I, figs. X, Y; Hein and Nowlan, 1998, p. 181, pl. 3, figs. 17–19.

Description — Elements have two morphotypes (a and e) with variants within two symmetry transition series. The a elements are subround to round, subsymmetrical to symmetrical, slender and wider based; varying degree of lateral compression, slight to strong, transitional forms compressed on one side. Oval to round in cross section, erect to reclined long cusp, deep basal cavity extends to point of curvature of cusp, cusp tip filled with white matter. Smooth surfaces, with or lacking keeled anterior and (or) posterior margin, basal margin drawn out slightly into anterior and posterior extensions, other specimens have relatively flat basal margin.

Asymmetrical, compressed (e) elements, cusp long and laterally bent, inner or concave side is carinate, giving slight lateral flare to inner side of base. Degree of compression along inner postero-lateral margin varies through a symmetry transition.

Discussion — The genus, as established by Miller (1980), is described as a bimembrate apparatus of non-geniculate, smooth elements, comprising a rounded, drepanodiform and a compressed asymmetrical, scandodiform element referred to by Bagnoli et al. (1987) as a p and q element, respectively. Some elements in which the tip of the cusp was broken have not been included because they could be *Proconodontus* sp. if white matter is lacking. Nicoll (1992) indicated that *Eoconodontus* could have a septimembrate apparatus but did not describe it. *Eoconodontus* is ancestral to lineages of *Cordylodus*, and broken specimens of *C. primitivus* and *C. proavus* can easily be mistaken for *Eoconodontus*.

Occurrence — *Eoconodontus* Zone to *Iapetognathus* Zone, Kechika Formation.

Material — 104 a elements, 22 e elements.

Repository — GSC 119040, 119041, 119042.

Genus *Iapetognathus* Landing in Fortey et al. 1982

Type species — *Pravognathus aengensis* Lindström, 1955

Discussion — The genus *Iapetognathus* was first established by Landing (in Fortey et al. 1982). Nicoll et al. (1999) define *Iapetognathus* as comprising ramiform elements with one or sometimes two lateral, denticulated processes. The cusp is compressed laterally or antero-posteriorly and may have anterior or lateral keels or carinae. The cusp becomes posteriorly recurved above the point of origin of the process. They describe *Iapetognathus* as a multielement taxon containing S and P elements and discuss its importance in defining the Cambrian–Ordovician boundary. *Iapetognathus* can be distinguished from *Cordylodus* by its denticulated processes, which are in a lateral position rather than posterior.

Iapetognathus aengensis (Lindström, 1955)

Plate 7, figs. 4, 5

Pravognathus aengensis Lindström, 1955, p. 585, pl. 5, figs. 10–13.

Iapetognathus aengensis (Lindström). Nicoll et al., 1999, pp. 44–46, pl. 1.1–1.5, 2.1–2.4, 3.1–3.4 only (contains synonymy through 1999).

Discussion — The specimens in this study are similar to those described by Nicoll et al. (1999). The Sb element is asymmetrical, its cusp recurved posteriorly. The posterior margin of the cusp is characterized by a prominent keel extending from the apex of the basal cavity to the edge of the basal margin. The deep basal cavity extends along the denticulated anterior inner-lateral process, and posterior shelf. The narrowing of the process anterior to the cusp and the keeled posterior margin are characteristics of the Sb elements. The Sc element has a wide cavity which does not narrow, and the keeled posterior margin extends to the basal margin with only a slight development of a posterior shelf. The Sd element has a prominent posterior keel and shelf, such that a posterior extension is created and it has a larger proximal denticle than in other elements. As in the Sc element, the process does not narrow anteriorly. White matter is located in the cusp tip and within the denticles.

Occurrence — *Iapetognathus* Zone, Kechika Formation.

Material — 13 Sb elements, 3 Sc elements, 3 Sd elements.

Repository — GSC 119067, 119084.

Iapetognathus fluctivagus Nicoll et al., 1999

Plate 7, figs. 6–8

Iapetognathus fluctivagus Nicoll et al., 1999, pp. 46–48, pl. 6.1–6.5, pl. 7.1–7.4 (Sb and Sc elements).

Discussion — Only two elemental morphotypes were recognized in this collection. The Sb element has an anterior outer lateral process that differentiates it from *I. aengensis*, which has an inner lateral process. The process is also strongly attached to the cusp, as its base does not constrict under the anteriormost denticle. The cusp has a keel extending to the basal margin and is laterally compressed. The keel twists outward and extends to the posterior edge of the base, which distinguished it from the Sc element. The Sc element has a reduced posterior keel. The elements lack a prominent keel and node on the anterior margin as described by Nicoll et al. (1999). The cusp tip and denticles contain white matter.

Occurrence — *Iapetognathus* Zone, Kechika Formation.

Material — 3 Sb elements, 2 Sc elements.

Repository — GSC 119081, 119082, 119083.

Iapetognathus sprakersi Landing et al., 1996

Plate 7, figs. 9, 10

Iapetognathus sprakersi Landing, Westrop, and Knox, 1996, p. 672, figs. 5.1–5.3; Nicoll et al., 1999, pp. 50–51, pl. 21.1–21.4 (Sb and Sc elements).

Discussion — *I. sprakersi* is characterized by a thin and weak attachment of the outer lateral process to the cusp and shallow basal cavity. The cusp is longer than in other elements in the collection and has an outer-lateral carina which is developed differently in the S elements. Only two morphotypes were observed, the Sb elements, which have a short outer-lateral carina, and the Sc elements, which lack the carina.

Occurrence — *Iapetognathus* Zone, Kechika Formation.

Material — 2 Sb elements, 11 Sc elements.

Repository — GSC 119085, 119086.

Iapetognathus sp.

Plate 7, fig. 11

Description — Cusp is strongly antero-posteriorly compressed and inclined posteriorly. It is broader than in other species described and lacks a keeled posterior margin, but has slight keel on anterior margin and strongly keeled lateral margins. Denticulated lateral process arises from outer anterior margin, and denticles are compressed in plane of process and are directed antero-laterally. Basal cavity is broad and deep, extending to point of recurvature of cusp, its tip is filled with white matter. Basal cavity narrows slightly under process and under a posterior-lateral extension of basal margin.

Discussion — The element is more robust than species and cannot be ascribed to an existing species.

Occurrence — *Iapetognathus* Zone, Kechika Formation.

Material — 1 element.

Repository — GSC 119066.

Family FRYXELLODONTIDAE Miller, 1980

Genus *Dzikodus* Zhang, 1998

Dzikodus tablepointensis (Stouge, 1984)

Plate 20, figs. 1–7

Remarks — Zhang (1998) erected the genus *Dzikodus* to encompass those specimens of *Polonodus* which have paired stelliplanate and unpaired pastiniplanate elements. *Polonodus* Dzik includes platforms which are stelliscaphate and pastiniscaphate. A series of ramiform elements and an oistodontiform element have been identified by Löfgren (1990).

Occurrence — *Dzikodus tablepointensis* Zone, Ospika Formation.

Material — 11 S elements, 12 P elements.

Repository — GSC 119438 to 119444.

Genus *Kallidontus* n. gen.

Type species — *Kallidontus serratus* n. gen. n. sp.

Diagnosis — Conodonts with apparatus of conical S elements and platform P elements. Elements have deeply excavated cusps which have thin walls, and cusp tip filled with white matter. Coniform elements range from simple costate forms to forms with blunt, rounded denticle along the posterior process. Sa elements have denticulate lateral cusp margins.

Platform elements have a variety of morphologies in which the nature of denticulation and ridge development varies. All P elements are broad with thin walls which bear faint to strongly developed transverse ridges. They are deeply excavated with white matter in cusp tip and denticle tips. Large Pb elements share characteristics of being large and caplike in which the base length is two times the element height. Elements are antero-posteriorly compressed and bear weak to strongly developed, equally spaced transverse ridges parallel to base, extending from base to tip, extending across anterior and posterior cusp faces. Pa elements are of two forms: one has broadly laterally expanded cusp wall which bears transverse ridges and nodose lateral ridges. Anterior cusp face convex; posterior cusp face convex. Nodose lateral processes curve in opposite directions, giving the basal margin an S-shape in oral view. In some forms, these lateral processes bifurcate distally. Cusp tip short and blunt. The second form is laterally compressed and has rounded anterior element face with nodose lateral margins. Posterior cusp face concave with transverse ridges. Cusp tip is one-third height of total element height and recurved posteriorly.

Etymology — Kalli means beautiful; dont means tooth.

Occurrence — *Scolopodus subrex* Zone to *Oepikodus communis* Zone, Kechika and Skoki formations.

Kallidontus nodosus n. gen. n. sp.

Plate 8, figs. 1–15

Diagnosis — A new species of *Kallidontus* in which the Pb element has strongly developed transverse ridges terminating laterally as strong nodes on one lateral process and as transverse nodes on the other lateral process. Coniform S elements have strongly developed nodose posterior and lateral processes.

Description — Apparatus of conical S elements and large, nodose, ridged P elements.

S elements are conical and deeply excavated. Cusp

tip filled with white matter and recurves posteriorly. Anterior cusp face rounded. Lateral cusp faces smooth and crossed by transverse ridges. Posterior process extended and denticulated from base to cusp tip. Base is wide and basal cavity deep. Basal margin slightly arched. Sa element (Pl. 8, fig. 10) bears nodose posterior process and denticulated lateral processes which curve posteriorly and are joined by transverse ridges. One morphotype observed, which may be an S element, is conical and bears costae on the anterior cusp face (Pl. 8, fig. 11). Element is stout with small cusp tip, recurved posteriorly and filled with white matter. Lateral cusp faces bear costae, and element has concentric ridges from base to tip.

Pb elements deeply excavated with thin, ridged, element walls. Cusp tip is recurved posteriorly. Anterior cusp face convex; posterior cusp face concave. Lateral processes strongly nodose and joined by strong transverse ridges from base to cusp tip. Ridges terminate laterally as strong nodes on one lateral process and as transverse nodes on the other lateral process (Pl. 8, fig. 4). In smaller Pb elements (?juvenile), the transverse ridges terminate along lateral processes which are equally developed and do not have strong transverse nodes.

Pa elements are stout and broad based. Element expands in anterior and posterior direction and constricts laterally under lateral processes. Strong transverse ridges are concentric about element and arise as nodose ridges along lateral process (Pl. 8, fig. 6). Second morphotype is one in which the element is compressed laterally and expands anteriorly and posteriorly. Anterior process is ridged, with ridges extending concentrically around elements. Posterior process bifurcates distally as two lateral processes which are ornamented by nodose denticles (Pl. 8, fig. 15).

Etymology — Named after the nodular appearance of ridges.

Discussion — *K. nodosus* differs from *K. serratus* in the development of strong nodes rather than blunt denticles along the lateral margins of the Pb elements and posterior process of S elements. The distinct feature of all elements of *K. nodosus* is the presence of well-developed transverse ridges, which are weak in *K. serratus* and barely visible in *K. princeps*.

Occurrence — *Acodus kechikaensis* Zone, Skoki Formation.

Material — 8 S elements, 10 P elements.

Repository — GSC 119352 to 119356 and GSC 119366 to 119374.

Kallidontus cf. *K. nodosus* n. gen. n. sp.

Plate 10, figs. 15–20; Plate 24, fig. 11

Discussion — A few elements from the upper Kechika and lower Skoki Formation are similar to S and P elements of *K. nodosus*. The P elements are elongate in which element height is equal to base length. Lateral cusp margins bear small denticles which are highly fused and extend from base to cusp tip. Denticles are much smaller than those of *K. serratus* and differ from the ridges developed along the lateral margins of *K. nodosus*. Cusp tip is bulbous. The P elements are more similar to *K. nodosus* than to *K. serratus* in the development of two nodose ridges on the posterior element face. Postero-lateral cusp margin is also ridged (Pl. 10, figs. 18, 20).

The S elements are conical with elongate, recurved cusp tip and deeply excavated basal cavity. Posterior and lateral cusp margins are sharp. These are similar to those of early forms of *K. serratus*. An ?Sa element (Pl. 10, fig. 15) has base expanded laterally with nodose lateral cusp margins and posterior cusp margin. Posterior cusp face is concave.

This species occurs stratigraphically higher than *K. nodosus*, and as it has some differences, mainly in the finer nodes and denticles of the P element, it is referred to as *K.* cf. *K. nodosus*.

Occurrence — *Acodus kechikaensis* Zone, Skoki Formation.

Material — 20 S elements, 4 P elements.

Repository — GSC 119332, 119333, 119334, 119375, 119376, 119377, and GSC 119509.

Kallidontus princeps n. gen. n. sp.

Plate 9, figs. 1–7

Diagnosis — A new species of *Kallidontus* with an apparatus of coniforms and platforms with thin walls joined to sharp processes. Coniforms are triangular in lateral view with three processes. Platform is broad with smooth walls which bear faint transverse ridges.

Description — Apparatus of deeply excavated coniforms (S elements) and platforms (P elements) which have thin walls. S elements are triangular in lateral view and have deep basal cavities which extend almost to cusp tip. Asymmetrical elements have one strong lateral costa extending from base to cusp tip. Anterior and posterior cusp margins sharp. Base extends posteriorly in some elements. Cusp is stout in some forms, as long as base is wide, and cusp margins extend below basal margin (Pl. 9, fig. 7).

Symmetrical (?Sa) element has strong posterior costa and two lateral costae. Cusp is slender and lateral margins project slightly below basal margin.

Platform element expands laterally and is caplike. Height of element is one-half length of base. Anterior cusp face convex; posterior cusp face unevenly concave and asymmetrical. One lateral process is longer than the other. Cusp walls bear faint transverse ridges. Base is wide and basal cavity is deep.

Etymology — Princeps means first.

Discussion — *K. princeps* appears to be the first species in this lineage, as it has the simplest morphology without the development of elaborate architecture of cusp walls and processes. The P elements bear weak transverse ridges, which become more visible in *K. serratus* and strongly developed in *K. nodosus*. The S elements in *K. princeps* have straight cusp margins, which in *K. serratus* become ornamented with nodose serrations and in *K. nodosus* are developed into nodose denticles.

Occurrence — *Scolopodus subrex* Zone to *Acodus kechikaensis* Zone, Skoki Formation.

Material — 113 S elements, 4 P elements.

Repository — GSC 119335 to 119339.

Kallidontus serratus n. gen. n. sp.

Plate 9, figs. 8–17; Plate 10, figs. 1–14

Diagnosis — A new species of *Kallidontus* in which coniforms have thin walls and three processes which have transverse ridges and nodes along one to three lateral processes. Platforms have one to three serrated to nodose processes and broad bases.

Description — Apparatus of conical S elements and platform P elements.

S elements are conical with three processes (two lateral, one posterior) connected by thin element walls. Elements are stout to elongate, with sharp lateral cusp margins that extend below basal margin. Elements are erect to proclined with weak curvature of distal portion of albid cusp tip. Lateral margins are straight edged in some elements and weakly serrated in others (Pl. 9, fig. 10). Anterior cusp face is flat to gently convex. Posterior process bears four to six rounded denticles or nodes which are short and blunt. Posterior process extends below basal margin in some elements and projects as a sharp process, directed downward. In others, the posterior process is denticulate right to end of process and expands posteriorly but does not project below basal margin (Pl. 10, fig. 1). All elements observed are asymmetrical because of unequal development of lateral processes (Pl. 10, figs. 3–5, posterior view).

Platform elements are broad based and comprise elements with a main cusp and lateral and posterior processes which extend as denticulated sheathes.

Pb element is diagnostic. It is an almost symmetrical, large platform element with two lateral denticulate processes. Element antero-posteriorly compressed. Cusp tip is slightly curved posteriorly and filled with white matter. Anterior face is convex; posterior face is broadly concave. Element has equally spaced transverse ridges parallel to the base, extending from the base to the cusp on posterior and anterior cusp faces. Lateral cusp margins bear at least seven large, blunt denticles, along each margin, which arise where transverse ridge meets lateral edge. Denticles give serrated appearance to cusp margins in posterior view (Pl. 9, fig. 8). Denticles filled with white matter. Basal cavity is deep and extends under lateral processes. Basal outline is straight.

In late Pb elements (Pl. 10, fig. 9), a posterior denticulate process is added. Element broadly expanded, and anterior and posterior cusp faces bear transverse ridges. Anterior cusp face convex; posterior cusp face convex in region between posterior process and lateral processes. Lateral processes and posterior process bear at least five broad denticles.

Pa elements are broad-based, stout elements. Cusp tip recurved posteriorly and filled with white matter. Anterior of element bears two costae which diverge from cusp tip to base and are connected by transverse ridges (Pl. 9, fig. 13). In lateral view, anterior element margin appears rounded (Pl. 9, fig. 14). Posterior process extends as denticulate process, which in some elements bifurcates distally (Pl. 9, fig. 15). Lateral element faces ornamented by transverse ridges. Element bulges laterally in anterior portion of element, but narrows posteriorly giving basal outline an S-shape (Pl. 9, fig. 15). A second form of Pa element is one in which element bulges to anterior and posterior rather than laterally. Cusp tip recurves posteriorly. Lateral processes curve in opposite directions, giving the basal margin an S-shape (Pl. 10, fig. 8). Lateral cusp margins bear a ridge extending from cusp to base. Transverse ridges cross anterior and posterior cusp face to join lateral ridges.

Etymology — Named for the serrated edges of the P elements.

Discussion — *K. serratus* appears to have arisen from *K. princeps* by the elaboration of a denticulated sheath in coniform and platform elements. Late forms of the large, broad Pb element develop a posterior process in addition to the lateral processes. There are two Pa elements which differ in the nature of expansion of the element (lateral versus antero-posteriorly) and in the bifurcation of the posterior process.

Occurrence — *Acodus kechikaensis* Zone, Skoki Formation.

Material — 126 S elements, 32 P elements.

Repository — GSC 119340 to 119351 and GSC 119357 to 119365.

New genus C, new species

Plate 29, figs. 14–17

Description — Two morphotypes (S and P) recognized which are similar to those of *Kallidontus* n. gen. S elements are simple cones with laterally compressed cusp. Anterior cusp margin rounded to sharply rounded and posterior cusp margin sharp. Inner lateral cusp face flat to concave, outer lateral face convex. Cusp erect and unornamented. Antero-basal corner sharply rounded. Base posteriorly extended. Basal margin straight; basal outline oval and expanded to outer lateral side. Basal cavity deep; cusp walls thin.

P element is diagnostic and a caplike platform. Element expands laterally and is wider than cusp height. Anterior side of element rounded with irregular surface and one prominent keel extending from base to lateral cusp margin. Posterior side of element concave. Element surface is irregular. Basal margin is irregular.

Occurrence — *Jumudontus gananda* Zone, Skoki Formation.

Material — 3 S elements, 1 P element.

Repository — GSC 119586, 119587, 119588, 119589.

Family PROCONODONTIDAE Lindström, 1970

Genus *Proconodontus* Miller, 1969

Emend. Szaniawski and Bengtson, 1998

Type species — *Proconodontus muelleri* Miller, 1969

Proconodontus muelleri Miller, 1969

Plate 12, figs. 20, 21

Proconodontus muelleri Miller, 1980, pp. 29–30, pl. 1, fig. 7, text-fig. 4C (contains synonymy to 1980); Landing, 1983, figs. 11G, H, p. 1180; Müller and Hinz, 1991, p. 56, pl. 42, figs. 1, 3–9 (includes synonymy to 1991); Hein and Nowlan, 1998, p. 180, pl. 3, fig. 9; Szaniawski and Bengtson, 1998, pp. 17–18, pl. 2, figs. 4–13 only, text figs. 4g–i only (includes synonymy to 1998).

Discussion — As described by Miller (1980), the elements of *P. muelleri* are similar to *E. notchpeakensis*, but the basal cavity extends to the tip, and elements have prominent posterior and anterior keels. Szaniawski and Bengtson (1998) described a multielement apparatus of simple cones with deep basal cavities, thin crowns, oval to round cross sections and keeled anterior and posterior margins. They describe an apparatus of three morphotypes, "comparatively slender and rounded in cross section, slightly wider and compressed from one side, and nearly geniculate," the latter of which are rare (Szaniawski and Bengtson 1998, p. 17). The compressed and nearly geniculate elements are within a transition series. The elements in this study include only specimens with intact distal cusps. The nearly geniculate elements were not observed. The rounded specimens studied are laterally compressed; asymmetrical elements are compressed along the posterior inner margin and have a slight lateral curvature of the cusp. All elements have keeled anterior and posterior margins.

Unlike *E. notchpeakensis*, the asymmetrical elements do not have an inner carina.

The *Proconodontus* lineage could be the ancestral stock for the Drepanodontidae as discussed by Szaniawski and Bengtson (1998), and appears to have given rise to one species of *Cordylodus*, *C. andresi*. From the material in this study, *C. hastatus* may also have arisen from *Proconodontus* because of strong similarity of its cusp morphology.

Occurrence — *Eoconodontus* Zone, Kechika Formation.

Material — 3 a elements, 4 e elements.

Repository — GSC 119044 and 119045.

Order BELODELLIDA Sweet, 1988

Family ANSELLIDAE Fåhræus and Hunter, 1985

Genus *Ansella* Fåhræus and Hunter, 1985

Ansella jemtlandica (Löfgren, 1978)

Plate 19, figs. 1, 2

Remarks — *Ansella jemtlandica* was described in detail by Löfgren (1978) and additional elements by Zhang (1998). The M elements were not recognized in the sparse material in this collection.

Occurrence — *Paroistodus originalis* Zone to *Amorphognathus tvaerensis* Zone, Skoki Formation and Road River Group.

Material — 8 S elements, 1 P element.

Repository — GSC 119453 and 119454.

Ansella longicuspica Zhang, 1998

Plate 23, figs. 2–4

Remarks — The elements of *A. longicuspica* differ from those of *A. jemtlandica* in having a shallower basal cavity which extends to about half of the length of the cusp rather than to the point of curvature of the cusp as in *A. jemtlandica*.

Occurrence — *Paroistodus originalis* Zone, Skoki Formation.

Material — 3 M elements, 2 S elements, 1 P element.

Repository — GSC 119613, 119614, 119615.

Family BELODELLIDAE Khodalevich and Tschernich, 1973

Genus *Stolodus* (Lindström, 1955)

Stolodus stola (Lindström, 1955)

Plate 14, figs. 26–29

Remarks — Elements are long, narrow, delicate conodonts which have a deeply excavated cusp ornamented by strong costae connecting a thin basal sheath. The cusp tip is small and erect to recurved. The apparatus plan for *S. stola* is as illustrated for the bellodellids in Sweet (1988) and Dzik (1994). However, a quinquimembrate plan as described by Albanesi (1998) includes a bicostate (a), tricostate (b), quadricostate (c), compressed (e), and possibly an f element. Of these morphotypes, a, b, and c elements have been identified among this collection.

Occurrence — *Paracordylodus gracilis* Zone, Kechika Formation

Material — 48 elements.

Repository — GSC 119246, 119247, 119248, 119249.

Genus *Walliserodus* Serpagli, 1967

Walliserodus costatus Dzik, 1976

Plate 22, fig. 17

Occurrence — *Dzikodus tablepointensis* Zone, Ospika Formation.

Material — 4 elements.

Repository — GSC 119464.

Walliserodus ethingtoni (Fåhræus, 1966)

Plate 22, figs. 19–22

Occurrence — *Paroistodus horridus* Zone, Ospika Formation.

Material — 35 elements.

Repository — GSC 119460, 119461, 119462, 119463.

?*Walliserodus nakholmensis* (Hamar, 1966)

Plate 22, fig. 18

Remarks — A few elements are tentatively assigned to *W. nakholmensis* based on the morphology of the P element.

Occurrence — *Paroistodus originalis* Zone, Ospika Formation.

Material — 10 elements.

Repository — GSC 119459.

Order PROTOPANDERODONTIDA Sweet, 1988

Family ACANTHODONTIDAE Lindström, 1970

Genus *Acanthodus* Furnish, 1938

Acanthodus uncinatus Furnish, 1938

Plate 1, figs. 1, 2

Occurrence — *Rossodus manitouensis* Zone, Kechika Formation.

Material — 59 a elements, 6 c elements, 48 e elements.

Repository — GSC 119279 and 119280.

Family CLAVOHAMULIDAE Lindström, 1970

Genus *Granatodontus* Wang, 1986

Type species — *Hirsutodontus*? *ani* Wang in Chen and Gong, 1986

Discussion — The genus was erected for small simple cones which are similar to *Hirsutodontus* but covered with a surface sculpture of fine warts, rather than sculpture of larger nodes confined to the basal region of the element.

Granatodontus cf. *G. ani* (Wang, 1986)

Plate 7, fig. 12

cf. *Hirsutodontus*? *ani* Wang in Chen and Gong., 1986, pp. 92–93, pl. 26, figs. 4–6.

cf. *Granatodontus ani* (Wang) Chen and Gong, 1986, pp. 148–149, pl. 26, fig. 8, pl. 27, figs. 1–5, 13, pl. 31, fig. 10, text-fig. 53 only (contains synonymy to 1986).

cf. *Hirsutodontus ani* (Wang) Mei, 1993, p. 18, pl. 1, fig. 7.

Discussion — A few specimens of small, simple cones with a deep basal cavity and surface sculpture of fine warts are reported. There appears to be two morphotypes. A symmetrical cap-shaped coniform element has a wide base and deep basal cavity. The asymmetrical form has a posterior carina and some compression along the posterior cusp face, but is similarly cap-shaped and covered in fine, warty sculpture.

Occurrence — *Iapetognathus* Zone to *Scolopodus sub-rex* Zone, Kechika Formation.

Material — 8 elements.

Repository — GSC 119046.

Family CORNUODONTIDAE Stouge, 1984

Genus *Macerodus* Fåhræus and Nowlan, 1978

Emend. Ji and Barnes, 1994

Type species — *Macerodus dianae* Fåhræus and Nowlan, 1978

Discussion — Ji and Barnes (1994) defined the genus as comprising a (subrounded curved) and e (compressed) elements characterized by an extremely long base and deep basal cavity.

Macerodus? n. sp. A

Plate 11, figs. 16–18

Diagnosis — A new species is tentatively assigned to *Macerodus* based on an apparatus of two morphotypes (a and e) that have a crescent-shaped outline of element with stout, broad, curved cusp.

Description — Apparatus of two morphotypes: a is subround and e is compressed. The a element is stout with broad, long base. Cusp is recurved posteriorly, giving element crescent shape. Anterior and posterior cusp margins are sharply rounded. Raised, rounded carina arises at base in medial-lateral position and fades at point of curvature of cusp. Base has triangular outline and is flared under carina.

The e element is laterally compressed with sharp cusp margins. Strong antero-lateral costa arises at base and merges with posterior cusp margin at point of curvature of cusp. Basal outline is oval.

Discussion — The specimens observed are close to *M. crassatus* but differ in the degree of asymmetry because of twisting of the cusp. The base is also not as wide as those illustrated by Ji and Barnes (1994).

Occurrence — *Rossodus manitouensis* Zone, Kechika Formation.

Material — 2 a elements, 1 e element.

Repository — GSC 119159 and 119160.

Macerodus? n. sp. B

Plate 11, figs. 12–15

Diagnosis — A new species tentatively assigned to *Macerodus* with apparatus of two morphotypes (a and e) in which elements bear one high, prominent costa on one lateral cusp face in medial or antero-lateral position.

Description — Apparatus of two morphotypes: a is subround and e is compressed. The a element is stout with broad, long base. Cusp is recurved, antero-posteriorly compressed, and flexed posteriorly. Anterior cusp margin rounded; posterior cusp margin sharp. Cusp bears costa which arises from basal margin at antero-lateral position and remains strong along length of cusp but fades near cusp tip. Costa has its highest point at basal margin, giving basal outline quadrate shape. Cusp is compressed along postero-lateral and antero-lateral flanks of costa.

The e element is laterally compressed. Cusp is broad and recurved. Cusp bears strong lateral costa which arises basally and joins anterior cusp margin distally. Posterior and anterior cusp margins sharp. Base is long and wide and has oval outline with flare posterior to costae.

Discussion — This new species shares similarities with and is likely derived from *Macerodus*? n. sp. A. It shows stronger development of the carina or costae in the a and e elements.

Occurrence — *Rossodus manitouensis* Zone, Kechika Formation.

Material — 3 a elements, 2 e elements.

Repository — GSC 119161, 119162, 119163.

Family DREPANOISTODONTIDAE Fåhræus and Nowlan, 1978

Genus *Drepanoistodus* Lindström, 1971

Emend. Ji and Barnes, 1994

Type species — *Oistodus forceps* Lindström, 1955

Discussion — The apparatus of *Drepanoistodus* was defined by Ji and Barnes (1994) as having three elemental morphotypes: a (subround, drepanodiform), c (suberect symmetrical), and e (compressed oistodontiform), forming two symmetry transition series.

Drepanoistodus amoenus (Lindström, 1955)

Plate 5, figs. 19–21

Drepanodus amoenus Lindström, 1955, p. 588, pl. 2, fig. 26 only.

Paroistodus amoenus (Lindström). van Wamel, 1974, pp. 78–79, pl. 7, figs. 8–11.

Drepanodus aff. *D. amoenus* (Lindström). Repetski, 1982, p. 19, pl. 5, fig. 8.

Discussion — *D. amoenus* has been synonymized with *P. numarcuatus* by some authors, but shows several differences. The base of the elements is much wider than in *Paroistodus*, and the basal edges are rounded. The elements in this collection are exactly as illustrated by van Wamel (1974) in which the a element bears a keeled posterior edge. There is no recessive basal margin as in *Paroistodus*, and the basal cavity is much deeper. The e element is geniculate, and the apparatus plan is like that of *Drepanoistodus* rather than *Drepanodus*.

Occurrence — *Acodus kechikaensis* Zone, Kechika Formation, to *Oepikodus communis* Zone, Skoki Formation.

Material — 60 a elements, 6 e elements.

Repository — GSC 119395, 119396, 119397.

Drepanoistodus basiovalis (Sergeeva, 1963)

Plate 23, figs. 8, 9

Remarks — The a elements of *D. basiovalis* and *D. forceps* are similar, but their c element varies slightly in the flare of the basal cavity and extension of the antero-basal corner (more flared and extended in *D. forceps*).

Occurrence — *Oepikodus communis* Zone, Skoki Formation, to *Phragmodus undatus* Zone, Road River Group.

Material — 17 a elements, 1 c element.

Repository — GSC 119571 and 119572.

Drepanoistodus bellburnensis Stouge, 1984

Plate 19, figs. 7–9

Remarks — Stouge (1984) described the species as having a compressed element which has a short base that is extended posteriorly. The compressed elements observed in this collection have a flexed posterior extension and rounded antero-basal margin. The c element is distinguished from other species by the posterior extension of the base.

Occurrence — *Paroistodus originalis* Zone, Ospika Formation.

Material — 3 a elements, 1 c element, 1 e element.

Repository — GSC 119417, 119418, 119419.

Drepanoistodus concavus (Branson and Mehl, 1933)

Plate 6, figs. 1–4

Occurrence — *Scolopodus subrex* Zone, Kechika Formation.

Material — 334 a elements, 38 c elements, 86 e elements.

Repository — GSC 119391, 119392, 119393, 119394.

Drepanoistodus expansus (Chen and Gong, 1986)

Plate 6, figs. 5–8

Drepanodus expansus Chen and Gong, 1986, pp. 136–137, pl. 48, figs. 2, 6, 9, pl. 50, figs. 1, 7, 8, 12, 13; pl. 52, fig. 13, text-fig. 44 (contains synonymy to 1986).

Discussion — Chen and Gong (1986) illustrate asymmetrical and subsymmetrical elements with a characteristic strongly flared base. Elements from this collection are assigned to this species, but have an apparatus plan of *Drepanoistodus*. The c element is diagnostic (Pl. 6, fig. 6) and is exactly as Chen and Gong (1986) illustrate and describe. The subsymmetrical element they describe as having a base which bulges on the inner lateral side is the a element.

Occurrence — *Rossodus manitouensis* Zone, Kechika Formation.

Material — 32 a elements, 16 c elements, 33 e elements.

Repository — GSC 119381, 119382, 119383, 119384.

Drepanoistodus forceps (Lindström, 1955)

Plate 19, figs. 10–12

Occurrence — *Oepikodus communis* Zone, Kechika and Skoki formations, to *Paroistodus originalis* Zone, Ospika Formation.

Material — 28 a elements, 9 c elements, 10 e elements.

Repository — GSC 119411, 119412, 119413.

Drepanoistodus minutus n. sp.

Plate 6, figs. 9–12

Drepanoistodus n. sp. Pyle and Barnes (2001), Pl. 1, fig. 10.

Diagnosis — A species of *Drepanoistodus* in which elements are characterized by small size and small base with slender, extremely elongate cusp.

Description — Apparatus of three morphotypes (a, c, e) of simple, unornamented coniforms characterized by small size and small base with slender, extremely elongate cusp.

The a element is subsymmetrical to symmetrical with elongate, erect to reclined cusp. Cusp margins sharply rounded. Base is short and expands weakly to posterior. Basal outline round; basal margin straight. Antero-basal corner is straight as cusp meets base at right angle. Basal cavity is shallow with apex located centrally.

The c element is diagnostic. Cusp is erect, long, and laterally compressed with rounded anterior and posterior cusp margins. Base is small and flared equally both posteriorly and anteriorly. Antero-basal corner concave just above basal margin. Basal margin is sinuous, arching under antero-basal corner.

The e element is compressed with sharp posterior margin and rounded anterior cusp margin. Asymmetrical element with compression along posterior margin. Antero-basal corner is straight; basal margin is straight. Base expands weakly to posterior.

Etymology — Named after its small size.

Discussion — This small, delicate apparatus is distinguished from other species of *Drepanoistodus* by the c element and the slender nature of the a and e elements, all of which have a small base.

Occurrence — *Rossodus manitouensis* Zone, Kechika Formation.

Material — 34 a elements, 14 c elements, 15 e elements.

Repository — GSC 119388, 119389, 119390.

Drepanoistodus nowlani Ji and Barnes, 1994

Plate 6, figs. 13–15

Occurrence — *Drepanoistodus nowlani* Zone, Kechika Formation.

Material — 71 a elements, 2 c elements, 9 e elements.

Repository — GSC 119385, 119386, 119387.

Drepanoistodus pervetus Nowlan, 1985

Plate 6, figs. 16–18

Occurrence — *Iapetognathus* Zone to *Rossodus manitouensis* Zone, Kechika Formation.

Material — 1311 a elements, 90 c elements, 890 e elements.

Repository — GSC 119378, 119379, 119380.

Drepanoistodus cf. *D. pitjanti* Cooper, 1981

Plate 19, figs. 13–16

Remarks — Cooper (1981) described the elements as robust and hyaline with one to five lateral costae near the posterior cusp margin. The elements in this collection are similar but exhibit a much more compressed antero-basal corner. The compressed e element bears the most pronounced posterior keeled cusp margin and a longitudinal postero-lateral groove.

Occurrence — *Dzikodus tablepointensis* Zone, Ospika Formation.

Material — 10 a elements, 1 c elements, 2 e elements.

Repository — GSC 119420, 119421, 119422, 119423.

Drepanoistodus suberectus (Branson and Mehl, 1933)

Plate 19, figs. 17–19

Occurrence — *Paroistodus originalis* Zone, Ospika Formation.

Material — 3 a elements, 1 c elements, 8 e elements.

Repository — GSC 119414, 119415, 119416.

Drepanoistodus n. sp. A

Plate 6, figs. 19–21

Description — A species of *Drepanoistodus* which is characterized by gaping base which opens widely to both inner and outer lateral sides of element. Apparatus of three morphotypes (a, c, e).

The a element is compressed with broad, erect cusp. Posterior and anterior cusp margins sharp. Lateral cusp faces smooth. Postero-basal margin sharp; antero-basal margin rounded. Basal margin has inner lateral flare and flares anteriorly.

The c element is suberect and symmetrical with compressed cusp (broken, so curvature not determined). Base opens widely laterally.

The e element is geniculate and compressed. Antero-basal margin rounded; postero-basal margin sharp. Base opens widely laterally and has prominent inner lateral flare.

Occurrence — *Acodus kechikaensis* Zone, Kechika Formation.

Material — 7 a elements, 3 c elements, 7 e elements.

Repository — GSC 119398, 119399, 119400.

Drepanoistodus n. sp. B

Plate 23, fig. 10

Drepanodus sp. 3 Serpagli, 1974, p. 45, pl. 10, figs. 7a, b, pl. 21, fig. 15.

Discussion — Serpagli (1974) described the suberect symmetrical element which has been observed in the present collection. The cusp is stout and bears an anticusp which is posterior to a characteristic notch in the short base.

Occurrence — *Oepikodus communis* Zone, Skoki Formation.

Material — 1 c element.

Repository — GSC 119573.

Drepanoistodus n. sp. C

Plate 23, figs. 11–13

Description — A new species of *Drepanoistodus* which is characterized by its compressed e element which has short base, extended posteriorly and anteriorly, with narrow basal opening and inner lateral flare. Apparatus of three elemental morphotypes (a, c, e). The a element is laterally compressed and geniculate. Cusp bears weak lateral carina which extends from cusp tip onto base. Antero-basal corner rounded; postero-basal edge sharp. The c element is symmetrical and has slender cusp and concavity in anterior cusp margin just above base. Base has inner lateral flare and extends posteriorly and anteriorly. The e element is diagnostic. It is laterally compressed and has narrow basal opening with inner lateral flare to basal margin. Antero- and postero-basal corners extend as flanges. Anterior cusp margin straight.

Occurrence — *Oepikodus communis* Zone, Skoki Formation.

Material — 1 a element, 3 c elements, 2 e elements.

Repository — GSC 119574, 119575, 119576.

Drepanoistodus n. sp. D

Plate 23, figs. 14–16

Description — A new species of *Drepanoistodus* which is close to *Drepanoistodus* n. sp. C, but has more open and flared base. The e element is diagnostic and has sinuous, flared base. Cusp is erect. Base expands anteriorly and opens anteriorly. Base has inner lateral flare. The a elements have open bases, round in profile, which also open anteriorly under anterior extension. Cusp is erect and laterally compressed. Cusp margins are sharply rounded. Anterior cusp margin concave above antero-basal corner.

Occurrence — *Jumudontus gananda* Zone, Skoki Formation.

Material — 2 a elements, 1 element.

Repository — GSC 119577, 119578, 119579.

Drepanoistodus n. sp. E

Plate 23, figs. 17–19

Description — A new species of *Drepanoistodus* with two morphotypes recognized (a and e). The e element is diagnostic and has a large, circular base which opens posteriorly and tapers anteriorly. Base has a thickened rim. Cusp is slender and erect. Cusp bears fine striae on lateral cusp faces and is laterally compressed. The a element is laterally compressed. Cusp ornamented by fine striae on medial part of lateral faces and one stronger costae which extends up cusp. Cusp is concave posterior to costae. Cusp margins sharp. Base is large and unornamented with notch in anterior basal corner.

Occurrence — *Jumudontus gananda* Zone, Skoki Formation.

Material — 8 a elements, 3 e elements.

Repository — GSC 119580, 119581, 119582.

Genus *Laurentoscandodus* Landing, Westrop, and Knox, 1996

Type species — *Oistodus*? *triangularis* Furnish, 1938

Discussion — Landing et al. (1996) established the genus to include elongate, microstriated, albid coniforms. They described subsymmetrical to symmetrical drepanodiforms (a elements), acodiforms (b elements), and suberectiforms (c elements). Compressed oistodiform (e) elements were also recognized. The morphotypes described are similar to those of *Drepanoistodus*, but there are four, not three, element morphotypes (a, b, c, and e) that are similar to early forms of *Drepanoistodus*; the e element is non-geniculate. The most distinct morphotype is the b element; all elements have a base that is triangular in lateral view, and some have longitudinal microstriae on the cusp which distinguish them from *Drepanoistodus*. Based on the present collection, two species are recognized that differ primarily in the shape of the basal outline. These two species occur within the *Rossodus manitouensis* Zone in Sections 1 and 13 and may represent morphological variants of *L. triangularis*, which, as observed in large collections from the Midwestern U.S., contains a range in variation of several elements.

Laurentoscandodus sinuosus n. sp.

Plate 11, figs. 1–5

Diagnosis — A new species of *Laurentoscandodus* characterized by the a element, which has a sinuous, flared base which expands laterally, anteriorly, and posteriorly.

Description — Apparatus of four morphotypes (a, b, c, e) in which elements are robust with elongate, longitudinally striated cusps and broad, expanded bases with shallow basal cavities and sinuous basal outline.

The a element is subround subsymmetrical to symmetrical. Cusp is erect and laterally compressed. Anterior and posterior cusp margins are sharp. Inner and outer lateral faces bear carina on base, giving lateral flare to base. Element is concave on either side of carinae. Subsymmetrical variants have basal flare on inner side only. Basal outline is sinuous. Basal cavity is large and gaping. Postero-basal margin straight; antero-basal margin convex.

The b element is laterally compressed. Anterior and posterior cusp margins are sharp. Postero-lateral margin bears a longitudinal groove and is compressed along length of posterior cusp margin, which is keeled and laterally deflected. Base flares laterally on inner and outer sides. Antero-lateral basal margin rounded; postero-basal margin straight.

The c element is suberect and symmetrical. Cusp is evenly laterally compressed along its length. Base expands slightly laterally and extends further posteriorly than anteriorly. Antero-basal corner is slightly concave.

The e element is non-geniculate and laterally compressed. Cusp is reclined and laterally flexed. Outer lateral surface is smoothly rounded with faint compression on postero-basal edge of cusp. Inner lateral surface bears even compression along cusp, but base is compressed on either side of inner lateral basal flare. Basal cavity cross section is oval.

Etymology — Named for sinuous nature of basal outline.

Discussion — This species is similar to *L.* cf. *L. triangularis* but is characterized by the flared base and more pronounced lateral deflection of the posterior keel on the b and e elements. The symmetrical element is close to that of *D. pervetus*, but can be distinguished by the longitudinal striae on the elongate cusp and the triangular profile of the base.

Occurrence — *Rossodus manitouensis* Zone, Kechika Formation.

Material — 106 a elements, 50 b elements, 13 c elements, 14 e elements.

Repository — GSC 119100, 119101, 119102, 119103.

Laurentoscandodus cf. *L. triangularis* (Furnish, 1938)

Plate 11, figs. 6–9

cf. *Oistodus*? *triangularis* Furnish, 1938, p. 330, pl. 42, fig. 22, text-fig. 1P.

cf. *Laurentoscandodus triangularis* (Furnish). Landing, Westrop, and Knox, 1996, pp. 672–674, figs. 8.12–8.29, 9.14, 9.18–9.21 (contains synonymy to 1996).

cf. *Drepanoistodus*? *pervetus* Nowlan, 1985, pp. 112–113, figs. 5.54–5.55, 6.2, 6.3 (only).

Discussion — An apparatus of four morphotypes in which all elements have a large, triangular base and weak lateral rounded costae is similar to that described by Landing et al. (1996). The oistodiform (e) element in the present collection has a much more flared and expanded base and tapered cusp than that illustrated by Landing et al. (1996). The suberect c element bears a faint longitudinal lateral carina and has an elongate antero-basal margin which is flattened. The a and b elements differ in having a posterior keel on the b element, and also a faint longitudinal carina and postero-lateral groove.

Occurrence — *Rossodus manitouensis* Zone, Kechika Formation.

Material — 37 a elements, 34 b elements, 9 c elements, 7 e elements.

Repository — GSC 119087, 119088, 119089, 119090.

Genus Paltodus *Pander, 1856*

Type species — *Paltodus subaequalis* Pander, 1856

?*Paltodus jemtlandicus* Löfgren, 1978

Plate 12, figs. 1–5

aff. *Paltodus*? *jemtlandicus* Löfgren, 1978, p. 65, pl. 4, figs. 1–3, 6, 7; Ethington and Clark, 1982, p. 75, pl. 8, figs. 10, 19.

Diagnosis — A species which is tentatively assigned to *Paltodus*, characterized by the compressed M element which has a short base with a sharp postero-basal corner and broad cusp. Apparatus of three morphotypes (S, M, P) of simple cones with triangular bases and cusps which bear lateral costae.

Description — M element is oistodiform and has broad cusp. Base is short with round antero-basal outline and sharp, straight postero-basal margin. Base flared on inner lateral face and opens laterally.

P element is compressed antero-posteriorly and bears a faint rounded carina on the posterior face. Anterior face rounded. Lateral cusp margins sharp. Base is short and has posterior flare.

S elements are erect with small triangular base. Strongly to weakly costae on outer lateral face. Costae arise above the base and fade distally. Inner lateral face smooth. Base poorly defined with rounded antero-basal margin and sharp to sharply rounded postero-basal margin.

Discussion — The M element has been illustrated previously, but the full apparatus has not been reconstructed. The assignment to *Paltodus* is questioned, but seems appropriate based on the triangular shape of the base of all elements.

Occurrence — *Scolopodus subrex* Zone, Kechika Formation.

Material — 161 M elements, 781 S elements, 88 P elements.

Repository — GSC 119290, 119291, 119292, 119293, 119294.

Genus *Paroistodus* Lindström, 1971

Type species — *Oistodus parallelus* Pander, 1856

Discussion — The apparatus of *Paroistodus* was reconstructed by Löfgren (1997) as containing seven elemental morphotypes. The genus is characterized by its geniculate M element and non-geniculate S and P elements which are all extremely laterally compressed and have a recessive antero-basal margin in all but the earliest species, *P. numarcuatus*. Löfgren (1997) makes subtle distinctions between S elements based on a large and well-preserved collection. Albanesi (1998) illustrated an apparatus of five morphotypes and discussed the evolution of *P. horridus* in which the dolabrate S and P elements are more easily distinguished. Albanesi and Barnes (2000) document the evolutionary transitions of *P. originalis* to *P. horridus*.

Paroistodus horridus (Barnes and Poplawski, 1973)

Plate 20, figs. 17–20

Cordylodus horridus Barnes and Poplawski, 1973, pp. 771–772, pl. 2, figs. 16–18.

Paroistodus horridus (Barnes and Poplawski). Albanesi, 1998, pp. 140–141, pl. 8, figs. 1–6, pl. 15, figs. 14–15, text-figs. 17–18 (contains synonymy to 1998); Albanesi and Barnes, 2000, p. 496, figs. 3, 4.

Discussion — Albanesi (1998) recognized two subspecies and illustrated five morphotypes in the apparatus. The material in the present collection is assigned to *P. horridus* sensu stricto. Five morphtypes are recognized and have been adequately described previously. The elements show elongate processes in the S and P elements, and the P elements are characterized by an alternating lateral deflection of the denticles.

Occurrence — *Paroistodus horridus* Zone, Ospika Formation.

Material — 19 M elements, 27 S elements, 26 P elements.

Repository — GSC 119455, 119456, 119457, 119458.

Paroistodus originalis (Sergeeva, 1963)

Plate 21, figs. 1–3

Occurrence — *Paroistodus originalis* Zone, Skoki and Ospika formations.

Material — 33 M elements, 102 S elements, 9 P elements.

Repository — GSC 119465, 119466, 119467.

Paroistodus parallelus (Pander, 1856)

Plate 21, figs. 4–6

Occurrence — *Acodus kechikaensis* Zone, Kechika Formation to *Paroistodus originalis* Zone, Ospika Formation.

Material — 33 M elements, 102 S elements, 9 P elements.

Repository — GSC 119414, 119415, 119416.

Paroistodus proteus (Lindström, 1955)

Plate 12, figs. 10–12

Occurrence — *Paroistodus proteus* Zone to *Oepikodus communis* Zone, Kechika Formation.

Material — 92 M elements, 523 S elements, 79 P elements.

Repository — GSC 119401, 119402, 119403, 119666.

Paroistodus n. sp.

Plate 26, figs. 1, 2

Description — Apparatus consists of some elements which are similar to *Paroistodus* in having inverted basal cavity and resemble S and P elements of genus. Antero-basal margin is recessive, and posterior margin and oral margin of base are thinly keeled. Base is expanded laterally, and cusp bears fine striae which do not extend on base. There are varying degrees of curvature of cusp and in lateral compression of elements.

Occurrence — *Jumudontus gananda* Zone, Skoki Formation.

Material — 4 elements.

Repository — GSC 119560 and 119561.

New genus A, new species

Plate 18, figs. 17–21

Diagnosis — Apparatus of S, M, and P elements in which the robust, albid elements have compression along the postero-lateral cusp margin, and P elements are diagnostic in having a keeled anterior cusp margin and inner lateral flare on the base.

Description — S elements are coniforms with erect cusps. Cusp is laterally compressed along the posterior. Posterior cusp margin is sharp. Anterior cusp margin is rounded. Base is short, about one-tenth the length of the cusp. Antero-basal corner is round; postero-basal corner is sharp. Base flared on inner lateral side.

M element is geniculate with a rounded antero-basal margin and sharp postero-basal margin. Angle between base and cusp is 30°. Cusp laterally compressed along its length. Basal cavity open downward.

P element is laterally compressed with raised and keeled anterior cusp margin. Posterior cusp margin sharp. Base flares on both lateral cusp faces. Another morphotype has less flared basal margin and less developed anterior keel.

Discussion — Elements of this new genus bear some resemblance to coniforms such as *Paltodus* and *Drepanoistodus*, so the genus is tentatively placed within the Protopanderodontidae. However, no elements resembling a symmetrical c element of the *Drepanoistodus* apparatus were observed, and the P elements are different from those of *Paltodus* in the strong development of the anterior keel on the cusp.

Occurrence — *Scolopodus subrex* Zone, Kechika Formation.

Material — 7 M elements, 9 S elements, 19 P elements.

Repository — GSC 119404, 119405, 119406, 119407, 119408.

Family ONEOTODONTIDAE Miller, 1980

Genus *Semiacontiodus* Miller, 1969

Semiacontiodus nogamii (Miller, 1969)

Plate 14, figs. 23–25

Occurrence — *Iapetognathus* Zone to *Polycostatus falsioneotensis* Zone, Kechika Formation.

Material — 35 a elements, 18 b elements, 4 c elements, 19 e elements.

Repository — GSC 119208, 119209, 119210.

Semiacontiodus cf. *S. cornuformis* (Sergeeva, 1963)

Plate 14, figs. 21, 22

Remarks — Specimens occur in the upper Kechika, and have long slender cusps that bear conspicuous fine striae on all cusp faces. Base flares to anterior and posterior, but most prominently on inner lateral side which distinguishes this species from *S. cornuformis*.

Occurrence — *Scolopodus subrex* Zone, Kechika Formation.

Material — 43 elements.

Repository — GSC 119226 and 119227.

Genus *Teridontus* Miller, 1980

Teridontus gracillimus Nowlan, 1985

Plate 15, figs. 12–14

Remarks — The specimens observed are in agreement with the description by Ji and Barnes (1994), however, a suberect symmetrical element was observed in this collection, which also has a small base and weak posterior carina.

Occurrence — *Iapetognathus* Zone to *Polycostatus falsioneotensis* Zone, Kechika Formation.

Material — 5 a elements, 7 b elements, 1 c element, 4 e elements.

Repository — GSC 119201, 119202, 119203.

Teridontus nakamurai (Nogami, 1967)

Plate 15, figs. 17–19

Occurrence — *Cordylodus proavus* Zone to *Polycostatus falsioneotensis* Zone, Kechika Formation.

Material — 195 a elements, 65 b elements, 12 c element, 36 e elements.

Repository — GSC 119198, 119199, 119200.

Teridontus obesus Ji and Barnes, 1994

Plate 15, figs. 20–22

Occurrence — *Iapetognathus* Zone, Kechika Formation.

Material — 21 a elements, 13 b elements, 10 c elements, 8 e elements.

Repository — GSC 119204, 119205, 119206.

Genus *Utahconus* Miller, 1980

Type species — *Paltodus utahensis* Miller, 1969

Emended diagnosis — The apparatus of *Utahconus* contains three to four elemental morphotypes characterized by compression along the cusp and round to triangular basal outline. The a elements are rounded and subsymmetrical simple cones with compression along the posterior cusp margin and a rounded basal outline in lateral view. The symmetrical c elements are antero-posteriorly compressed with sharp lateral cusp margins and a rounded basal cavity. The e elements are variable but have an evenly compressed cusp along its length. The b elements are present in *U. utahensis* but not in *U. longipinnatus* and are laterally compressed morphotypes.

Discussion — Ji and Barnes (1994) described the apparatus of *U. longipinnatus* as having three morphotypes in which the a elements are unicostate, asymmetrical subround with lateral and (or) posterior costae, and the e elements are bicostae and postero-laterally compressed. A bimembrate apparatus plan containing a and e elements occurs in early species as originally described by Miller (1980). Suberect symmetrical c elements which have two lateral costa and one faint posterior carina are recognized by Ji and Barnes (1994). The e elements are variable and comprise three morphotypes in *U. utahensis*.

Utahconus longipinnatus Ji and Barnes, 1994

Plate 17, figs. 1–3

Utahconus longipinnatus Ji and Barnes, 1994, pp. 66–67, pl. 25, figs. 1–4, 7, 8 only, text-fig. 38A (see synonymy within).

Scalpellodus longipinnatus (Ji and Barnes). Landing et al., 1996, pp. 675–676, figs. 7.9, 7.10, 7.12–7.19, 7.24, 9.22, 9.23 only.

Discussion — The elements in this study are in agreement with the Ji and Barnes (1994) description, except for the c (suberect symmetrical) element in which the basal cavity is oval. The c element of *U. longipinnatus* is rare, and has a rounded base and smooth anterior cusp face. Specimens of *U. longipinnatus* which occur with *Variabiloconus* are distinguished from it by lacking grooves and a prominent posterior carina and wide base. *U. longipinnatus* has a triangular to round basal cross section and is characterized by strong antero-posterior compression in bicostate forms and antero- or postero-lateral compression in unicostate forms. The apparatus lacks the compressed e elements in the *U. utahensis* apparatus but shares the characteristics of the genus such as the compression of the cusp and the rounded base. The elements of *U. longipinnatus* are recognized by their robustness and elongate cusps.

Occurrence — *Rossodus manitouensis* Zone, Kechika Formation.

Material — 450 a elements, 42 c elements, 266 e elements.

Repository — GSC 119164, 119165, 119166.

Utahconus utahensis (Miller, 1980)

Plate 16, figs. 17–23

Paltodus utahensis Miller, 1969, p. 436, text-fig. 5F, pl. 63, figs. 33–40.

Utahconus utahensis (Miller). Miller, 1980, pp. 35–36, figs. 3B, C, F, G, pl. 2, figs. 1, 2 (contains synonymy to 1980); Wang, 1986, p. 242, pl. 2, figs. 1, 2, pl. 6, fig. 6, pl. 8, fig. 19, pl. 11, figs. 1, 8, pl. 13, figs. 19, 20; Wang in Chen and Gong, 1986, pp. 195–197, pl. 44, figs. 2–6, 10–17, pl. 51, figs. 2, 10, 13–14 only, text-fig. 82 (in part); Orndorff, 1988, p. A16, pl. 2, figs. 22, 23; Landing, 1993, p. 6, figs. 4.11–4.15.

Utahconus aff. *U. utahensis* (Miller). Nowlan, 1985, pp. 117–118, figs. 5.45–5.52, 9.1, 9.3.

Description — Apparatus of four elemental morphotypes of small, simple cones with shallow, rounded basal cavities and compressed cusps. The a, b, and c elements form the first symmetry transition series, and three compressed e elements form the second series.

The a elements are subsymmetrical forms which vary in degree of postero-lateral compression. Unicostate forms have rounded anterior cusp margin and sharply keeled posterior cusp margin. Cusp is compressed along inner lateral face; outer lateral face is rounded. Basal cavity is right triangle in lateral view, apex anterior. Postero-basal margin is rounded. Basal cross section round to oval. Bicostate forms bear sharp anterior and posterior cusp margins and develop inner lateral carina.

The b element is laterally compressed with sharply rounded antero-lateral and postero-lateral cusp margins. Basal cross section is oval; basal cavity is right triangle in lateral view. Posterior basal corner is drawn out slightly.

The c element is symmetrical and erect with long, slender cusp. Cusp curves posteriorly. Anterior cusp face is rounded and smooth. Posterior cusp face is concave and bears faint costae from cusp tip to base. Lateral cusp margins are sharp. Basal cavity is shallow, apex is centrally located. Basal cross section is round. Base is expanded slightly posteriorly to produce posterior carina.

Three compressed, non-geniculate e elements vary from erect to proclined. All have laterally compressed cusps with sharp posterior cusp margins. The most common are those with even compression above the basal portion (Pl. 16, fig. 18). Anterior and posterior cusp margins are sharp. Basal cavity is extremely narrow, oval, and shallow and is right triangle in lateral view. Another variant (Pl. 16, fig. 17) has postero-lateral compression and bears inner lateral carina. Anterior cusp margin rounded. Posterior cusp margin is sharp. Base is more rounded and extremely shallow.

The third e element (Pl. 16, fig. 23) is non-geniculate with shallow basal cavity. Cusp is compressed postero-laterally and is flexed laterally. Posterior cusp margin is sharp, anterior margin is sharply rounded. Anterior and posterior basal margin sharp; basal cavity is oval; inner lateral cusp face flared basally. Basal cavity shallow and narrow and close to anterior cusp margin.

Discussion — The apparatus of *U. utahensis* has previously been described as bimembrate by Miller (1980) who illustrates unicostate (a) and bicostate (e) elements. This collection has allowed the multielement reconstruction of an apparatus which contains the ancestor to *Rossodus*. There is an abundance of compressed elements in the apparatus, which perhaps suggests there were more pairs of the three e elements. Two pairs were subsequently lost in *U. longipinnatus*.

Occurrence — *Iapetognathus* Zone to *Rossodus manitouensis* Zone, Kechika Formation.

Material — 334 a elements, 141 b elements, 90 c elements, 843 e elements.

Repository — GSC 119105, 119106, 119107, 119108, 119109, 119110, 119111.

Family PROTOPANDERODONTIDAE Lindström, 1970

Genus *Anodontus* Stouge and Bagnoli, 1988

Anodontus longus Stouge and Bagnoli, 1988

Plate 23, fig. 1

Remarks — Stouge and Bagnoli (1988) described the apparatus of *Anodontus* as forming a symmetry transition series of subrounded antero-posteriorly compressed elements, asymmetrical elements, and laterally compressed elements. In this collection, a few subround (a) elements and compressed (e) elements were observed.

Occurrence — *Jumudontus gananda* Zone to *Tripodus laevis* Zone, Skoki Formation.

Material — 6 a elements, 5 e elements.

Repository — GSC 119511.

Genus *Colaptoconus* Kennedy, 1994

Emend. Ji and Barnes, 1994

Type species — *Scolopodus quadraplicatus* Branson and Mehl, 1933

Discussion — The genus contains up to five morphotypes of simple cones which have coarse carinae and deep grooves as described by Ji and Barnes (1994). Kennedy (1980) originally assigned symmetrical and asymmetrical cones to *Glyptoconus*, but provided the replacement name, *Colaptoconus*, because *Glyptoconus* was a junior homonym of a gastropod taxon (Kennedy 1994).

Colaptoconus bolites (Repetski, 1982)

Plate 3, figs. 1–4

Occurrence — *Scolopodus subrex* Zone, Kechika Formation.

Material — 75 a elements, 3 c elements, 2 e elements.

Repository — GSC 119299, 119300, 119301, 119302.

Colaptoconus floweri (Repetski, 1982)

Plate 3, figs. 5–8

Occurrence — *Scolopodus subrex* Zone, Kechika Formation.

Material — 127 a elements, 10 b elements, 8 c elements, 2 e elements.

Repository — GSC 119295, 119296, 119297, 119298.

Colaptoconus greypeakensis n. sp.

Plate 3, figs. 9–14

Diagnosis — A new species of *Colaptoconus* in which apparatus consists of four morphotypes (a, b, c, and e) of strongly costate and grooved simple cones with wide and expanded bases.

Description — The a element bears strong costae on lateral cusp faces and grooved posterior cusp face. Anterior cusp face rounded. Cusp is erect and twice as long as base is wide. Lateral cusp faces may bear smaller costae in grooves flanking major costae and arise basally and fade distally at point of curvature of cusp. Elements are symmetrical to asymmetrical, resulting from twisting and compression of cusp. Base is wide and round to oval in outline. Base expands under lateral costae which extend to base.

The b element is transitional and postero-laterally compressed. Antero-lateral and posterior cusp margins are sharp. Inner lateral cusp face bears strong central costae flanked by two grooves. Basal profile sinuous, as costae project below basal margin as sharply rounded flanges.

The c element is suberect and symmetrical. Lateral cusp margins sharp. Anterior cusp face rounded; posterior cusp face concave with two major costae flanked by two deep grooves. Central costae are separated by groove which contains smaller costae that arise basally and fade distally at point of curvature of cusp. Base is wide and expands laterally. Basal margin flares under posterior costae.

The e element is laterally compressed and nongeniculate. Antero-lateral and posterior cusp margins are sharp. Inner lateral cusp face bears one major costae flanked by a secondary postero-lateral costa which fades distally. Base is oval in outline and expanded laterally.

Etymology — Named after Grey Peak, the section locality where this new species is most abundant.

Discussion — The apparatus plan is similar to that of *C. floweri*, but the elements are more robust with stronger costae and an expanded base.

Occurrence — *Scolopodus subrex* Zone, Kechika Formation.

Material — 114 a elements, 31 b elements, 3 c elements, 18 e elements.

Repository — GSC 119303, 119304, 119305, 119306, 119307.

Colaptoconus sp.

Plate 3, figs. 15, 16

Discussion — Limited material is assigned to *Colaptoconus* based on the rounded lateral costae and concave posterior cusp face that bears strong, rounded posterior carina. The symmetrical c elements have a very shallow cavity, and the base is rounded and expanded antero-posteriorly. Anterior cusp face is rounded with a weak carina from tip to base. Elements are spear-head shaped. Smith (1991) assigned symmetrical elements to a form genus, *Aloxoconus*. Herein, the interpretation of Albanesi (1998) is followed by assigning the symmetrical elements to *Colaptoconus* sp. rather than to *Polycostatus*, which has a deeper basal cavity, or to the form genus of *Aloxoconus*.

Occurrence — *Rossodus manitouensis* Zone, Kechika Formation.

Material — 74 elements.

Repository — GSC 119183, 119184.

Genus *Eucharodus* Kennedy, 1980

Eucharodus parallelus (Branson and Mehl, 1933)

Plate 23, fig. 20

Occurrence — *Scolopodus subrex* Zone to *Oepikodus communis* Zone, Kechika and Skoki formations.

Material — 130 elements.

Repository — GSC 119510.

Genus *Juanognathus* Serpagli, 1974

Type species — *Juanognathus variabilis* Serpagli, 1974

Discussion — Serpagli (1974) originally defined the genus as having an apparatus of elements within a transition series in which elements bear two lateral keel-like costae on a twisted cusp in which one costa projects beyond the base. Albanesi (1998) defined the apparatus as quinquimembrate, containing a, b, c, e, and f elements, however f elements were not observed in this collection.

Juanognathus jaanussoni Serpagli, 1974

Plate 24, figs. 5–8

Juanognathus jaanussoni Serpagli, 1974, pp. 50–51, pl. 11, figs. 8a–12c, pl. 23, figs. 1a–5b (contains synonymy to 1974); Lehnert, 1995, pp. 92–93, pl. 3, figs. 2, 3 (contains synonymy to 1995); Albanesi, 1998, p. 125, pl. 5, figs. 4–7, 9, text-fig. 13 (only).

Discussion — Albanesi (1998) proposed a quinquimembrate apparatus, however, the f element is not observed in this collection. The other elements are similar to those previously described, and some have a basal body attached. *J. jaanussoni* differs from *J. variabilis* in that only the antero-lateral costa projects beyond the base as a process. The elements are also more slender than those of *J. variabilis*. The compressed element may be similar to the f element illustrated by Albanesi (1998) and has a similar morphology to the e element of *J. variabilis* but lacks the prominent postero-lateral costa.

Occurrence — *Oepikodus communis* Zone to *Jumudontus gananda* Zone, Skoki Formation.

Material — 61 a/b elements, 4 c elements, 4 e elements.

Repository — GSC 119556, 119557, 119558, 119559.

Juanognathus variabilis Serpagli, 1974

Plate 24, figs. 1–4

Juanognathus variabilis Serpagli, 1974, pp. 49–50, pl. 11, figs. 1a–7c, pl. 22, figs. 6–17, text-fig. 8; Albanesi, 1998, p. 126, pl. 5, figs. 10–14 (contains synonymy to 1998).

Reutterodus andinus Serpagli, 1974, pp. 78–81, pl. 17, figs. 7a, b, 9a–d; pl. 28, figs. 4, 8, 9a, b, text-fig. 19 (left) only; Stouge and Bagnoli, 1988, p. 138, pl. 10, fig. 8; Rao and Tortello, 1998, p. 150, pl. 3, fig. 9 (coniform element).

?*Reutterodus andinus* Serpagli. Ethington and Clark, 1982, p. 91, pl. 10, fig. 18.

Reutterodus andinus? Serpagli. Repetski, 1982, pl. 18, fig. 7, pl. 19, figs. 1–3.

Discussion — The apparatus is close to that illustrated by Albanesi (1998) but has some differences in that an f element is not included in the apparatus. The most important aspect of this apparatus reconstruction is the inclusion of the conelike element of *Reutterodus andinus* as the compressed e element of *J. variabilis*. Serpagli (1974) reported that the conelike element of *R. andinus* occurred alone, without the denticulated elements of the apparatus. The other reported occurrences listed in the synonymy are also isolated, but do occur in beds with *J. variabilis*. This element is very similar in shape and preservation to those of *J. variabilis* and seems more likely to fit in its apparatus than with the denticulated elements of *R. andinus*. As Ross et al. (1997) defined a zone based on the occurrence of this distinctive element, the proposal that it is part of the *J. variabilis* apparatus has implications to the zonation of the Middle Arenigian.

Occurrence — *Oepikodus communis* Zone to *Jumudontus gananda* Zone, Skoki Formation.

Material — 227 a/b elements, 39 c elements, 119 e elements.

Repository — GSC 119552, 119553, 119554, 119555.

Genus *Parapanderodus* Stouge, 1984

Type species — *Protopanderodus asymmetricus* Barnes and Poplawski, 1973

Discussion — The apparatus plan of *Parapanderodus* was established by Stouge (1984) as comprising slender, simple cones which have a posterior groove and are finely striated. Ji and Barnes (1994) described the apparatus plan as having two to four morphotypes (a, b, c, and e), and Albanesi (1998) described f elements in the apparatus of *P. paracornuformis*.

Parapanderodus striatus (Graves and Ellison, 1941)

Plate 25, figs. 17–22

Drepanodus striatus Graves and Ellison, 1941, p. 11, pl. 1, figs. 3, 12.

Parapanderodus striatus (Graves and Ellison). Smith, 1991, pp. 49–52, figs. 28a–f, 29a–d, 30 (contains

synonymy to 1991); Ji and Barnes, 1994, pp. 49–50, pl. 21, figs. 1–10, text-fig. 31A; Lehnert, 1995, pp. 106–107, pl. 1, fig. 15, pl. 2, figs. 7–10, pl. 8, figs. 12, 13, pl. 9, fig. 8, pl. 12, fig. 12, pl. 20B, figs. 10, 11; Albanesi, 1998, p. 117, pl. 7, fig. 27.

Semiacontiodus potrerillensis Albanesi, 1998, pp. 134–135, pl. 12, figs. 1–7, 34, text-fig. 1 (contains synonymy to 1998).

Discussion — The elements of *P. striatus* bear fine longitudinal striae which extend from the cusp tip to the basal margin. The cusp is elongate, and elements vary through a symmetry transition. The a and b elements show a range of morphologies in which some bear longitudinal grooves on the lateral cusp faces in addition to the posterior sulcus. The apparatus reconstruction by Smith (1991), who observed fused clusters of the elements, is followed. The compressed elements are like those illustrated as t elements by Smith (1991) rather than those illustrated as e elements by Ji and Barnes (1994).

Occurrence — *Scolopodus subrex* Zone, Kechika Formation, to *Tripodus laevis* Zone, Skoki Formation.

Material — 84 a elements, 14 b elements, 2 c elements, 7 e elements.

Repository — GSC 119512, 119513, 119514, 119515, 119516, 119517.

Genus *Planusodus* n. gen.

Type species — *Planusodus gradus* n. gen. n. sp.

Diagnosis — Coniform elements which are laterally compressed with broad cusp that bears lateral costa on each lateral cusp face. Elements have thin, keeled posterior and anterior cusp margins. Basal cavity, in lateral view, has concave anterior margin and apex which lies close to anterior cusp margin at point of curvature of cusp. Cusp tip filled with white matter. Apparatus of four morphotypes (a, b, c, and e elements) with a plan similar to that of *Protopanderodus*.

Etymology — Plano means flat.

Planusodus gradus n. gen. n. sp.

Plate 26, figs. 9–18

Diagnosis — Elements are broad, flat, costate cones with sharp anterior and posterior cusp margins. Basal cavity is shallow and in lateral view has concave anterior margin and apex which lies close to anterior cusp margin at point of curvature of cusp.

Description — Apparatus of four morphotypes (a, b, c, and e elements) comprising simple cones which bear a lateral costa on each cusp face. All morphotypes have a basal cavity with diagnostic shape described above.

The a element is subsymmetrical and laterally compressed. Lateral cusp faces bear a lateral costa near medial position, flanked by antero-lateral groove from base to tip. Anterior cusp margin sharply keeled comprising two keels: one arises on outer lateral cusp face near antero-basal corner, the other arises on inner lateral face at antero-basal corner. Anterior keels are closely spaced along anterior cusp margin basally and merge near cusp tip. Posterior cusp margin is an extremely thin keel. Base is short with straight basal margin and oval basal outline. Anterior basal corner is rounded. Base expands weakly to posterior. In late form (Pl. 26, figs. 14, 15), costae are sharper and cusp margins more strongly developed and thinly keeled. Cusp also is recurved rather than erect, and the base extends more posteriorly compared to early forms.

The b element shares all characteristics described for the a element above but is asymmetrical because the lateral costae are in postero-lateral position rather than medial.

The c element shares all characteristics of the a elements but is symmetrical, with both lateral costae in a medial position on the lateral cusp faces (Pl. 26, fig. 16).

The e element is more compressed than other morphotypes. Cusp is recurved. Anterior and posterior cusp margins are sharply keeled. Lateral cusp faces each bear weakly developed carinae which extend from base to cusp tip. Basal margin is straight and antero-basal corner is rounded. In late form (Pl. 26, figs. 17, 18), cusp margins are more strongly keeled. Lateral carinae are more prominent and flanked by compression on antero- and postero-lateral cusp face. Base bears inner lateral flare under carina.

Etymology — Gradus means step, referring to steplike appearance of sharp posterior keel meeting the lateral costae.

Discussion — This new genus has a protopanderodontid apparatus plan and has therefore been placed in the Family Protopanderodontidae. The early and late forms may represent separate species, but more material is needed to ascertain this. The elements are different from those of *Protopanderodus* in the shape of the basal cavity and nature of the thinly keeled cusp margins. *Planusodus* also does not have the characteristic notch in the base found in most species of *Protopanderodus*.

Occurrence — *Oepikodus communis* Zone to *Jumudontus ganada* Zone, Skoki Formation.

Material — 36 a/b elements, 23 c elements, 16 e elements.

Repository — GSC 119590, 119591, 119592, 119593, 119594, 119595, 119596.

Genus *Polycostatus* Ji and Barnes, 1994

Type species — *Acodus oneotensis* Furnish, 1938

Discussion — The apparatus of *Polycostatus* consists of three morphotypes characterized by multicostate cusps. The a (subround) elements have two lateral and one posterior costa, the c (suberect symmetrical) elements have a characteristic posterior carina with a median groove, and the e elements (postero-laterally compressed) vary in having or lacking a posterior costa.

Polycostatus falsioneotensis Ji and Barnes, 1994

Plate 12, figs. 13–15

Remarks — As defined by Ji and Barnes (1994), the apparatus comprises robust, multicostate elements with a deep basal cavity. The genus is distinguished from *Variabiloconus* by being multicostate, and the c elements are much more compressed antero-posteriorly.

Occurrence — *Polycostatus falsioneotensis* Zone, Kechika Formation.

Material — 12 a elements, 4 c elements, 2 e elements.

Repository — GSC 119191, 119192, 119193.

Polycostatus cf. *P. sulcatus* (Furnish, 1938)

Plate 12, figs. 16–19

cf. *Scolopodus sulcatus* Furnish, 1938, p. 334, pl. 41, figs. 14, 15.

cf. *Polycostatus sulcatus* (Furnish). Ji and Barnes, 1994, pp. 52–53, pl. 15, figs. 20–29, text-fig. 32A (contains synonymy to 1994).

Discussion — *P. sulcatus*, as defined by Ji and Barnes (1994), contains three morphotypes, a (subround), c (suberect, symmetrical), and e (compressed), of robust, multicostate coniform elements. The elements are distinguished by having more than four major costae, and the c element is distinct in having an anterior costa. Elements in this material are close to those described by Ji and Barnes (1994), but the c element lacks a broad posterior carina and rather has a groove flanked by two sharp longitudinal carinae and two strong lateral keels. The posterior face is concave, and the posterior part of the base has a rounded, liplike expansion. The e element has a more defined base than those illustrated by Ji and Barnes (1994) and bears fainter costae. The a elements do not have as an expanded base but are sharply costate. The apparatus may have a symmetrical b element with two strong costae on the anterior lateral margin. The anteriormost costa on the inner and outer lateral cusp face are sharp basally but smoothly curve into the anterior cusp margin.

Occurrence — *Rossodus manitouensis* Zone, Kechika Formation.

Material — 11 a elements, 5 b elements, 2 c elements, 4 e elements.

Repository — GSC 119194, 119195, 119196, 119197.

Genus *Protopanderodus* Lindström, 1971

Protopanderodus calceatus Bagnoli and Stouge, 1996

Plate 21, figs. 10–12

Remarks — Although not all of the elemental morphotypes are present in this collection, those that are unicostate to bicostate with a small, hyaline base are assigned to *P. calceatus*. The species differs from *P. varicostatus* by having fewer costae and by having a well-developed antero-lateral groove.

Occurrence — *Dzikodus tablepointensis* Zone, Ospika Formation.

Material — 38 a/b elements, 5 c elements, 2 e elements, 1 f element.

Repository — GSC 119481, 119482, 119483.

Protopanderodus cooperi (Sweet and Bergström, 1962)

Plate 21, figs. 13, 14

Remarks — *P. cooperi* is characterized by having one strong costa on one or two lateral sides and by its basal profile. The aboral margin of the base bears a notch near the antero-basal corner, which is compressed.

Occurrence — *Dzikodus tablepointensis* Zone, Ospika Formation.

Material — 21 a/b elements, 14 c elements, 3 e elements, 1 f element.

Repository — GSC 119479 and 119480.

Protopanderodus gradatus Serpagli, 1974

Plate 21, figs. 15–17

Occurrence — *Oepikodus communis* Zone to *Paroistodus originalis* Zone, Skoki and Ospika formations.

Material — 79 a/b elements, 21 c elements, 17 e elements.

Repository — GSC 119475, 119476, 119477.

Protopanderodus leonardii Serpagli, 1974

Plate 27, figs. 5–7

Remarks — *P. leonardii* has an apparatus containing simple cones with a short base and less deflection of the anterior edge of the element compared to other species of *Protopanderodus*. The elements have rounded to sharp costae which do not reach the basal margin.

Occurrence — *Oepikodus communis* Zone to *Paroistodus originalis* Zone, Skoki and Ospika formations.

Material — 87 a/b elements, 29 c elements, 17 e elements, 2 f elements.

Repository — GSC 119624, 119625, 119626.

Protopanderodus rectus (Lindström, 1955)

Plate 27, fig. 8

Occurrence — *Oepikodus communis* Zone, Skoki Formation.

Material — 8 a elements, 2 c elements, 3 e elements.

Repository — GSC 119627.

Protopanderodus robustus (Hadding, 1913)

Plate 21, figs. 19–22

Occurrence — *Paroistodus originalis* Zone, Ospika Formation.

Material — 26 a/b elements, 2 c elements, 3 e elements.

Repository — GSC 119471, 119472, 119473, 119474.

Protopanderodus varicostatus (Sweet and Bergström, 1962)

Plate 21, fig. 23

Remarks — *P. varicostatus* has been synonymized with *P. gradatus* by Albanesi (1998), but as illustrated in Dzik (1994), the former has a wider base with more alate external margins and a more sinuous basal profile. Not all of the elements of the apparatus are present in this collection, but the a–b elements are distinctive.

Occurrence — *Paroistodus originalis* Zone to *Dzikodus tablepointensis* Zone, Ospika Formation.

Material — 8 elements.

Repository — GSC 119478.

Genus Scolopodus Pander, 1856

Emend. Ji and Barnes, 1994

Type species — *Scolopodus sublaevis* Pander, 1856

Discussion — Pander's (1856) form species *Scolopodus* was redescribed by Fåhræus (1982), and Baltoscandian forms are mainly from the *Oepikodus evae* Zone and higher. Lindström (1955, 1971) and van Wamel (1974) also emended the genus, but the species they figured occur much higher than the first occurrence in this collection. Herein, the emended diagnosis of Ji and Barnes (1994) is followed in which costate, simple cones comprise an apparatus consisting of a, b, and e elements associated with rarer c and f elements. The elements may have an anterior or posterior keel such as in the compressed morphotypes, but have a rounded cross section and bear fine to coarse striae.

Scolopodus amplus n. sp.

Plate 28, figs. 1–7

Diagnosis — A new species of *Scolopodus* in which multicostate, robust, hyaline elements have extremely large, expanded bases.

Description — An apparatus of four morphotypes (a, b, c, e) and perhaps and ?f element in which the multicostate, robust, hyaline elements have an extremely large, expanded base. Basal margin is well defined by smooth, thickening around margin, producing a rim. Costae are strongly developed and directed posteriorly.

The a element is rounded and symmetrical. Cusp is recurved and bears coarse costae that extend from cusp tip but do not reach basal margin. Postero-lateral cusp face is smooth. Antero-basal corner rounded. Base is expanded posteriorly and basal cavity is wide.

The b element is similar to the a element but is asymmetrical because of twisting of cusp. Rounded base is not as extended posteriorly but is circular and opens widely to posterior.

The c element is suberect and symmetrical. Posterior cusp face bears three coarse costae separated by deep grooves. Anterior cusp face is smooth. Wide base is expanded laterally and has oval cross section.

The e element is laterally compressed and nongeniculate. Cusp bears three coarse costae on inner lateral face and multiple costae on outer antero-lateral face. Base is expanded posteriorly and has oval cross section.

The ?f element is suberect and asymmetrical because of twisting of cusp. It is multicostate and has a narrow, oval base which is slightly expanded posteriorly.

Etymology — Named for its large size (amplus = large).

Discussion — This new species of *Scolopodus* differs from *S. subrex* and *S. krummi* in having additional morphotypes (c and ?f) in its apparatus. *S. amplus* has coarser costae and expanded base in contrast to finer costae and smaller base in *S. subrex* and *S. krummi*.

Occurrence — *Oepikodus communis* Zone, Kechika and Skoki formations.

Material — 12 a/b elements, 2 c elements, 4 e elements, 1 f element.

Repository — GSC 119518, 119519, 119520, 119521, 119522, 119523.

Scolopodus krummi (Lehnert, 1995)

Plate 14, figs. 12–14

Scolopodus cornutiformis Branson and Mehl. Ethington and Clark, 1965, p. 200, pl. 1, fig. 12.

Scolopodus aff. *cornutiformis* Branson and Mehl. Barnes and Poplawski, 1973, p. 786, pl. 1, figs. 9–10.

Scolopodus cf. *S. cornutiformis* Branson and Mehl. Tipnis et al., 1978, pl. 2, figs. 14, 16.

Scolopodus rex Lindström. Serpagli, 1974, pp. 70–71, pl. 17, figs. 1a–3b, pl. 28, fig. 10; Repetski, 1982, pp. 50–51, pl. 23, fig. 6.

Tropodus? *krummi* Lehnert, 1995, pp. 130–131, pl. 5, figs. 7, 9, 13–19.

Scolopodus krummi (Lehnert), Albanesi, 1998, pp. 131–132, pl. 12, figs. 18–26, text-fig. 15.

?*Walliserodus ethingtoni* (Fåhræus). Ethington and Clark, 1982, pp. 116–117, pl. 13, fig. 16 only (e element).

Scolopodus rex paltodiformis Lindström. Repetski, 1982, p. 51, pl. 23, figs. 8, 10, 11 (e element).

Discussion — *Scolopodus krummi* contains costate elements in which the costae arise above the basal margin and are finer than in *S. subrex*. The base is less pronounced than in *S. subrex* and the cusp is more tapered. As in *S. subrex*, three morphotypes were observed: a, b, and e.

Occurrence — *Acodus kechikaensis* Zone to *Oepikodus communis* Zone, Kechika and Skoki formations.

Material — 126 a elements, 110 b elements, 37 e elements.

Repository — GSC 119317, 119318, 119319.

Scolopodus subrex Ji and Barnes, 1994

Plate 14, figs. 15–20

Scolopodus rex Lindström. Lee, 1970, p. 334, pl. 8, figs. 8, 9; Lee, 1975, p. 89, pl. 2, fig. 13, text-fig. 4I.

Scolopodus cf. *S. rex* Lindström. Ji and Barnes, 1996, p. 883, figs. 14.9–14.18.

Scolopodus subrex Ji and Barnes, 1994, pp. 58–59, pl. 19, figs. 1–15, text-fig. 35A.

Discussion — *Scolopodus subrex* has been defined by Ji and Barnes (1994) as containing up to five elemental morphotypes and is the name for the Midcontinent Realm form of *Scolopodus*, which occurs much earlier than *S. rex* in Baltoscandia. The main differences are the more robust nature compared to *S. rex*, reported by Lindström (1955, 1971), van Wamel (1974), and Löfgren (1994), which occur in the *Prioniodus elegans* to *Paroistodus originalis* zones. The elements in this collection are distinctive and characterized by their strong, coarse costae which extend from the basal margin to the cusp tip, small but defined base, and slender cusp. However, of the five morphotypes illustrated by Ji and Barnes (1994) only the a and b elements are observed, and a non-geniculate e element with a compressed and twisted cusp is present. There appear to be early and late forms of the species in which the early forms have less numerous and distinct costae.

Occurrence — *Scolopodus subrex* Zone, Kechika Formation.

Material — 454 a elements, 137 b elements, 41 e elements.

Repository — GSC 119284, 119285, 119286, 119287, 119288, 119289.

Genus *Striatodontus* Ji and Barnes, 1994

Type species — *Striatodontus prolificus* Ji and Barnes, 1994

Discussion — Ji and Barnes (1994) defined the genus as having up to five morphotypes of simple cones which bear longitudinal striae and well-developed posterior grooves.

Striatodontus prolificus Ji and Barnes, 1994

Plate 15, figs. 1–4

Occurrence — *Rossodus manitouensis* Zone, Kechika Formation and Road River Group of Cassiar Terrane.

Material — 79 a elements, 21 c element, 17 e elements.

Repository — GSC 119171, 119172, 119173, 119174, and 119665.

Striatodontus strigatus n. sp.

Plate 15, figs. 5–9

Striatodontus n. sp. Pyle and Barnes (2001), Pl. 1, fig. 14.

Diagnosis — A new species of *Striatodontus* in which robust simple cones are covered by fine striae and bear two prominent longitudinal grooves. Apparatus of at least three morphotypes: a, b, and c elements.

Description — Apparatus of at least three morphotypes, a, b, and c elements, of slender cones with slightly expanded bases and cusps which are covered by fine striae and bear two prominent longitudinal grooves on inner lateral or posterior cusp face.

The a element is subsymmetrical and rounded with proclined cusp. Anterior cusp face is rounded and covered in fine longitudinal striae. Posterior cusp face is covered in striae and bears two longitudinal grooves on either side of posterior costa, which is striated basally. Basal outline is round and basal cavity is deep.

The b element is erect and laterally compressed with fine striae covering whole element. Two faint longitudinal grooves are located antero- and postero-laterally.

The symmetrical c element has erect, gently tapered cusp. Anterior cusp face is smoothly rounded and finely striated. Posterior cusp face bears two longitudinal grooves and posterior costa which is striated basally. Basal outline is round.

The e element is laterally compressed and has proclined cusp. Cusp is finely striated on both lateral faces. Outer lateral cusp face bears one stronger medial stria. Anterior and posterior margins of cusp are sharp. Base has oval outline and is expanded posteriorly.

Etymology — Strigatus means furrowed or grooved.

Discussion — The elements observed are similar in the overall morphology to *S. prolificus* but are more robust and have more prominent striae covering the elements and deeper longitudinal grooves.

Occurrence — *Rossodus manitouensis* Zone, Kechika Formation.

Material — 122 a elements, 14 b elements, 28 c elements, 15 e elements.

Repository — GSC 119167, 119168, 119169, 119170.

Genus *Stultodontus* Ji and Barnes, 1994

Stultodontus costatus (Ethington and Brand, 1981)

Plate 15, figs. 10, 11

Occurrence — *Rossodus manitouensis* Zone, Kechika Formation.

Material — 9 elements.

Repository — GSC 119189 and 119190.

?*Stultodontus* sp.

Plate 15, figs. 15, 16

Description — Stout coniform elements which have longitudinal costae are questionably assigned to *Stultodontus*. Two morphotypes (a and e) have short cusps which are slightly longer than base is wide. Anterior cusp margin round; posterior cusp margin sharply rounded. Base is short but extends posteriorly as flattened flange. The e element is similar to the a element but lacks strong longitudinal costae and is more laterally compressed.

Occurrence — *Scolopodus subrex* Zone, Kechika Formation.

Material — 17 a elements, 5 e elements.

Repository — GSC 119409 and 119410.

Genus *Variabiloconus* Landing et al., 1986

Emend. Ji and Barnes, 1994

Type species — *Paltodus bassleri* Furnish, 1938

Discussion — Four morphotypes are recognized as described by Ji and Barnes (1994). The distinguishing characteristics of this genus are the posterior carina, wide base, sharp costae and grooves. The a elements are subround with a wide posterior carina bounded by grooves. The transitional b elements have an offset carina and unequal grooves because of lateral twisting of the cusp. The symmetrical c elements have a wide base and two posterior grooves. The e elements are laterally compressed and have an antero-lateral groove. Landing et al. (1996) revised the genus to include forms such as *Teridontus gracillimus*, which is rejected from the definition of the multielement apparatus in which the elements are costate and grooved and much more robust than *Teridontus*.

Variabiloconus bassleri (Furnish, 1938)

Plate 17, figs. 5–12

Paltodus bassleri Furnish, 1938, p. 331, pl. 42, fig. 1; Ethington and Clark, 1971, p. 72, pl. 2, figs. 2, 4, 6; Repetski, 1982, p. 37, pl. 14, fig. 12.

Variabiloconus bassleri (Furnish). Ji and Barnes, 1994, pp. 67–68, pl. 25, figs. 14–26, text-fig. 38B (contains multielement synonymy to 1994); Landing et al., 1996, p. 678, figs. 7.25–7.33 only; Smith and Clark, 1996, p. 370, pl. 1, figs. 9–11; Parsons and Clark, 1999, p. 7, figs. 5.10, 5.11.

Variabiloconus neobassleri Ji and Barnes, 1994, pp. 68–69, pl. 25, figs. 9–13, text-fig. 38C.

Discussion — In early forms of *V. bassleri*, the costae, grooves, and posterior carinae are not as pronounced as in later forms, such as *V. neobassleri* Ji and Barnes which is herein synonymized because it is considered to show morphologic variation within the same species. Specimens co-occur with *Utahconus longipinnatus* and are somewhat similar but are characterized by the development of postero- and antero-lateral grooves, wide posterior carina, and wide base. The two species are likely derived from the same ancestor.

A later form is recognized in the material and has an apparatus containing four morphotypes, but the elements have a more defined base and deeper grooves on the cusp. The a elements are symmetrically to subsymmetrically rounded coniforms with a more deeply grooved posterior to postero-lateral cusp face, which has a sharply rounded to sharp posterior carina and one or two rounded lateral costae. The anterior cusp face is smoothly rounded. The b elements are more compressed along the posterior cusp margin and have displaced lateral costae. The symmetrical c element is similar to that of earlier forms but has a more defined rounded posterior costa which expands onto the base as a carina. The compressed e elements bear faint lateral costae and are compressed along the length of the element in most specimens, although some have a base which expands slightly postero-laterally.

Occurrence — *Rossodus tenuis* Zone to *Rossodus manitouensis* Zone, Kechika Formation.

Material — 550 a elements, 416 b elements, 83 c elements, 324 e elements.

Repository — GSC 119179, 119180, 119181, 119182, 119185, 119186, 119187, 119188.

Genus *Ventricodus* n. gen.

Type species — *Paltodus spurius* Ethington and Clark, 1964

Ventricodus spurius (Ethington and Clark, 1964)

Plate 17, figs. 13–16

Paltodus spurius Ethington and Clark, 1964, p. 695, pl. 114, figs. 3, 10, text-fig. 2B; Ethington and Clark, 1971, p. 72, pl. 1, fig. 12; Greggs and Bond, 1971, p. 1468, pl. II, figs. 1, 2, 2a; Ethington and Clark, 1982, p. 76, pl. 8, figs. 9, 13; Repetski, 1982, pp. 37–38, pl. 15, figs. 2, 4, 6; Taylor and Landing, 1982, p. 189, text-fig. 5H; An et al., 1983, p. 113, pl. 9, figs. 25–27; Miller and Dockum, 1983, p. 411, figs. 2A, B.

Paltodus aff. *P. spurius* Ethington and Clark. Orndorff, 1988, p. 23, pl. 2, fig. 4.

Paltodus sp. cf. *P. spurius* Ethington and Clark. Mound, 1968, p. 415, pl. 4, figs. 14–17.

"*Paltodus*" *spurius* Ethington and Clark. Harris et al., 1995*b*, p. 24, pl. 1, fig. T.

Diagnosis — A new genus in which the apparatus comprises four morphotypes (a, b, c, e) characterized by simple coniforms with sharply recurved, elongate cusps, which are bulbous or bulging midway along their length, constrict basally, and bear longitudinal grooves.

Description — An apparatus of four morphotypes. The a element is symmetrical with reclined to recurved cusp. Lateral cusp faces each bear longitudinal groove which arises above base. Cusp commonly expands distally and is constricted basally. Posterior cusp margin is sharp. Anterior cusp margin rounded and sharpens distally. Base is expanded anteriorly and posteriorly. Basal cross section is rounded; basal cavity shallow.

The b element is an asymmetrical morphotype in which there is unequal development of longitudinal groove on lateral cusp faces. Element is more compressed on inner lateral side.

Rare, symmetrical c element bears bulbous cusp. Cusp bulges distally and is constricted basally. Posterior cusp margin is sharp; anterior cusp margin is round. Cusp lacks longitudinal grooves. Basal cross section is round; base is expanded.

The e element is geniculate with strongly recurved, laterally compressed cusp that is constricted basally. Anterior and posterior cusp margins are sharp. Cusp has maximum width basally and tapers distally. Angle between base and cusp is about 30°. Base is flared posteriorly and has rounded antero-basal margin.

Etymology — Ventrico (ventricosus) means "bulging out" in reference to the shape of the cusp.

Discussion — Ethington and Clark (1964, 1982) have described the a elements, but the multielement apparatus has not been reconstructed previously. The c and e elements are rare but have the characteristic constriction of the cusp near the base. This taxon is given a new generic name because although it has been grouped into *Paltodus* by early workers, the *Paltodus* apparatus has been demonstrated to be different and characterized by triangular-based elements.

Occurrence — *Rossodus manitouensis* Zone, Kechika Formation.

Material — 208 a elements, 166 b elements, 5 c elements, 1 e element.

Repository — GSC 119175, 119176, 119177, 119178.

Order PANDERODONTIDA Sweet, 1988

Family PANDERODONTIDAE Lindström, 1970

Genus *Panderodus* Ethington, 1959

Panderodus unicostatus (Branson and Mehl, 1933)

Plate 20, figs. 15, 16

Occurrence — *Distomodus staurognathoides* Zone, Pesika Formation.

Material — 30 elements.

Repository — GSC 119488 and 119489.

Order PRIONIODONTIDA Dzik, 1976

Family ACODONTIDAE Dzik, 1994

Genus *Acodus* Pander, 1856

Type species — *Acodus erectus* Pander, 1856

Discussion — The apparatus of *Acodus* is sexi- to septimembrate and has geniculate M and P elements in late forms. Early species of *Acodus* lack strongly geniculate M and P elements but are characterized by S elements with well-developed costae. The S elements among some species such as *A. kechikaensis* n. sp., *A. deltatus*, and *A. longibasis* are similar, but the P elements are diagnostic.

Acodus deltatus Lindström, 1955

Plate 1, figs. 3–10

Acodus deltatus Lindström, 1955, p. 544, pl. 3, fig. 30; Albanesi, 1998, p. 146, pl. 1, figs. 1–5 (contains synonymy to 1998).

Discussion — The apparatus is seximembrate and contains a transition series of S elements with variable positions of the costae. The M element is non-geniculate in early forms and is laterally compressed with a flexed cusp. The P element is non-geniculate and bears strong outer lateral costae. A symmetrical Sa element appears to be lacking in early forms, and other S elements show a great deal of variability in the number of lateral costae.

Occurrence — *Acodus deltatus* Zone to *Acodus kechikaensis* Zone, Kechika Formation.

Material — 278 M elements, 1318 S elements, 288 P elements.

Repository — GSC 119232, 119233, 119234, 119235, 119236, 119237, 119238, 119239.

Acodus kechikaensis n. sp.

Plate 1, figs. 11–17

Acodus? aff. *A. emanuelensis* McTavish. Ethington and Clark, 1982, pp. 19–20, pl. 1, figs. 11–13 (Sb, Sa, P) only, text-fig. 5 (in part).

Diagnosis — Apparatus of six element morphotypes characterized by robust elements with elongate cusps and ramiform complex in which adentate processes are basally adjoined by sheath which is concave in interprocess region of element.

Description — Sa is symmetrical with posterior process and two lateral processes. Processes are joined by sheath which extends between all three and is concave in between. Anterior cusp face is flat but basal sheath region is concave. Posterior cusp face concave except for prominent posterior process. Basal cavity triangular, apex located under posterior process. Basal margin is flat.

Sb element is asymmetrical and bears three processes; posterior process is most prominent and extends below basal margin. Processes join cusp in smooth curve just above the point of curvature of cusp. Cusp is reclined. Anterior cusp face is rounded.

Sc element is laterally compressed and broad based. Anterior and posterior processes taper distally and extend below the basal margin. Cusp is erect. Inner postero- and antero-lateral cusp margins are slightly concave longitudinally. Antero-basal corner sharply rounded. Base extends posteriorly, and basal margin is slightly arched under anterior and posterior processes.

Sd element is asymmetrical, its cusp is erect and bears four processes. Processes taper distally, posterior process is most prominent.

M element is geniculate with posterior process joining the cusp at 30° angle. Posterior process is rounded along top margin. Cusp bears broad carina which extends to the midline of base and posterior process, giving lateral flare to basal region.

P element has erect cusp and is laterally compressed with flat outer lateral margin and flared inner lateral margin with carina on base. Antero-basal margin is concave. Outer lateral face is slightly concave along basal margin region where sheath extends between two sharp anterior and posterior cusp margins which taper distally.

Etymology — Named after the Kechika Formation.

Discussion — The adenticulate prioniodids described by McTavish (1973) included a trichonodelliform (Sa), tetraprioniodiform (Sd), oistodiform (M), and prioniodiform (P). Cooper (1981) illustrated five morphotypes in which he designated the tetraprioniodiform element as an Sb rather than an Sd. Therefore, the Sb element described herein has not yet been reported. Ethington and Clark (1982) illustrated an Sc element and noted that, unlike the Australian material, the base of their elements is not as long. This is also observed in this collection in that the Sc and P elements are not the same as illustrated by McTavish (1973) and Cooper (1981). The Australian specimens have more slender cusps. As Ethington and Clark (1982) indicate, there are several species of *Acodus* diversifying in this interval, and the elements among intermediate forms are very similar, particularly the S elements. However, this new species of *Acodus* is distinct and abundant, especially the P elements which bear a prominent inner lateral costa.

Occurrence — *Acodus kechikaensis* Zone, Kechika Formation.

Material — 271 M elements, 627 S elements, 301 P elements.

Repository — GSC 119250, 119251, 119252, 119253, 119254, 119255.

Acodus longibasis McTavish, 1973

Plate 1, figs. 18–24

Acodus deltatus longibasis McTavish, 1973, p. 40, pl. 1, figs. 16, 20, 22, 23, 25, 26, pl. 2, fig. 14, text-fig. 3I-L.

Emended diagnosis — A species of *Acodus* characterized by elongate posterior process of the S elements in which process is longer than lateral costae and anterior keel, and its free portion is as long as cusp. The M and P elements have elongate posterior extensions of the base.

Description — Pb elements are non-geniculate and compressed, broad based with erect cusp. Elements vary in the width of the base and extension of postero-basal margin. Cusp has keeled anterior and posterior margins and inner lateral keel. Lateral keel is most strongly developed on the base, and bifurcates at medial basal margin. Outer lateral cusp face slightly concave. Basal cavity fills wide base, apex at point of curvature of cusp. Pa has posteriorly drawn-out base and a slightly drawn-out and arched anterior basal margin. Cusp is proclined with keeled anterior and posterior cusp margins. Inner lateral keel not as strong and becomes weaker distally along cusp. Inner lateral cusp face has lateral expansion under keel on basal portion. Outer lateral cusp face is slightly concave.

M element is geniculate and distinguished by its long, straight posterior extension of the base. Cusp is compressed and reclined with concave anterior cusp margin and slight antero-basal extension.

Sa element is symmetrical with erect cusp. Cusp bears strong posterior keel and two lateral keels. Posterior cusp face bears rounded keel. Anterior cusp face is flat. Basal cavity confined under posterior keel; lateral keels are filled with white matter. Basal margin straight, basal cavity triangular.

Sb element is asymmetrical with three strong keels. Posterior keel is strongest and elongate, its free portion as long as cusp. Two lateral keels merge anteriorly above the point of curvature of cusp. Basal cavity is deep but narrow. Asymmetric variants have four keels; two lateral keels merge with anterior keel distally. Posterior keel is most prominent and elongate. Cusp is erect to reclined. Basal cavity is narrow and deep.

Sc element is laterally compressed with anterior and posterior keels. Posterior keel extends below the basal margin slightly. Cusp is proclined; element is broad based with deep basal cavity. Basal margin is slightly arched, inner lateral surface has slight basal expansion.

Sd element bears four costae and is similar to Sb element.

Discussion — McTavish (1973) originally described *Acodus deltatus longibasis* as a subspecies, but it is herein raised to species level because it is distinct. In his description, elements described as tetraprioniodiform and gothodiform are Sd and Sb elements (four-keeled and three-keeled), and the oistodiform and prioniodiform are M and P elements. The material from this collection suggests seven morphotypes are present in *Acodus longibasis.*

Occurrence — *Acodus kechikaensis* Zone, Kechika Formation.

Material — 34 M elements, 209 S elements, 63 P elements.

Repository — GSC 119239, 119240, 119241, 119242, 119243, 119244, 119245.

Acodus neodeltatus n. sp.

Plate 2, figs. 1–7

Diagnosis — A new species of *Acodus* in which elements are thin walled, delicate, and laterally compressed with strong development of keels (lateral, posterior, and anterior).

Description — Apparatus of three morphotypes (M, S, P).

S elements are characterized by the strong development of keels which are separated from each other by thin walls of element. Elements laterally compressed. Keels extend from base to cusp tip and are drawn out to basal margin as sharply rounded processes. No symmetrical Sa element was observed. Sb element has proclined cusp and long base, about half the height of the element. Anterior and posterior cusp margins sharp. Inner lateral cusp face bears strong postero-lateral costa. Antero- and postero-basal corners sharply rounded.

Sc element has sharp anterior and posterior cusp margins. Inner lateral cusp face bears weak costa on cusp above point of curvature of cusp. Outer lateral cusp face unornamented and flat. Base is long, about one-third height of element. Antero-basal corner sharp, as anterior costa extends below basal margin as projection. Base expands posteriorly, and postero-basal corner is sharply rounded and laterally compressed.

Sd element is similar to the Sb element but bears a lateral costa on each lateral cusp face in a medial position.

M element has expanded and sharply rounded

antero-basal margin and less than 20° angle between base and cusp.

P element has well-defined central keel on inner lateral face. Outer lateral cusp face flat. Anterior and posterior cusp margins are thin, sharp keels. Area between inner lateral keel and posterior keel is concave. Postero-basal margin drawn out into well-defined process. Basal corners sharply rounded. Basal margin slightly arched under lateral keel.

Etymology — "Neo" refers to younger or newer.

Discussion — Elements of this new species of *Acodus* differ from the others in their delicate nature and development of well-defined keels into processes. This new species appears to be the direct ancestor of *Oepikodus communis.*

Occurrence — *Acodus kechikaensis* Zone, Kechika Formation.

Material — 34 M elements, 304 S elements, 59 P elements.

Repository — GSC 119263, 119264, 119265, 119266, 119267.

Acodus oneotensis Furnish, 1938

Plate 2, figs. 8–12

Acodus oneotensis Furnish, 1938, p. 325, pl. 42, figs. 26–29, text-fig. 1N; Ethington and Clark, 1971, p. 72, pl. 1, figs. 3, 6, 8; Ethington and Clark, 1982, p. 20, pl. 1, fig. 16.

"*Acodus*" *oneotensis* Furnish, Wang, 1986, pp. 114–116, pl. 52, fig. 10, text-fig. 28.1 only.

Description — Apparatus of slender, strongly keeled coniform elements.

M element is non-geniculate and has outer and inner lateral keels from base to cusp tip. Base is expanded slightly, gradually tapers to tip, laterally compressed. Antero- and postero-basal corners are sharp; basal outline is oval. Some are strongly compressed with inner lateral keel weakly developed at base and faint distally up cusp. Outer lateral cusp face bears very faint keel. Base is extremely compressed. Basal cavity lies just posterior of midline of element.

P element is similar to the M element but has more expanded base. It is non-geniculate and compressed with a slight flare on the inner basal margin.

Sa element has four strong costae, all extending from base to cusp tip. Cusp is erect and curved posteriorly. Cusp and basal cross section are diamond shaped. Posterior cusp face has compression on both sides of the posterior costa. The base is expanded posteriorly and drawn out laterally.

Sb elements are characterized by four strong keels: two lateral, one anterior, and one posterior. Cusp is reclined to recurved and twisted in some specimens. Cusp cross section is diamond shaped. Anterior cusp face is smoothly rounded with a medial costa extending from base to tip. Posterior cusp face is concave with strong medial costa. Offset of the posterior and anterior costae occurs through symmetry transition series. Base is narrow and drawn out posteriorly and anteriorly, producing diamond-shaped cross section. Basal cavity is a right triangle in posterior view, apex lateral from midline, extends to point of curvature of cusp.

The Sc element is characterized by one strong lateral costa on outer face. Inner cusp face is flattened to slightly concave. Cusp is erect, laterally compressed, and bladelike with keeled anterior and posterior margins. Base is narrow and triangular, drawn out slightly under lateral keel. Basal cavity is triangular, its apex located centrally in lateral view.

Discussion — *Acodus oneotensis* is similar to *Acodus quentinenesis* n. sp. but is much more delicate and strongly costate. There is strong similarity among the M elements to e elements of *U. utahensis*, but M elements can be distinguished by the position of the apex of the basal cavity, which is extremely posterior in *U. utahensis* but just posterior of the midline in *A. oneotensis*.

Ji and Barnes (1994) assigned elements illustrated by Furnish (1938) and Ethington and Clark (1971) to *Polycostatus oneotensis* and include in their remarks that the species is widely cited and has been assigned to a number of genera. The interpretation of the apparatus herein, which includes the a elements originally illustrated by Furnish (1938), is that the species is an early representative of *Acodus*.

Occurrence — *Iapetognathus* Zone, Kechika Formation.

Material — 42 M elements, 252 S elements, 40 P elements.

Repository — GSC 119222, 119223, 119224, 119225.

Acodus quentinensis n. sp.

Plate 2, figs. 13–19

Diagnosis — The apparatus consists of S, M, and P elements characterized by bladelike cusps with prominent costae and broad base. Sa element bears weak, rounded anterior carina and strong posterior carina.

Description — The compressed M element is non-geniculate and has laterally compressed cusp, erect to reclined, sharp anterior and posterior margins. Antero-basal corner is drawn out slightly, postero-basal corner is rounded. Inner lateral face is costate basally, giving flare to the inner basal margin, becoming fainter distally. Outer lateral face is flat to slightly concave at the base. Basal cavity is triangular, apex is anterior under inner lateral costa, cavity is oval to round, tapers anterior to anterior cusp margin. Transitions occur in the degree of inner lateral costae and drawn-out postero-basal corner, and compression of the base. Most of the compressed element has erect, flexed cusp. Cusp margins are keeled posteriorly and anteriorly.

P element is non-geniculate and compressed. Inner lateral face bears short costa on the flare of the basal margin, extending faintly distally up to the cusp tip. Outer lateral face is smoothly round with faint costa extending from cusp tip to base. Outer lateral face of the base is concave. Base is slightly expanded posteriorly, basal cross section is oval, expanding laterally. Basal cavity is triangular, tip is anterior of the midline of cusp.

Sa element is symmetrical and has three strong costae, two lateral and one posterior. Anterior cusp face bears faint anterior costa. Posterior cusp face is concave with raised posterior costa, strongest at base. Cusp is erect and flexed posteriorly, compressed antero-posteriorly. Base is slightly rounded along anterior margin, drawn out slightly at lateral margins and flared where posterior costa joins basal margin. Base is wide, basal cavity is deep, extending to point of flexure of cusp, its apex is central.

The Sb elements are asymmetric variants in which cusp twists from midline, but all other characters are same as in Sa elements.

The Sc element is laterally compressed, lacks lateral costae, but is costate anteriorly and posteriorly. Cusp is erect, outer lateral basal margin is slightly round, inner lateral margin is flat. Cusp is gently flexed above apex of cavity, its cross section is oval, and cusp tapers anteriorly. Basal cavity is triangular in lateral view, the apex is anterior to the midline.

Etymology — Named after the Quentin Member of the Kechika Formation.

Discussion — *Acodus quentinensis* n. sp. is similar to *A. oneotensis*, but its elements have broader bases and the Sa lacks a strong anterior costa. The P element is close to one type of e element of *Rossodus tenuis* and *Utahconus utahensis*, however, the concavity of the outer lateral basal face and equality of the antero- and postero-basal expansion distinguishes it from both *R. tenuis* and *U. utahensis* in which the outer lateral basal face is flattened and the basal margin is drawn out posteriorly.

Occurrence — *Iapetognathus* Zone, Kechika Formation.

Material — 74 M elements, 171 S elements, 95 P elements.

Repository — GSC 119227, 119228, 119229, 119230, 119231.

Acodus warensis n. sp.

Plate 2, figs. 20–27

Diagnosis — A new species of *Acodus* in which base of all elements (M, S, P) is laterally expanded, and elements are stout and robust, with cusp as long as base is wide.

Description — Apparatus of three morphotypes (M, S, P).

S elements are laterally expanded, stout and robust, with cusp as long as base is wide. All bear three or four keeled processes. Sa element is symmetrical with thinly keeled lateral cusp margins which expand laterally at base. Cusp curved posteriorly. Posterior cusp face bears strong keel flanked by concave regions between keel and lateral keels. Anterior cusp face rounded. Basal margin flat. Basal cavity oval in posterior view, extends to, but not into, lateral extensions.

Sb element is diagnostic in bearing sharp anterior and posterior cusp margins and strong lateral keel which arises at center of base and curves up to cusp tip in medial position. Outer lateral face slightly concave. Anterior keel extends below basal margin as laterally compressed projection. Base expands posteriorly as sharply rounded extension.

Sc element has cusp which is as long as base is wide. Anterior and posterior cusp margins are sharp. Lateral cusp faces lack costae. Base is expanded laterally, anteriorly, and posteriorly. Basal margin concave, arched between anterior and posterior keels which extend below basal margin.

Sd element is similar to the Sb element but bears two lateral costae and is not as stout in lateral view, as anterior cusp face is more rounded.

M element has wide, short base, and cusp is as long as base is wide. Base is flared laterally. Oral margin rounded in lateral profile. Aboral margin also rounded, giving base oval shape. Antero-basal corner projects and tapers anteriorly such that it has a concavity on the oral and aboral surfaces of the projection.

P elements have wide bases, expanded laterally and posteriorly. Keel on inner lateral cusp face fades distally up cusp. Pa element has sharp keel on inner lateral base and is laterally compressed. Pb element has stronger inner lateral keel and wider base which is drawn out posteriorly and longer than in the Pa element.

Etymology — Named after the Ware map area.

Discussion — Compared to other species of *Acodus*, *A. warensis* n. sp. is distinct in its stout nature, expansion of the base, and shortening of the cusp compared to *A. longibasis* in which the keels are exaggerated and elements are elongate. The M elements are distinct in the concave oral and aboral margins of the antero-basal corner, which has straight margins in other species.

Occurrence — *Acodus kechikaensis* Zone, Kechika Formation.

Material — 3 M elements, 30 S elements, 9 P elements.

Repository — GSC 119256, 119257, 119258, 119259, 119260, 119261.

Genus *Diaphorodus* Kennedy, 1980

Type species — *Acodus delicatus* Branson and Mehl, 1933

Discussion — The apparatus consists of six morphotypes as illustrated by Albanesi (1998). The genus is distinguished by its P element. The multielement descriptions by Serpagli (1974) and Kennedy (1980) identified five morphotypes.

Diaphorodus russoi (Serpagli, 1974)

Plate 5, figs. 7–11

Acodus? *russoi* Serpagli, 1974, pp. 35–37, pl. 8, figs. 1a–5b, pl. 20, figs. 7–8, text-fig. 5 (contains synonymy to 1974); Repetski, 1982, p. 13, pl. 3, figs. 1–5.

Diaphorodus russoi (Serpagli). Albanesi, 1998, pp. 147–148, pl. 4, figs. 11–15, text-fig. 20 (contains synonymy to 1998).

Discussion — The elements of *D. russoi* have been adequately described previously, but the material in this collection supports a seximembrate apparatus plan proposed by Albanesi (1998). The elements are tiny and delicate, and characterized by the geniculate, modified pastinate P element. The S elements have tonguelike posterior processes as Serpagli (1974) described.

Occurrence — *Acodus kechikaensis* Zone, Kechika Formation.

Material — 6 M elements, 93 S elements, 8 P elements.

Repository — GSC 119273, 119274, 119275, 119276, 119277.

?*Diaphorodus* n. sp.

Plate 5, figs. 12–18

Description — A species of ?*Diaphorodus* with slender cusps and slender, tonguelike basal processes extending from keeled cusp margins. Apparatus of at least five morphotypes which have lateral keels and tonguelike basal processes.

Sa element is symmetrical and delicate. Cusp is erect and flexed posteriorly. Anterior cusp face is rounded and bears faint carina along the midline. Posterior cusp face is concave and bears keel on base which extends to point of flexure of cusp. Lateral cusp margins are sharp and extend as basal keels which are compressed antero-posteriorly and rounded distally. Base is narrow and expands under posterior keel. Basal cavity is shallow.

Sb element bears three sharp keels. Cusp is slender and erect. Anterior and posterior cusp margins are keeled and extend as compressed keels below basal margin. Outer lateral face bears keel which is strongest basally and extends below basal margin, and fades distally to tip of cusp. Inner lateral cusp face is concave. Basal cavity is shallow and triangular in lateral profile, its apex is located centrally beneath lateral keel.

Sd element bears four sharp keels and is asymmetrical. Anterior cusp face is rounded and bears rounded keel. Posterior cusp face is concave and bears strong keel which extends below basal margin as tonguelike process. Lateral cusp margins are sharply keeled. Basal cavity is small with diamond-shaped cross section.

P element is non-geniculate. Cusp is erect and bears outer lateral keel which extends below basal margin. Anterior and posterior cusp margins are sharp. Inner lateral cusp face is concave. Base is not well defined, its anterior and posterior margins are straight.

M element is geniculate and compressed. Cusp is concave on inner lateral side and convex on outer lateral side. Edge of base meets posterior edge of cusp at a 20° angle. Basal process is compressed and flexed inward; its upper surface is rounded. Base is curved toward process, and inner basal margin is flared slightly.

Discussion — The apparatus is complete except for an Sc element. The P element is similar to those in the *Diaphorodus* apparatus because it is oepikodiform, but non-geniculate unlike those in *D. russoi*, perhaps because it is an earlier species. Hence, the assignment of this new species of *Diaphorodus* is held in question at this stage.

Occurrence — *Rossodus manitouensis* Zone, Kechika Formation.

Material — 4 M elements, 8 S elements, 2 P elements.

Repository — GSC 119211, 119212, 119213, 119214.

Genus *Pteracontiodus* Harris and Harris, 1965

Pteracontiodus cryptodens (Mound, 1965)

Plate 27, figs. 21, 22

Occurrence — *Paroistodus originalis* Zone to *P. horridus* Zone, Skoki and Ospika formations.

Material — 3 elements.

Repository — GSC 119584 and 119585.

Genus *Triangulodus* van Wamel, 1974

Type species — *Scandodus brevibasis* (Sergeeva), sensu Lindström, 1971

Discussion — The diagnoses of *Trigonodus* Nieper, *Triangulodus* and *Scandodus* Lindström, *Tripodus* Bradshaw, and *Pterocontiodus* Harris and Harris have been problematic as discussed by Bauer (1987) and Stouge and Bagnoli (1990). Herein, *Triangulodus* is used for hyaline coniforms with keeled cusps which form a septimembrate apparatus. van Wamel (1974) described the apparatus as having scandodiform (P), oistodiform (M), drepanodiform (Sc), acodiform (Sb), and paltodiform and roundyaform (Sa) elements. The P elements of *Triangulodus* have two sharp edges, but those of *Pterocontiodus* bear three sharp edges. *Tripodus* includes elements which are albid but has a similar apparatus plan to both *Pterocontiodus* and *Triangulodus*.

Triangulodus akiensis n. sp.

Plate 22, figs. 8–16

Diagnosis — A species of *Triangulodus* in which S and P elements have keels which are elongate and extend below basal margin. M element has extended antero- and postero-basal corners.

Description — Pa and Pb elements have keeled anterior and posterior cusp margins which extend as antero- and postero-basal corners of element. Pa element is erect and basal margin is straight. Pb element is proclined and bears inner lateral flare on base. Basal margin arched under anterior and posterior keels.

M element has broad, low base which is as long as cusp is tall. Cusp is compressed but somewhat convex on outer surface. Basal margins are extended as pointed keels.

Sa element is symmetrical and bears extremely elongate posterior keel and two extended lateral keels. Anterior cusp margin is costate and rounded. In some Sa elements, cusp strongly recurved (almost 90° to base) and anterior margin projects (Pl. 22, figs. 10, 11). Anterior cusp margin is concave above basal margin and a thin sheath joins the two lateral processes. Sb element also bears two extended lateral keels, strong posterior keels which projects below basal margin and rounded anterior cusp margin, but is asymmetrical.

Sc is laterally compressed and has sharp anterior and posterior keels which extend below basal margin. Posterior keel thin and elongate and extends posteriorly. Basal margin straight between anterior and posterior extensions.

Sd element bears four keels — two lateral, one anterior, and one posterior. Region between keels is thinly sheathed basally. All keels project below basal margin.

Etymology — Named after the Akie River, south of the type section of the Ospika Formation.

Discussion — The elements bear a resemblance to *Acodus longibasis* McTavish, but have much more distinct keels and are hyaline and not albid. The apparatus reconstruction is similar to that shown by Stouge and Bagnoli (1990) for *Trigonodus* (= *Triangulodus*) *brevibasis* (Sergeeva).

Occurrence — *Dzikodus tablepointensis* Zone, Ospika Formation.

Material — 4 M elements, 43 S elements, 17 P elements.

Repository — GSC 119445, 119446, 119447, 119448, 119449, 119450, 119451, 119452.

Genus *Tripodus* Bradshaw, 1969

Tripodus laevis Bradshaw, 1969

Plate 28, fig. 8

Occurrence — *Tripodus laevis* Zone, Skoki Formation.

Material — 4 elements.

Repository — GSC 119583.

Genus *Tropodus* Kennedy, 1980

Type species — *Paltodus comptus* Branson and Mehl, 1933

Discussion — *Tropodus* has a septimembrate apparatus plan in which there is a symmetry transition of S elements bearing three to five costae, M elements with a characteristic arched and curved posterior process, and one or two P elements which have keeled anterior and posterior cusp margins which project below the basal margin as rounded flanges. The base of the element bears a process which fades distally. The genus was defined by Kennedy (1980), and Stouge and Bagnoli (1988) interpreted the apparatus as a prioniodontid-type containing M and P elements. Smith (1991) assigned the P elements to the genus *Chionoconus*, which has herein been synonymized with *T. sweeti*. The apparatus plan is similar to that of *Diaphorodus* in having a geniculate P element, but the costae of *Diaphorodus* are tonguelike.

Tropodus australis (Serpagli, 1974)

Plate 16, figs. 10–13

Walliserodus australis Serpagli, 1974, pp. 89–91, pl. 19, figs. 9–10, pl. 29, figs. 8, 10, 14.

Tropodus australis (Serpagli). Albanesi, 1998, p. 151, pl. 13, figs. 12–18 (contains synonymy to 1998).

Discussion — The elements of *T. australis* are similar in overall morphology to those of *T. sweeti*, but have less well-developed keels on the S elements. The M and P elements are distinct in having a rounded antero-basal margin, which becomes much sharper and keeled in *T. sweeti*. A final difference between the two species is the development of a keel on the lateral basal margin of the M and P elements of *T. sweeti*, which in *T. australis* is a rounded carina.

Occurrence — *Acodus kechikaensis* Zone to *Oepikodus communis* Zone, Kechika and Skoki formations.

Material — 35 M elements, 160 S elements, 32 P elements.

Repository — GSC 119313. 119314, 119315, 119316.

Tropodus comptus (Branson and Mehl, 1933)

Plate 16, figs. 14–16

Occurrence — *Scolopodus subrex* Zone, Kechika Formation.

Material — 4 elements.

Repository — GSC 119215, 119216, 119217.

Tropodus sweeti (Serpagli, 1974)

Plate 28, figs. 9–17

"*Paltodus*"? *sweeti* Serpagli, 1974, pp. 58–59, pl. 14, figs. 13a–14b, pl. 24, figs. 8–10, text-fig. 12.

Tropodus sweeti (Serpagli) Stouge and Bagnoli, 1988, p. 142, pl. 16, figs. 10–15 (contains synonymy to 1988); Albanesi, 1998, p. 152, pl. 13, figs. 23–30, text-fig. 23 (contains synonymy to 1998).

Chionoconus avanga, Smith, 1991, pp. 22–24, figs. 14a, b, c, d only.

Discussion — Some elements of *Tropodus sweeti* have been adequately described previously. The Pa element was originally described as "*Paltodus*" *sweeti* Serpagli and the M as "*Scandodus*" *americanus* Serpagli. The material from this collection includes a Pb element which is a geniculate form with a recurved cusp. The cusp bears a sharper lateral costa than the Pa element, and the costa extends to the cusp tip rather than just on the base as in the Pa. The longitudinal groove is deeper, close to the anterior margin of the cusp. The anticusp is more prominent, and the postero-basal margin is more drawn out and flattened.

The S elements were originally described as having three to five costae as represented by *Walliserodus australis* Serpagli; the five-costate form is the symmetrical Sc, the four-costate form is the Sd, and the three-costate form is the Sb element. Albanesi (1998) figured an Sa element of the apparatus which is present in this collection, a symmetrical, antero-posteriorly compressed element with strong posterior and lateral keels. The Sa element is a broad-based, strongly compressed erect element with strong posterior keel and two lateral keels. The anterior cusp face is flat but concave basally. The two lateral costae extend below the base. The basal cross section is triangular.

Some elements, particularly the tricostate Sb elements, in which the postero-lateral keel is drawn out, have incipient denticles on the keel. Juvenile forms of the Sb, Sc, and Sd have denticulate keels on the elements.

Occurrence — *Oepikodus communis* Zone, Skoki Formation.

Material — 64 M elements, 155 S elements, 98 P elements.

Repository — GSC 119534, 119535, 119536, 119537, 119538, 119539, 119540, 119541, 119542.

Family DISTOMODONTIDAE Klapper in Clark et al., 1981

Genus *Distomodus* Branson and Branson, 1947

Distomodus staurognathoides (Walliser, 1964)

Plate 19, figs. 3–6

Occurrence — *Distomodus staurognathoides* Zone, Pesika Formation.

Material — 3 M elements, 3 S elements, 3 P elements.

Repository — GSC 119484, 119485, 119486, 119487.

Family MULTIOISTODONTIDAE Harris, 1964

Genus *Multioistodus* Cullison, 1938

Type species — *Multioistodus subdentatus* Cullison, 1938

Multioistodus compressus? Harris and Harris, 1965

Plate 24, fig. 15

? *Neomultioistodus compressus* Harris and Harris, 1965, pp. 43–44, pl. 1, figs. 7a–c.

?*Multioistodus compressus* (Harris and Harris) Ethington and Clark, 1982, pp. 58–59, pl. 6, fig. 10 (only) (contains synonymy to 1981).

Discussion — An element from the Skoki Formation observed in this collection is similar to the acodiform element illustrated by Ethington and Clark (1982) and is an asymmetrical Sb element within the quinquimembrate apparatus plan illustrated in Sweet (1988). The denticles are compressed and flexed posteriorly. The cusp bears four costae, the lateral ones which join the sharp edge of the denticle in a smooth curve.

One specimen from the Chesterfield Member occurs with *M. subdentatus*, which is an Sc with an extremely slender, elongate cusp and posteriorly deflected denticle arising from the aboral basal margin. It differs from the Sc of *M. subdentatus* in the gracile nature of the element and the position of the denticle which arises higher on the base in *M. subdentatus*.

Occurrence — *Jumudontus gananda* Zone, Skoki Formation, to *Paroistodus originalis* Zone, Ospika Formation.

Material — 2 elements.

Repository — GSC 119601.

Multioistodus subdentatus Cullison, 1938

Plate 20, figs. 10, 11

Multioistodus subdentatus Cullison, 1938, p. 226, pl. 29, figs. 13a, b; Ethington and Clark, 1982, pp. 59–60, pl. 6, figs. 12–14 (contains synonymy to 1982); Craig et al., 1986, p. 10, pl. 1, figs. 33, 34; Bauer, 1989, p. 97, figs. 4.11–4.13.

Discussion — The apparatus is incompletely preserved, but a series of S elements can be recognized and possibly an oistodiform M, which has a denticle arising from the antero-basal corner. Ethington and Clark (1982) described four elemental morphotypes — cordylodiform (Sc), distacodiform (Sa), trichonodelliform (?P), and acodiform (Sb), and although the Sb was not illustrated, the elements in this collection are similar to the Sc and Sa.

Occurrence — *Jumudontus gananda* Zone, Skoki Formation, to *Paroistodus originalis* Zone, Ospika Formation.

Material — 11 elements.

Repository — GSC 119426 and 119427.

Family OISTODONTIDAE Lindström, 1970

Genus *Graciloconus* n. gen.

Type species — *Graciloconus concinnus* n. gen. n. sp.

Diagnosis — Extremely thin, delicate coniform elements with slender, elongate cusps that have sharply keeled anterior and posterior margins. Base is triangular with small, shallow basal cavity. Apparatus includes S elements, geniculate M elements and P elements. Cusp is filled with white matter.

Etymology — Gracile + cone refers to thin, delicate nature of the elements.

Graciloconus concinnus n. gen. n. sp.

Plate 7, figs. 13–20

Diagnosis — A species of *Graciloconus* with apparatus of three morphotypes (S, M, P) in which S elements are diagnostic in having long, slender, extremely compressed cusps and triangular base in which the basal cavity is small and shallow.

Description — S elements are coniforms with erect cusps. Cusp is laterally compressed with sharp anterior and posterior cusp margins. In all S elements, base is short and triangular in lateral view. Basal cavity is small and confined to middle third of base; triangular shaped with concave anterior margin; apex lies just anterior to midline of cusp. Sa element has unornamented cusp faces, except for small inner lateral carina on base. Posterior cusp margin extends below basal margin as tonguelike projection. Antero-basal corner is rounded.

Sb element has smooth lateral cusp faces and sharp anterior and posterior cusp margins. Postero-basal corner is sharp but does not project below basal margin. Antero-basal margin rounded. No inner carina on base.

Sc element has erect cusp in which sharp anterior and posterior cusp margins project below basal margin. Basal carina is small and basal margin is arched. Sc has broadest base of all S elements.

?Sd element is similar to the Sa but with a more pronounced basal carina and faint lateral costa from base to cusp tip. Antero- and postero-basal corners extend as tonguelike projections below basal margin.

M element is geniculate with slender, elongate, tapered cusp (Pl. 7, fig. 20) and long tonguelike projection of posterior basal margin (Pl. 7, fig. 19). Antero-basal margin drawn out and compressed. Basal cavity shallow and small. Inner basal margin flared.

P element similar to S elements but has less sharp cusp margins and rounded basal margins. It is not as laterally compressed and bears an inner lateral flare on basal margin which expands the base more than in S elements.

Etymology — Concinnus means elegant.

Discussion — This new genus and species is tentatively placed within the Oistodontidae based on its apparatus structure which is similar to that of *Oistodus*. The gracile nature of the elements with posteriorly expanded bases could warrant placement among the Paracordylodontidae, however, *G. concinnus* is adenticulate.

Occurrence — *Scolopodus subrex* Zone, Kechika Formation.

Material — 54 M elements, 163 S elements, 15 P elements.

Repository — GSC 119324 to 119331.

Genus *Jumudontus* Cooper, 1981

Jumudontus gananda Cooper, 1981

Plate 24, figs. 9, 10

Occurrence — *Jumudontus gananda* Zone, Skoki Formation.

Material — 3 P elements.

Repository — GSC 119622 and 119623.

Genus *Oistodus* Pander, 1856

Oistodus bransoni Ethington and Clark, 1982

Plate 25, fig. 7

Remarks — *O. bransoni* was described by Ethington and Clark (1982) as having a short and stubby posterior extension of the base, which is a feature displayed by the few elements in this collection which do not have characteristics of *O. lanceolatus* nor *O. multicorrugatus*.

Occurrence — *Oepikodus communis* Zone, Skoki Formation.

Material — 3 elements.

Repository — GSC 119529.

Oistodus lanceolatus Pander, 1856

Plate 25, figs. 8–11

Remarks — The apparatus is quinquimembrate and comprises hyaline elements with a symmetry transition of elements which lack a strong lateral costa (Sc), have one or two strong lateral outer costa(e) (Sb), or are symmetrical (Sa). The P element has been described by Bagnoli et al. (1988).

Occurrence — *Acodus kechikaensis* Zone, Kechika Formation, to *Paroistodus originalis* Zone, Skoki and Ospika formations.

Material — 20 M elements, 34 S elements, 5 P elements.

Repository — GSC 119530, 119531, 119532, 119533.

Oistodus multicorrugatus Harris, 1962

Plate 25, figs. 12–16

Remarks — The apparatus has been adequately described by Stouge and Bagnoli (1988) as comprising S, M, and P elements which have one or more costae on the base. The elements have a variable number of costae. The elements originally illustrated by Harris (1962) have many costae, but those illustrated by Stouge and Bagnoli (1988) include acostate M elements with one strong lateral carina, and costate to multicostate S and P elements, as well as a range in the development of the costae.

Occurrence — *Oepikodus communis* Zone to *Paroistodus originalis* Zone, Skoki Formation.

Material — 12 M elements, 51 S elements, 6 P elements.

Repository — GSC 119524, 119525, 119526, 119527, 119528.

Oistodus sp.

Plate 11, fig. 19

Discussion — Oistodiform elements which have a slender, twisted cusp and broad base with sinuous basal outline are not assigned to a known species of *Oistodus*. The lateral cusp faces lack costae, and basal margins are rounded, which distinguish it from *O. lanceolatus* and *O. multicorrugatus*.

Occurrence — *Acodus kechikaensis* Zone, Kechika Formation.

Material — 3 elements.

Repository — GSC 119278.

Genus *Rossodus* Repetski and Ethington, 1983

Emend. Ji and Barnes (1994)

Type species — *Rossodus manitouensis* Repetski and Ethington, 1983

Discussion — Ji and Barnes (1994) described the apparatus of *Rossodus* as containing four elemental morphotypes (a, b, c, and e), based on the apparatus description by Repetski and Ethington (1983). Albanesi (1998) emended the genus by describing a later form, *R. barnesi*, which has seven elemental morphotypes. The material in this present study shows that early species such as *R. tenuis* (Miller), which have four elemental morphotypes, descended from *Utahconus utahensis* Miller. The evolutionary transition from *Utahconus* to *Rossodus* has been noted by Repetski and Ethington (1983) and Ji and Barnes (1994). All of the species of *Rossodus* in this collection are characterized by elements that have bladelike cusps filled with white matter, which bear rounded anterior costae. There are up to three types of compressed morphotypes within each apparatus in which one is geniculate and the other two are non-geniculate.

Rossodus kwadachaensis n. sp.

Plate 13, figs. 16–20

Diagnosis — A species of *Rossodus* with laterally expanded base on c element and characterized by flared base on a, b, and e elements.

Description — The a elements are stout with inner postero-laterally expanded base and carina. Antero-lateral carina is weak. Cusp is erect. Antero-basal corner rounded. Anterior and posterior cusp margins are sharp. Base expanded under anterior carina and sharper basally. Base is rounded under posterior carina.

The c element is symmetrical and stout with erect cusp which is curved posteriorly at tip. Anterior cusp face is rounded and bears faint, rounded carina. Posterior cusp face is concave. Base is expanded laterally and is triangular in posterior view. Basal cavity is triangular, apex at midline. Basal cavity expands under posterior carina.

The b element is laterally compressed with posteriorly expanded base. Cusp bears antero-lateral carina. Basal cavity apex is close to anterior cusp margin. Anterior and posterior cusp margins are sharp. Basal cavity extremely compressed and slitlike.

The e element is geniculate with a 45° angle between cusp and base. Posterior base is flangelike and compressed, joins expanded base. Carina on inner lateral face. Faint outer lateral carina.

Etymology — Named after Kwadacha Wilderness Park.

Discussion — This species is close to *R. manitouensis* and *R. tenuis* but has an expanded base and poorly developed posterior carina.

Occurrence — *Rossodus manitouensis* Zone, Kechika Formation.

Material — 12 a elements, 4 b elements, 3 c elements, 5 e elements.

Repository — GSC 119136, 119137, 119138, 119139.

Rossodus manitouensis Repetski and Ethington, 1983

Plate 13, figs. 1–6

Rossodus manitouensis, Repetski and Ethington, 1983, pp. 289–301, figs. 1–5 (contains synonymy to 1983); Ji and Barnes, 1994, p. 56, pl. 17, figs. 1–9 (contains synonymy to 1994); Ji and Barnes, 1996, p. 882, figs. 13.13–13.20.

Discussion — The apparatus of *R. manitouensis* was described adequately by Ji and Barnes (1994) as containing four morphotypes (a, b, c, and e). In this collection, two additional compressed elements are recognized. The compressed elements comprise two non-geniculate forms and one geniculate form. The non-geniculate elements have a broad, laterally compressed cusp which is flexed laterally. Cusp margins are sharply keeled, the outer and inner lateral surfaces bear faint costae. The inner basal margin is carinate, more pronounced in one type in which the postero-basal margin is drawn out. The base is expanded and basal cross section is oval, tapering posteriorly. Ji and Barnes (1994) described the compressed geniculate element.

Occurrence — *Rossodus manitouensis* Zone, Kechika Formation.

Material — 2297 a elements, 671 b elements, 123 c elements, 814 e elements.

Repository — GSC 119118 to 119123.

Rossodus muskwaensis n. sp.

Plate 13, figs. 7–15

New genus A, n. sp. A Repetski, 1982, pp. 56–57, pl. 28, figs. 1–4.

Diagnosis — A species of *Rossodus* with extremely shallow and constricted basal cavity and broad, blade-like cusp.

Description — Apparatus of four morphotypes (a, b, c, e) in which all elements have an anterior or antero-lateral, rounded carina which extends from cusp tip to base. All elements have broad, gently tapering cusp.

The a elements are asymmetrical. Anterior cusp face rounded with rounded, laterally displaced carina. Posterior cusp face flat to slightly concave. Basal margin flat on inner surface with slight expansion. Basal cavity expands on outer lateral surface and is constricted by lateral keels extending on either side of central basal cavity.

The b element has proclined cusp with two lateral keels and anterior carina. Anterior basal margin is extended down. Basal cavity is laterally constricted, with elliptical outline. Element is compressed laterally with inner and outer lateral expansion of base.

The c element is symmetrical, with erect cusp. Anterior cusp face bears round, medial carina, and cusp tapers gradually. Posterior cusp face flat to slightly concave. Lateral cusp margins keeled. Basal margin straight, expands anteriorly and posteriorly. Basal cavity oval in lateral view, its apex shallow and at midline.

There are three types of compressed (e) elements. Two are non-geniculate with broad proclined cusp which tapers gradually. Basal two-thirds of cusp is same width as base, tapers gently. Cusp is keeled anteriorly and posteriorly. Antero-lateral cusp face bears rounded carina from cusp tip to base. Postero-lateral cusp face is concave and twisted posteriorly. Basal cavity oblique, antero-basal margin is drawn out. Basal cavity apex is anterior to midline in postero-lateral view. More common, second compressed (e) element (Pl. 13, fig. 13) has broad, gently tapered, proclined cusp. Anterior cusp face bears carina; lateral cusp margins are sharp. Posterior cusp face concave. Base is straight with extremely shallow, small, circular cavity located centrally. Basal cavity is right triangle in posterolateral view, its apex anterior of midline. Base has slight inner lateral expansion.

Most compressed (e) element (Pl. 13, fig. 7) is geniculate with broad cusp and shallow basal cavity. Base meets cusp at less than 45°. Anterior and posterior cusp margins are sharp. Antero-lateral carina is faint on slightly rounded antero-lateral face. Base is flat with very faint inner flare.

Etymology — Named for the Muskwa River.

Discussion — Elements differ from *R. manitouensis* in the constriction of the base and in the straight broad cusps of all morphotypes. The compressed geniculate elements of the two species are close, but differ in that *R. manitouensis* has a well-developed inner lateral basal expansion and sharper, more tapered cusp compared to the broad cusp of *R. muskwaensis* n. sp.

Occurrence — *Rossodus manitouensis* Zone, Kechika Formation.

Material — 241 a elements, 53 b elements, 20 c elements, 64 e elements.

Repository — GSC 119124 to 119130.

Rossodus sheffieldensis n. sp.

Plate 14, figs. 1–6

Rossodus manitouensis Repetski and Ethington. Ji and Barnes, 1996, p. 882, figs. 13.14 (b element), 13.16 (a element), 13.19, 13.20 (e element).

Diagnosis — A species of *Rossodus* in which four morphotypes (a, b, c, e) bear strong posterior carina extending from base to cusp tip and have lateral or antero/postero-lateral keels which extend below basal margin.

Description — An apparatus of four elemental morphotypes (a, b, c, e) characterized by strong carina or keels. The a elements are part of symmetry transition series of elements with four strong keels (two lateral, one anterior, one posterior). Cusp is erect but curved posteriorly. Anterior cusp face bears strong rounded carina. Posterior cusp face is concave but with rounded posterior keel from cusp to base. Base is shallow and expands posteriorly and anteriorly under carinae. Base constricted under sharp lateral keels which extend below basal margin.

The b element is asymmetrical with laterally compressed cusp that is more compressed along postero-lateral margin. Cusp bears strong lateral keels which extend below basal margin. Anterior cusp face bears rounded carina that sharpens distally and extends below basal margin. Posterior cusp face bears two longitudinal grooves which flank a broad, rounded carina. Base is flared posteriorly and anteriorly under carinae.

The c element is symmetrical with same characteristics as a element but has less extension of lateral keels. It is characterized by strong, rounded posterior carina from base to cusp tip. Cusp is slightly recurved posteriorly.

The e element is compressed and geniculate. Cusp bears an inner and outer lateral rounded carina and sharp antero-lateral and posterior cusp margins. Base bears median carina extended down from cusp which is flanked by two longitudinal grooves on inner lateral surface.

Etymology — Named after Mt. Sheffield.

Discussion — Overall, the apparatus is similar to that of *R. manitouensis*, but the elements are much more robust with the strong development of the posterior or postero-lateral carinae and longitudinal grooves. These features also distinguish *R. sheffieldensis* from *R. muskwaensis*, which has reduced development of carinae. All elements are also characterized by the strong development of lateral keels which extend below the basal margin in all but the c element. The e element is distinguished by its strong carina on the antero- and postero-lateral cusp faces.

Occurrence — *Rossodus manitouensis* Zone, Kechika Formation.

Material — 54 a elements, 12 b elements, 4 c elements, 11 e elements.

Repository — GSC 119131 to 119135.

Rossodus subtilis n. sp.

Plate 14, figs. 7–11

Diagnosis — A new species of *Rossodus* with apparatus of four morphotypes (a, b, c, e). Elements are small and compressed laterally or posteriorly with thin, flat, blade-like cusp, rounded basal margins, and shallow basal cavity.

Description — The a elements are subsymmetrical and rounded. Cusp has two strong postero-lateral and antero-lateral margins. Postero-lateral face is concave; antero-lateral face is convex and bears a rounded carina. Basal length is half cusp height.

The b element is laterally compressed. Cusp is erect, extremely thin, and bears faint anterior carina. Basal cavity is right triangle in lateral view, its apex located at midline. Postero-basal corner is drawn down below basal margin.

The c element is symmetrical. Cusp is antero-posteriorly compressed with sharp lateral cusp margins. Anterior cusp face bears distinct carina from base to tip and is gently convex. Posterior cusp face is strongly concave. Base is two-thirds height of cusp. Basal cavity is confined to central part of element and expands only slightly under posterior carina.

The e element is geniculate. Cusp is close to postero-basal margin, almost fused, at a 20° angle to base. Postero-lateral face is concave. Antero-lateral face bears faint, broad carina.

Etymology — Subtilis means thin, referring to delicate nature of all elements.

Discussion — This species is distinguished from all other species of *Rossodus* by its small size and thin nature of elements. It is most similar to *R. kwadachaensis*, but elements are much thinner with less broad bases.

Occurrence — *Rossodus manitouensis* Zone, Kechika Formation.

Material — 12 a elements, 5 b elements, 4 c elements, 4 e elements.

Repository — GSC 119143, 119144, 119145, 119146.

Rossodus tenuis (Miller, 1980)

Plate 13, figs. 21–26

Oistodus minutus Miller, 1969, p. 443, pl. 66, figs. 5–7 only.

Utahconus tenuis (Miller). Miller, 1980, pp. 36–37, pl. 2, figs. 5–7, text-fig. 4T; Taylor and Landing, 1982, p. 189, text-fig. 5L; Wang, 1986, pp. 241–242, pl. 2, fig. 3.

Rossodus tenuis (Miller). Ji and Barnes, 1994, p. 56, pl. 17, figs. 10–19.

Description — Apparatus is characterized by a and c elements, which have rounded anterior costae characteristic of the genus, and basal cavity outline is round in lateral view.

Three types of compressed e elements are long, slender, tiny coniform elements with shallow basal cavities. Two are geniculate, one is non-geniculate. Non-geniculate element (Pl. 13, fig. 26) has proclined cusp which is postero-laterally compressed. Anterior cusp margin is rounded and straight in lateral profile. Posterior basal margin is smoothly round. Inner antero-lateral margin bears faint carina and is flared along basal margin. Basal cavity is shallow, its apex under carina.

Second form (Pl. 13, fig. 22) has laterally compressed cusp with sharp anterior and posterior margins. Cusp is flexed laterally. Posterior basal corner is drawn out; angle between base and cusp is almost 90°. Basal margin is expanded on inner lateral side. Anterior basal margin smoothly joins cusp. Anterior cusp margin is straight. Basal margin outline is oval, drawn out anteriorly and posteriorly with slight expansion of inner lateral basal margin.

The most compressed, geniculate element (Pl. 13, fig. 21) is delicate with laterally compressed cusp with sharp anterior and posterior margins. Inner face bears rounded carina extending along midline to basal margin. Inner anterior base is compressed, and slight carina gives expansion on middle of inner base. Base meets cusp at 45° angle, and inner antero-basal corner is rounded.

The c element is symmetrical with erect cusp which bears anterior and posterior carinae. Base expands posteriorly. Posterior cusp face is concave, gradually tapers distally. Basal outline round and expanded anteriorly.

The a element is asymmetrical. Base is wide and rounded but tapers laterally and antero-laterally. Cusp margins are strongly keeled laterally and antero-laterally. Cusp is long and slender, compressed anteriorly and posteriorly.

The b element is laterally compressed with sharp anterior and posterior cusp margins. Cusp bears faint anterior and posterior carina. Base is drawn out posteriorly, postero-basal margin flared. Basal cavity is shallow right triangle in lateral view, its apex near the anterior.

Discussion — *R. tenuis* was partially described by Ji and Barnes (1994), and as indicated in the synonymy, some of its elements have been illustrated by others, characteristically the larger geniculate element. The material in this present study shows the apparatus is more complex than previously described and is intermediate between *U. utahensis* and *R. manitouensis* by further elaboration of the costae on the cusp. The compressed elements are close to those of *U. utahensis* but are distinguished by the extension of the base and more central location of the basal cavity apex. The presence of a geniculate M element distinguished *R. tenuis* from *U. utahensis*. All morphotypes of *R. tenuis* are similar to *R. manitouensis* but small in size with more slender cusp and base and more rounded basal outline.

Occurrence — *Rossodus tenuis* Zone to *Rossodus manitouensis* Zone, Kechika Formation.

Material — 810 a elements, 213 b elements, 120 c elements, 350 e elements.

Repository — GSC 119112 to 119117.

Genus *Tricostatus* Ji and Barnes, 1994

Type species — *Tricostatus glyptus* Ji and Barnes, 1994

Tricostatus cf. *T. glyptus* Ji and Barnes, 1994

Plate 15, figs. 23–26

cf. *Tricostatus glyptus* Ji and Barnes, 1994, p. 66, pl. 17, figs. 20–25.

Discussion — An apparatus of three morphotypes is observed, however, the a elements bear four costae, not three, and are characterized by a strong posterior costa. The e elements are close to a compressed element of *R. tenuis*, but bear a postero-lateral costa on the basal expansion.

Occurrence — *Rossodus manitouensis* Zone, Kechika Formation.

Material — 61 a elements, 16 c elements, 5 e elements.

Repository — GSC 119147, 119148, 119149, 119150.

Tricostatus infundibulum n. sp.

Plate 16, figs. 1–4

Diagnosis — A new species of *Tricostatus* with apparatus of three morphotypes (a, b, c) which are characterized by prominent rounded posterior or postero-lateral costa and rounded basal carina which looks like an inverted funnel.

Description — Apparatus of three morphotypes (a, c, e) which have prominent posterior or postero-lateral costa and round basal carina, which together look like inverted funnel.

The a elements are subsymmetrical with elongate, proclined to erect cusp. Antero-lateral cusp face bears faint carina from cusp to base. Postero-lateral cusp face is concave with sharp costa which fades distally and grades basally into basal carina. Anterior and posterior cusp margins are sharp. Base is oval and shallow, expanded postero-laterally.

The c element is symmetrical with erect cusp which is recurved posteriorly. Posterior cusp face is concave and bears sharp rounded costa which smooths basally into basal carina. Lateral cusp margins are sharp. Anterior cusp face bears faint costa. Base is oval and expands posteriorly.

The e element is compressed and non-geniculate. Cusp bears posterior costa, and anterior cusp face is rounded. Base extends postero-laterally and flares posteriorly.

Etymology — Infundibulum means funnel-shaped, referring to shape of rounded posterior costae which joins rounded basal carina.

Discussion — The elements are similar to those of *T. glyptus* but have a much more prominent costa and expanded base.

Occurrence — *Rossodus manitouensis* Zone, Kechika Formation.

Material — 23 a elements, 2 c elements, 8 e elements.

Repository — GSC 119140, 119141, 119142.

Tricostatus terilinguis n. sp.

Plate 16, figs. 5–9

Diagnosis — A new species of *Tricostatus* with an apparatus of four morphotypes (a, b, c, e) in which elements are characterized by a small, round basal cavity. The c element is diagnostic in bearing a tonguelike flange which is a projection of the anterior carina.

Description — Apparatus of four elemental morphotypes: a (subround), b (drepanodiform), c (suberect, symmetrical), and e (compressed oistodiform). The elements are extremely compressed, elongate, and bladelike and characterized by a small, round basal cavity.

The a element is compressed laterally with bladelike cusp and weakly developed anterior carina. Cusp is proclined to erect and broad, narrowing only slightly distally from base, flexed postero-laterally. Anterior face is smoothly rounded; carina is round and bulbous at its base, terminating at a point higher than that on posterior face. Posterior face is concave and weakly carinate; carina fades distally along cusp. Lateral cusp margins are sharply keeled, one longer than the other, producing tapered base. Base is small and shallow, broadened under anterior and posterior carinae.

The b element is extremely compressed antero-posteriorly and bladelike. Cusp has faint anterior and posterior carinae and sharply keeled lateral cusp margins. Cusp is elongate, narrowing slightly distally, flexed slightly posteriorly. Basal cavity is extremely compressed and shallow.

The c element is symmetrical, elongate, compressed laterally. Cusp has rounded anterior and posterior carinae and sharp lateral cusp margins. Anterior cusp face is carinate, from base to tip, carina extends below basal margin as a flange. Posterior face flattened, rounded carina is raised, extends from base to tip. Basal cavity is small, circular, and shallow.

The e element is oistodiform. Cusp bears faint carina on inner lateral face, posterior of midline. Outer lateral face is weakly carinate along midline. Cusp forms gentle acute angle to base. Basal margin joined by inner lateral carina.

Etymology — Named for the straight (teri-), tonguelike (linguis) flange of the c element.

Discussion — *T. terilinguis* n. sp. appears to be closely related to *R. manitouensis* in that it shares characteristic features of the genus such as the anterior carina of the a, b, and c elements. The e element is close to that of *R. manitouensis*, but is more compressed and the cusp joins the base less acutely. Compared to other species of *Tricostatus*, *T. terilinguis* has less developed postero-lateral costae. The c element, although rare, is diagnostic in its anterior flange. The a element is also distinct in having a flangelike projection of posterior margin.

Occurrence — *Rossodus manitouensis* Zone, Kechika Formation.

Material — 11 a elements, 4 b elements, 5 c elements, 2 e elements.

Repository — GSC 119153, 119154, 119155, 119156.

Family PARACORDYLODONTIDAE Bergström in Clark et al., 1981

Genus *Fahraeusodus* Stouge and Bagnoli, 1990

Fahraeusodus marathonensis (Bradshaw, 1969)

Plate 23, figs. 21–24

Occurrence — *Acodus kechikaensis* Zone, Kechika Formation, to *Oepikodus communis* Zone, Skoki Formation.

Material — 7 M elements, 47 S elements, 16 P elements.

Repository — GSC 119597, 119598, 119599, 119600.

Genus *Oelandodus* van Wamel, 1974

Oelandodus elongatus van Wamel, 1974

Plate 11, figs. 20–23

Occurrence — *Paracordylodus gracilis* Zone, Kechika Formation.

Material — 14 M elements, 97 S elements, 19 P elements.

Repository — GSC 119320, 119321, 119322, 119323.

Genus *Paracordylodus* Lindström, 1971

Paracordylodus gracilis Lindström, 1955

Plate 12, figs. 6–9

Occurrence — *Paracordylodus gracilis* Zone, Kechika Formation.

Material — 34 M elements 54 S elements, 34 P elements.

Repository — GSC 119268, 119269, 119270, 119271.

Genus *Protoprioniodus* McTavish, 1973

Type species — *Protoprioniodus simplicissimus* McTavish, 1973

Discussion — The description of the apparatus of *Protoprioniodus* McTavish by Stouge and Bagnoli (1988) as comprising M, S, and P elements is herein followed. Most elements are extremely delicate and small. Analogies with the apparatus plans of *Paracordylodus* and *Oistodus* are clarified using the M, S, P notation, however, the apparatus plans of some species, such as *P. nyinti* Cooper, may contain highly specialized P elements.

Protoprioniodus aranda Cooper, 1981

Plate 22, figs. 1–5

Protoprioniodus aranda, Cooper, 1981, pp. 175–176, pl. 29, figs. 1, 6, 7, 10, 12; Ethington and Clark, 1982, pp. 86–87, pl. 9, figs. 24–30; Repetski and Alberstadt, 1989, p. 234, figs. 8A, B; Pohler, 1994, p. 66, pl. 6, figs. 10–12; Harris et al, 1995*b*, p. 27, pl. II, fig. R.

Discussion — Apparatus is characterized by the P element which bears a short cusp and anterior process and a curved posterior process with a prominent notch between the posterior process and the cusp. Some P elements have a well-developed lateral carina and a ledge along the lower basal margin. The upper margin of the posterior process is rounded. The M element described by Cooper (1981) is more likely the P element, similar to that of *P. cowheadensis* as described by Stouge and Bagnoli (1988). The S elements comprise a symmetrical Sa element and asymmetrical Sb element, both characterized by the conspicuous notch between the curved posterior process and the cusp. The anterior and lateral processes are sharp and short.

Occurrence — *Jumudontus gananda* Zone, Skoki Formation, to *Paroistodus originalis* Zone, Ospika Formation.

Material — 14 M elements, 22 S elements, 15 P elements.

Repository — GSC 119428, 119429, 119430, 119431, 119432.

Protoprioniodus cowheadensis Stouge and Bagnoli, 1988

Plate 27, figs. 11–13

Protoprioniodus cowheadensis Stouge and Bagnoli, 1988, p. 137, pl. 14, figs. 1–5 (contains synonymy to 1990); Albanesi, 1998, pp. 169–170, pl. 10, figs. 19–21.

Discussion — Stouge and Bagnoli (1988) described the apparatus as containing M, S, and P elements, however, only an Sa, Sb, and Sc were recovered in this small collection. The S elements have a characteristic bladelike posterior process which has an arched aboral margin and anticusp comprising the anterior basal margin.

Occurrence — *Oepikodus communis* Zone, Skoki Formation.

Material — 7 S elements.

Repository — GSC 119635, 119636, 119637.

Protoprioniodus nyinti Cooper, 1981

Plate 27, figs. 14–20

Protoprioniodus nyinti Cooper, 1981, pp. 176–178, pl. 29, figs. 1–8, 11, 12.

Discussion — A species with highly specialized P elements which Cooper (1981) described as an arched bar, swollen on one side and concave on the other. The swollen face bears a longitudinal ledge along the basal margin. The M elements illustrated by Cooper (1981) are likely a second type of P element with fusion of the cusp to the base along much of its length. The S elements differ from those described by Cooper (1981) in that they have strong, curved lateral costae and wider, arched posterior processes. An M element is included in the apparatus which is not illustrated by Cooper (1981) in which the arched posterior process and lateral cusp bear a rounded costae along their length. The cusp and process are thinly keeled and have curved posterior and upper margins, respectively.

Occurrence — *Oepikodus communis* Zone, Skoki

Formation, to *Paroistodus originalis* Zone, Ospika Formation.

Material — 20 M elements, 51 S elements, 58 P elements.

Repository — GSC 119628 to 119634.

Protoprioniodus n. sp.

Plate 27, figs. 9, 10

Description — An incomplete apparatus of S and P elements which bear strong costate "bar" along the base and posterior process and costae on the cusp.

P element is similar to that of *P. nyinti* but has robust base which bears costate bar, comprising two rounded costae which converge posteriorly and are separated by groove. Fusion is strong between cusp and base, and cusp bears strong rounded costae.

S element also has strong fusion of the keeled base and cusp. It also has two costae on the base which meet at the strong lateral keel and diverge posteriorly along the base.

Discussion — The elements occur with *P. nyinti* and are similar, but the presence of the grooved bar on the base suggests this is a new species. However, there is not enough material to reconstruct the apparatus.

Occurrence — *Jumudontus gananda* Zone, Skoki Formation.

Material — 4 elements.

Repository — GSC 119638 and 119639.

Family PERIODONTIDAE Lindström, 1970

Genus *Microzarkodina* Lindström, 1971

Type species — *Prioniodina flabellum* Lindström, 1955

Microzarkodina n. sp.

Plate 24, figs. 12–14

Description — P elements are different from species previously described because they bear a large anterior denticle and small posterior denticles. As so few elements were recovered and no S elements, specific assignment is not attempted. Compared to *M. flabellum*, this new species has a larger anterior denticle and smaller posterior process and denticles.

Occurrence — *Jumudontus gananda* Zone, Skoki Formation.

Material — 8 M elements, 6 P elements.

Repository — GSC 119543, 119544, 119545.

Genus *Periodon* Hadding, 1913

Periodon aculeatus Hadding, 1913

Plate 21, figs. 7–9

Occurrence — *Oepikodus communis* Zone, Skoki Formation to *Dzikodus tablepointensis* Zone, Ospika Formation.

Material — 216 M elements 562 S elements, 167 P elements.

Repository — GSC 119433, 119434, 119435.

Periodon flabellum (Lindström, 1955)

Plate 26, figs. 3–6

Occurrence — *Oepikodus communis* Zone, Skoki Formation, to *Paroistodus originalis* Zone, Ospika Formation.

Material — 69 M elements 98 S elements, 18 P elements.

Repository — GSC 119618, 119619, 119620, 119621.

Periodon selenopsis (Serpagli, 1974)

Plate 26, figs. 7, 8

Occurrence — *Oepikodus communis* Zone to *Jumudontus gananda* Zone, Skoki Formation.

Material — 8 M elements, 7 P elements.

Repository — GSC 119116 and 119617.

Family PRIONIODONTIDAE Bassler, 1925

Genus *Oepikodus* Lindström, 1955

Type species — *Oepikodus smithensis* Lindström, 1955

Discussion — The apparatus plan of *Oepikodus* was emended by van Wamel (1974) as multielemental and comprises a series of S elements in which the Sa position is not readily distinguished, which distinguishes it from *Prioniodus*. There are two P elements and one longiramiform M element.

Oepikodus communis (Ethington and Clark, 1964)

Plate 24, figs. 16–19

Occurrence — *Oepikodus communis* Zone, Kechika and Skoki formations.

Material — 110 M elements, 752 S elements, 273 P elements.

Repository — GSC 119494, 1194495, 119496, 119497.

Oepikodus evae (Lindström, 1955)

Plate 25, figs. 1–3

Occurrence — *Oepikodus evae* Zone, Skoki Formation.

Material — 13 M elements, 67 S elements, 27 P elements.

Repository — GSC 119503, 119504, 119505.

Oepikodus intermedius (Serpagli, 1974)

Plate 25, figs. 4–6

Occurrence — *Oepikodus communis* Zone to *Jumudontus gananda* Zone, Kechika and Skoki formations.

Material — 27 M elements, 142 S elements, 61 P elements.

Repository — GSC 119500, 119501, 119502.

Oepikodus n. sp.

Plate 24, figs. 20, 21

Description — A new species of *Oepikodus* characterized by P elements with large denticles of equal size which are fused at base. Cusp bears lateral adenticulate process which is deflected anteriorly and arises from cusp at point of curvature of cusp. Process extends up to cusp tip as a strong keel.

Discussion — This new species of *Oepikodus* requires more material to determine the apparatus structure, but the nature of the denticulation distinguishes it from other species. The denticles are fewer in number and larger than in *O. evae*, *O. intermedius*, and *O. communis*.

Occurrence — *Jumudontus gananda* Zone, Skoki Formation.

Material — 4 P elements.

Repository — GSC 119498 and 119499.

Genus *Prioniodus* Pander, 1856

Type species — *Prioniodus elegans* Pander, 1856

Prioniodus adami Stouge and Bagnoli, 1988

Plate 26, figs. 19–21

Occurrence — *Oepikodus communis* Zone, Skoki Formation.

Material — 7 S elements, 2 P elements.

Repository — GSC 119564, 119565, 119566.

Prioniodus elegans Pander, 1856

Plate 12, figs. 22–25

Occurrence — *Prioniodus elegans* Zone, Kechika and Ospika formations.

Material — 2 M elements, 11 S elements, 1 P element.

Repository — GSC 119309, 119310, 119311, 119312.

?*Prioniodus* n. sp.

Plate 27, figs. 1–4

Description — An incomplete apparatus containing S, M, and P elements is tentatively assigned to *Prioniodus*. The denticles are highly fused. The Sb elements have a lateral adenticulate process and a ledge along the posterior process. The Sd elements have a stronger lateral process and extension of the anterior cusp margin. The denticulate posterior process bears fused denticles on an arched process. Compared to *P. adami* and *P. elegans*, the arched posterior process with a ledge is the distinguishing feature.

Occurrence — *Oepikodus communis* Zone to *Jumudontus gananda* Zone, Skoki Formation.

Material — 2 M elements, 15 S elements, 2 P elements.

Repository — GSC 119567 to 119570.

Family RHIPIDOGNATHIDAE Lindström, 1970

Genus *Bergstroemognathus* Serpagli, 1974

Bergstroemognathus extensus (Graves and Ellison, 1941)

Plate 23, figs. 5–7

Remarks — The apparatus of *B. extensus* was described as multielemental by Serpagli (1974), and Albanesi (1998) illustrates a quinquimembrate apparatus plan containing Sa, Sb, Sc, M, and P elements. As Stouge and Bagnoli (1988) noted, the nature of the denticulation of the anterior process of the M element can vary between early and late members of the species, which is observed in early and late forms in this present collection.

Occurrence — *Oepikodus communis* Zone, Skoki Formation.

Material — 19 M elements, 10 S elements.

Repository — GSC 119506, 119507, 119508.

Order PRIONIODINIDA Sweet, 1988

Family CHIROGNATHIDAE Branson and Mehl, 1944

Genus *Erraticodon* Dzik, 1978

Erraticodon balticus Dzik, 1978

Plate 20, figs. 8, 9

Occurrence — *Paroistodus originalis* Zone, Ospika Formation.

Material — 1 S element, 1 P element.

Repository — GSC 119436 and 119437.

Genus *Spinodus* Dzik, 1976

Spinodus spinatus (Hadding, 1913)

Plate 22, figs. 6, 7

Occurrence — *Paroistodus originalis* Zone to *P. horridus* Zone, Ospika Formation.

Material — 4 S elements.

Repository — GSC 119424 and 119425.

Family PRIONIODINIDAE Bassler, 1925

Genus *Oulodus* Branson and Mehl, 1933

Oulodus sp. A

Plate 20, figs. 12, 13

Remarks — The specimens are treated in open nomenclature because too few specimens are available for a determination. Sa, Sb, and Sc elements can be recognized.

Occurrence — *Distomodus staurognathoides* Zone, Pesika Formation.

Material — 9 S elements.

Repository — GSC 119490 and 119491.

Oulodus sp. B

Plate 20, fig. 14

Remarks — Specimens from the Kwadacha Formation are assigned to *Oulodus*, but too few specimens were recovered to make a determination of species.

Occurrence — Kwadacha Formation.

Material — 5 S elements.

Repository — GSC 119492.

Family NOV. 5 Aldridge and Smith, 1993

Genus *Pseudooneotodus* Drygant, 1974

Pseudooneotodus beckmanni (Bischoff and Sannemann, 1958)

Plate 21, fig. 18

Remarks — *P. beckmanni* shows variation in its basal outline from round to oval and bears a single apex which is reclined posteriorly (Cooper 1977). The specimens in this collection are laterally compressed and have an oval basal outline.

Occurrence — *Distomodus staurognathoides* Zone, Pesika Formation.

Material — 2 elements.

Repository — GSC 119493.

Taxa of unknown affinity

Order UNKNOWN

Family NOV. 3 Aldridge and Smith, 1993

Genus *Loxodus* Furnish, 1938

Type species — *Loxodus bransoni* Furnish, 1938

Loxodus latibasis Ji and Barnes, 1994

Plate 11, figs. 10, 11

Loxodus latibasis Ji and Barnes, 1994, p. 46, pl. 12, figs. 14–17, text-fig. 29B.

Discussion — As described by Ji and Barnes (1994), the apparatus is unimembrate, with left and right elements. The basal portion is large, occupying one-half to two-thirds of the element. The denticulated portion comprises vaguely discrete denticles which are inclined posteriorly and confluent along their bases. The elements are laterally compressed, and the denticulated portion is flexed along a line between the denticle row and basal portion, which is a distinguishing character from *L. bransoni*, which shows no flexure in the element. A difference noted in this collection is that the anterior margin is concave at the point of flexure. Some elements have a small node which is anteriorly directed, originating from the point of concavity in the anterior margin. The anterior denticle is small and erect to slightly flexed posteriorly. The basal cavity is narrow and deep.

Occurrence — *Rossodus manitouensis* Zone, Kechika Formation.

Material — 17 dextral elements, 14 sinistral elements.

Repository — GSC 119072 and 119073.

Order UNKNOWN

New genus B, new species

Plate 29, figs. 10–13

Diagnosis — Apparatus of at least three coniform morphotypes (a, b, e) in which the a elements are diagnostic in bearing an elongate, erect cusp, small base, and prominent antero-lateral keel.

Description — The a element is laterally compressed with erect cusp. Cusp strongly keeled anteriorly and posteriorly. Outer lateral face bears strong antero-lateral keel which arises one-third length of cusp from basal margin as anterior cusp margin diverges. Keel is deflected anteriorly. Divergence of anterior cusp margin gives basal outline a triangular profile. Base is deep.

The b element is laterally compressed with erect cusp and small base. Cusp is laterally compressed. Anterior and posterior cusp margins sharply rounded. Cusp twisted laterally. Base open laterally and is round in lateral view.

The e element is geniculate and compressed. Cusp is flexed inward and outer lateral cusp face is keeled. Antero-basal margin is rounded.

Discussion — As so few elements were recovered, the assignment of this new genus is unknown. More material is needed to reconstruct the whole apparatus.

Occurrence — *Oepikodus communis* Zone, Skoki Formation.

Material — 6 elements.

Repository — GSC 119546, 119547, 119548, 119549.

Coniform indet. A

Plate 18, fig. 1

Description — Simple coniform with proclined cusp, at least twice length of base. Cusp bears two strong costae, one along medial lateral cusp face, the other antero-lateral. Posterior and anterior cusp margins sharp. Base is square, and angle between postero-basal edge and cusp is greater than 100°.

Occurrence — *Rossodus manitouensis* Zone, Kechika Formation.

Material — 5 elements.

Repository — GSC 119104.

Coniform indet. B

Plate 18, figs. 2, 3

Description — Stout coniforms of two morphotypes which resemble elements of *Clavohamulus* and have shallow basal cavities. One morphotype (a) is stout with triangular cross section and widely rounded base. Posterior cusp face deeply concave; anterior cusp face rounded. Posterior basal margin flat; anterior basal margin rounded. The e element is stout with slender, erect, bulbous cusp. Base is large and round with thickened rim. Above basal margin at point where cusp emerges are four pockets or concave regions separated by a rounded costae extending off cusp onto basal rim.

Occurrence — *Rossodus manitouensis* Zone, Kechika Formation.

Material — 2 elements.

Repository — GSC 119151 and 119152.

Coniform indet. C

Plate 18, figs. 4, 5

Description — Stout coniforms which have a costate cusp and wide base with rounded rim. Cusp is erect and bears sharp costae flanked by deep concave regions which extend from cusp tip to basal rim.

Occurrence — *Rossodus manitouensis* Zone, Kechika Formation.

Material — 10 elements.

Repository — GSC 119157 and 119158.

Coniform indet. D

Plate 18, fig. 6

Description — Simple coniforms which are robust and stout with large, round, gaping basal cavity and sharp posterior cusp margin. Three morphotypes are recognized in which a and c elements comprise first symmetry transition series and compressed e elements the second series.

The a element is rounded and asymmetrical to nearly symmetrical. Cusp is proclined and has straight anterior cusp margin. Posterior cusp margin is sharp. Base is slightly defined but not expanded. Basal cavity is deep and wide, opening posteriorly.

The c element is symmetrical, with erect cusp and posteriorly sharpened margin, flanked by slight concavity along postero-lateral sides of element. Base is expanded with rounded cross section.

The e element is compressed and broad with oval cross section. Posterior cusp margin is sharp. As in c element, outer and inner lateral faces have a slight concavity along posterior margin. Cusp is erect, bears faint lineation along antero-lateral margin. Basal margin bears inner lateral flare or carina.

Occurrence — *Rossodus manitouensis* Zone, Kechika Formation.

Material — 5 elements.

Repository — GSC 119207.

Coniform indet. E

Plate 18, figs. 7–9

Description — Simple cones with erect to proclined cusp, compressed laterally, some more than others. Cusp margins sharp. Base has characteristic shape in being drawn out posteriorly and shallow with flare on inner lateral basal margin producing small flange. Outer lateral cusp and base are flat. All specimens are symmetrical.

Occurrence — *Iapetognathus* Zone, Kechika Formation.

Material — 13 elements.

Repository — GSC 119218, 119219, 119220.

Coniform indet. F

Plate 18, fig. 10

Description — An apparatus characterized by elongate, simple cones with compressed, twisted cusps so that the round basal cavity gapes on inner lateral cusp face. There are at least two morphotypes within two symmetry transition series. The a element is rounded and has a reclined cusp which is laterally compressed above base. Posterior and anterior cusp margins are sharp. Inner lateral cusp face bears carina. Base is expanded laterally. Outer lateral face is smooth and bears faint carina along its length. Basal outline is circular and same width as cusp. Basal cavity is deep.

Compressed e element is erect and compressed along whole element. Inner lateral cusp face bears a costa along anterior cusp margin. Outer lateral cusp face is smooth except for faint carina along anterior cusp margin. Basal outline is oval and constricted.

Occurrence — *Rossodus manitouensis* Zone, Kechika Formation.

Material — 17 elements.

Repository — GSC 119221.

Coniform indet. G

Plate 18, fig. 11

Description — Simple cones with a slender, recurved cusp which bears fine striae over all cusp faces. Base is poorly defined but opens widely and has rounded margins.

Occurrence — *Acodus kechikaensis* Zone, Kechika Formation.

Material — 19 elements.

Repository — GSC 119272.

Coniform indet. H

Plate 18, figs. 12–14

Description — Elements are extremely laterally compressed simple cones, lacking any costae or carinae on elongate cusp, with rounded cusp margins. Compressed e element has sharp posterior cusp margin with more compression along posterior cusp margin. Basal cavity is narrow and extends to point of curvature of cusp, its apex located centrally. Some elements have a long base and concave anterior cusp margin.

Occurrence — *Iapetognathus* Zone, Kechika Formation.

Material — 55 elements.

Repository — GSC 119281, 119282, 119283.

Coniform indet. I

Plate 18, figs. 15, 16

Description — Robust simple cones which bear faint, rounded postero-lateral costae on the cusp. One morphotype is laterally compressed with a broadly tapered cusp and slightly flared base. The other has a slender cusp and pronounced flare on inner lateral base.

Occurrence — *Scolopodus subrex* Zone, Kechika Formation.

Material — 2 elements.

Repository — GSC 119308, 120665.

Coniform indet. J

Plate 28, figs. 18, 19

Description — Elements are simple coniforms which are laterally compressed with bladelike cusps. Anterior and posterior cusp margins and basal corners sharp. Basal cavity shallow with apex located anterior to midline. Elements have an overall similarity to the compressed elements of *Utahconus utahensis*.

Occurrence — *Jumudontus gananda* Zone, Skoki Formation.

Material — 15 elements.

Repository — GSC 119550 and 119551.

Coniform indet. K

Plate 28, fig. 21

Description — Simple coniform elements with broad, posteriorly expanded base with triangular profile. Cusp is slender and bears one strong lateral costae and minor striae posterior to costae. Overall morphology is similar to *Walliserodus*.

Occurrence — *Oepikodus communis* Zone, Skoki Formation.

Material — 4 elements.

Repository — GSC 119562.

Coniform indet. L

Plate 28, fig. 20

Description — Some elements from the Skoki in which simple hyaline coniforms have a large, circular base and small, tapering cusp which bears one weak to moderately strong longitudinal groove from the cusp tip to almost the basal margin. No suberect symmetrical elements or compressed elements were observed, only a elements The elements have affinities with *Semiacontiodus* or *Acanthodus* and could be placed tentatively among the Protopanderodontidae.

Occurrence — *Jumudontus gananda* Zone, Skoki Formation.

Material — 6 elements.

Repository — GSC 119563.

Coniform indet. M

Plate 29, figs. 1–3

aff. *Aloxoconus* sp. A. Smith, 1991, p. 20, fig. 12c.

Description — Incomplete apparatus of hyaline elements which have rounded costae and rounded basal margins.

Symmetry transition of a elements in which symmetrical a element has rounded carinate posterior face and rounded anterior cusp margin. Lateral cusp margins are slightly costate. Asymmetrical elements have rounded inner lateral carina. Basal margin is rounded and base constricted.

Symmetrical suberect c element has sharp lateral cusp margins. Posterior cusp face is convex with central costa flanked by two rounded costae. Anterior cusp face is smooth. Base flares up under the central costa and has round basal outline.

Discussion — The elements are similar to those of Coniform indet. O, but have a constricted base and more rounded costae. The symmetrical element is more stout with costae instead of striae. The symmetrical element is somewhat similar to *Aloxoconus* sp. A of Smith (1991) but is broader and has more defined lateral cusp margins which flatten basally into keel-like projections. The overall morphology is similar, especially the shape of the basal cavity in posterior view on the c element.

Occurrence — *Tripodus laevis* Zone, Skoki Formation.

Material — 15 elements.

Repository — GSC 119602, 119603, 119604.

Coniform indet. N

Plate 29, figs. 4–6

Description — Apparatus of possibly four morphotypes in which hyaline elements are characterized by a elements which have a long, straight-edged posterior process.

The a element has laterally compressed cusp with sharper posterior cusp margin than anterior. Antero-basal corner is expanded and square. Postero-basal corner is drawn out into adenticulate process which has straight edges and is deflected below basal margin.

The symmetrical suberect c element is antero-posteriorly compressed. Base is small and cusp tapered. Anterior and posterior cusp faces bear faint costae. Lateral cusp margins are sharp.

The e element is geniculate and compressed with a sharp posterior cusp edge as in the other elements. The postero-basal corner has straight edges and joins the posterior cusp margin at a 20° angle. Cusp is broader than width of base.

Occurrence — *Tripodus laevis* Zone, Skoki Formation.

Material — 6 elements.

Repository — GSC 119607, 119608, 119609.

Coniform indet. O

Plate 29, figs. 7–9

Description — Incomplete apparatus of at least three morphotypes (a, c, e).

The a element is subround and asymmetrical with four costae. Anterior cusp margin is costate and outer lateral cusp face smoothly costate. Inner lateral cusp face is flat between anterior and inner lateral costae. Lateral costae fade distally up the cusp. Posterior cusp margin sharp. Base is flared laterally. Basal cavity deep.

The symmetrical c element is suberect and has sharp lateral cusp margins. Posterior cusp face convex and bears faint striae which extend from the base to the cusp tip. Anterior cusp face is convex and smoothly rounded. Base is expanded laterally and antero-posteriorly. Basal cross section is round.

A compressed e element is non-geniculate and has sharp anterior and posterior cusp margins and faint costae laterally. The base is expanded posteriorly. Basal cavity is deep and base is flared laterally.

Discussion — As there were only a few elements discovered, the apparatus may contain more elemental morphotypes, such as a symmetrical a element and a transitional b element.

Occurrence — *Tripodus laevis* Zone, Skoki Formation.

Material — 9 elements.

Repository — GSC 119610, 119611, 119612.

Platform indet.

Plate 28, figs. 22, 23

Description — A few small platform elements from the Skoki Formation have a small recurved cusp tip and wide, deep basal cavity. Posterior face of element concave; anterior face rounded. Both faces ornamented by a series of nodose ridges which extend from base and fade distally, except for the lateral ridges which are strongest.

Occurrence — *Oepikodus communis* Zone, Skoki Formation.

Material — 4 elements.

Repository — GSC 119605 and 119606.

References

Albanesi, G.L. 1998. Taxonomia de conodontes de las Secuencias Ordovicicas Del Cerro Potrerillo, Precordillera Central De San Juan, R. Argentina. *In* Bioestratigrafia, Biofacies y Taxonomia de Conodontes de las Secuencias Ordovicicas del Cerro Potrerillo, Precordillera Central De San Juan, R. Argentina. *Edited by* M.A. Hünicken. Academia Nacional de Ciencias, Cordoba, pp. 101–253.

Albanesi, G.L., and Barnes, C.R. 2000. Subspeciation within a punctuated equilibrium evolutionary event: phylogenetic history of the Lower–Middle Ordovician *Paroistodus originalis – P. horridus* complex (Conodonta). Journal of Paleontology, **74**: 492–502.

Aldridge, R.J., and Schönlaub, H.P. 1989. Conodonts. *In* A Global Standard for the Silurian System. *Edited by* C.H. Holland and M.G. Bassett. National Museum of Wales, Geological Series, Cardiff, pp. 274–279.

Aldridge, R.J., and Smith, M.P. 1993. Conodonta. *In* The Fossil Record 2. *Edited by* M.J. Benton. Chapman and Hall, London, pp. 563–572.

An, T., Zhang, F., Xiang, W., Zhang, Y., Xu, W., Zhang, H., Jiang, D., Yang, C., Lin, L., Cui, Z., and Yang, X. 1983. The Conodonts of North China and the Adjacent Region. Chinese Science Publishing House, Beijing, 223 pp.

Andres, D. 1988. Structures, apparatuses and phylogeny of primitive conodonts. Paleontographica (Abt. A), **200**: 105–152.

Austin, R.L. 1987. Conodonts: Investigative Techniques and Applications. Ellis Horwood Limited, Chichester, 416 pp.

Bagnoli, G., and Stouge, S. 1996. Lower Ordovician (Billingenian–Kunda) conodont zonation and provinces based on section from Horns Udde, north Öland, Sweden. Bollettino della Societa Paleontologica Italiana, **35**: 106–163.

Bagnoli, G., Barnes, C.R., and Stevens, R.K. 1987. Tremadocian conodonts from Broom Point and Green Point, western Newfoundland. Bolletino della Societa Paleontologica Italiano, **35**: 145–158.

Bagnoli, G., Stouge, S., and Tongiorgi, M. 1988. Acritarchs and conodonts from the Cambro–Ordovician Furuhall (Kopingsklint) section (Oland, Sweden). Rivista Italiana di Paleontologia e Stratigrafia, **94**: 163–248.

Barnes, C.R. 1988. The proposed Cambrian–Ordovician global boundary stratotype and point (GSSP) in Western Newfoundland, Canada. Geological Magazine, **125**: 381–414.

Barnes, C.R., and Poplawski, M.L. 1973. Lower and Middle Ordovician conodonts from the Mystic Formation, Quebec, Canada. Journal of Paleontology, **47**: 760–790.

Barnes, C.R., Kennedy, D.J., McCracken, A.D., Nowlan, G.S., and Tarrant, G.A. 1979. The structure and evolution of Ordovician conodont apparatuses. Lethaia, **12**: 125–151.

Barnes, C.R., Ji, Z., and Pohler, S.M.L. 1991. A review of Ordovician conodont paleontology. *In* Ordovician to Triassic Conodont Paleontology of the Canadian Cordillera. *Edited by* M.J. Orchard and A.D. McCracken. Geological Survey of Canada, Bulletin 417, pp. 27–39.

Bassler, R.S. 1925. Classification and stratigraphic use of the conodonts. Geological Society of America Bulletin, **36**: 218–220.

Bateson, W. 1886. The ancestry of the chordata. Quarterly Journal of Microscopical Science, **26**: 535–571.

Bauer, J.A. 1987. Conodonts and conodont biostratigraphy of the McLish and Tulip Creek formations (Middle Ordovician) of south-central Oklahoma. Oklahoma Geological Survey Bulletin, **141**: 1–58.

Bauer, J.A. 1989. Conodont biostratigraphy and paleoecology of Middle Ordovician rocks in eastern Oklahoma. Journal of Paleontology, **63**: 92–107.

Bengtson, P. 1988. Open nomenclature. Palaeontology, **31**: 223–227.

Bergström, S.M. 1971. Conodont biostratigraphy of the Middle and Upper Ordovician of Europe and eastern North America. Geological Society of America Memoir, **127**: 83–161.

Bergström, S.M. 1988. On Pander's Ordovician conodonts: distribution and significance of the *Prioniodus elegans* fauna in Baltoscandia. Senckenbergiana Lethaea, **69**(3/4): 217–251.

Bischoff, G., and Sannemann, D. 1958. Unterdevonische Conodonten aus dem Frankenwald. Notisblatt des Hessischen Landesamt für Bodenforschung, **86**: 87–110.

Bradshaw, L.E. 1969. Conodonts from the Fort Pena Formation (Middle Ordovician), Marathon Basin, Texas. Journal of Paleontology, **43**: 1137–1168.

Branson, E.B., and Branson, C.C. 1947. Lower Silurian conodonts from Kentucky. Journal of Paleontology, **21**: 549–556.

Branson, E.B., and Mehl, M.G. 1933. Conodont studies, numbers 1 and 2. University of Missouri Studies, **8**(1–2): 1–349.

Branson, E.B., and Mehl, M.G. 1944. Conodonts. *In* Index Fossils of North America. *Edited by* H.W. Shimer and R.R. Shrock. Wiley, New York, pp. 235–246.

Buggisch, W., and Repetski, J.E. 1987. Uppermost Cambrian (?) and Tremadocian conodonts from Handler Ridge, Robertson Bay Terrane, North Victoria Land, Antarctica. Geologisches Jahrbuch, **B66**: 145–185.

Cecile, M.P., and Norford, B.S. 1979. Basin to platform transition, Lower Paleozoic strata of Ware and Trutch map areas, northeastern British Columbia. *In* Current Research, Part A, Geological Survey of Canada, Paper 79-1A, pp. 219–226.

Cecile, M.P., Morrow, D.W., and Williams, G.K. 1997. Early Paleozoic (Cambrian to Early Devonian) tectonic framework, Canadian Cordillera. Bulletin of Canadian Petroleum Geology, **45**: 54–74.

Chen, J., and Gong, W. 1986. Conodonts. *In* Aspects of Cambrian–Ordovician Boundary in Dayangcha, China. *Edited by* J. Chen. China Prospect Publishing House, Beijing, pp. 93–223.

Clark, D.L., Sweet, W.C., Bergström, S.M., Klapper, G., Austin, R.L., Rhodes, H.T., Müller, K.J., Ziegler, W., Lindström, M., Miller, J.F., and Harris, A.G. 1981. Conodonta. Pt. W, Supplement 2. *In* Treatise on Invertebrate Paleontology. *Edited by* R.A. Robison. Geological Society of America and University of Kansas, Kansas, p. 202.

Cooper, B.J. 1977. Toward a familial classification of Silurian conodonts. Journal of Paleontology, **51**: 1057–1071.

Cooper, B.J. 1981. Early Ordovician conodonts from the Horn Valley Siltstone, Central Australia. Palaeontology, **24**: 147–183.

Cooper, R.A., and Nowlan, G.S. 1999. Proposed global stratotype section and point for base of the Ordovician System. Acta Universitatis Carolinae Geologica, **43**: 61–64.

Cope, E.D. 1889. Synopsis on the families of the Vertebrata. American Naturalist, **23**: 1–29.

Craig, W.W., Ethington, R.L., and Repetski, R.E. 1986. Guidebook to the conodont paleontology of uppermost Lower Ordovician through Silurian strata, northerneastern Arkansas. Annual Meeting of the South-Central and Southeastern Sections of the Geological Society of America, Memphis, 57 pp.

Cullison, J.S. 1938. Dutchtown fauna of southeastern Missouri. Journal of Paleontology, **12**: 219–228.

Davies, E.J.L. 1966. Ordovician and Silurian of the northern Rocky Mountains between Peace and Muskwa Rivers, British Columbia. Unpublished Ph.D. thesis, University of Alberta, Edmonton, Alberta, 139 pp.

Donoghue, P.C.J., Forey, P.L., and Aldridge, R.J. 2000. Conodont affinity and chordate phylogeny. Biological Reviews, **75**: 191–251.

Druce, E.C., and Jones, P.J. 1971. Cambro–Ordovician conodonts from the Burke River Structural Belt, Queensland. Australia Bureau of Mineral Resources Bulletin, **110**: 1–159.

Drygant, D.M. 1974. Simple conodonts from the Silurian and lowermost Devonian of the Volyno–Podolian area. Paleontologicheski y Sbornik, **10**: 64–69.

Dzik, J. 1976. Remarks on the evolution of Ordovician conodonts. Acta Paleontologica Polonica, **21**: 395–455.

Dzik, J. 1978. Conodont biostratigraphy and paleogeographical relations of the Ordovician Mojcza Limestone (Holy Cross Mts., Poland). Acta Palaeontologica Polonica, **23**: 51–72.

Dzik, J. 1994. Conodonts of the Mojcza Limestone. *In* Ordovician Carbonate Platform Ecosystem of the Holy Cross Mountains. *Edited by* J. Dzik, E. Olempscka, and A. Pisera. Palaeontologia Polonica, **53**: 43–128.

Epstein, A.G., Epstein, J.B., and Harris, L.D. 1977. Conodont color alteration — an index to organic metamorphism. United States Geological Survey Professional Paper, **995**: 1–27.

Ethington, R.L. 1959. Conodonts from the Ordovician Galena Formation. Journal of Paleontology, **33**: 257–292.

Ethington, R.L., and Austin, R.L. 1991. Conodonts of the Dounans Limestone, Highland Border Complex, Scotland. Journal of Micropaleontology, **10**: 51–56.

Ethington, R.L., and Brand, U. 1981. *Oneotodus simplex* (Furnish) and the genus *Oneotodus* (Conodonta). Journal of Paleontology, **55**: 239–247.

Ethington, R.L., and Clark, D.L. 1964. Conodonts from the El Paso Formation (Ordovician) of Texas and Arizona. Journal of Paleontology, **38**: 685–705.

Ethington, R.L., and Clark, D.L. 1965. Lower Ordovician conodonts and other microfossils from the Columbia Ice fields section, Alberta, Canada. Brigham Young University, Geological Studies, **12**: 185–205.

Ethington, R.L., and Clark, D.L. 1971. Lower Ordovician conodonts in North America. *In* Symposium on Conodont Biostratigraphy. *Edited by* W.C. Sweet and S.M. Bergström. Geological Society of America Memoir **127**: 63–82.

Ethington, R.L., and Clark, D.L. 1982. Lower and Middle Ordovician conodonts from the Ibex area, western Millard County, Utah. Brigham Young University, Geologic Studies, **28**: 1–160.

Fåhræus, L.E. 1966. Lower Viruan (Middle Ordovician) conodonts from the Güllhögen quarry, southern central Sweden. Sveriges Geologiska Undersökning, **60**: 1–40.

Fåhræus, L.E. 1982. Recognition and redescription of Pander's (1856) *Scolopodus* (form-) species — Constituents of multi-element taxa (Conodontiphora, Ordovician). Geologica et Palaeontologica, **16**: 19–22.

Fåhræus, L.E., and Hunter, D.R. 1985. Simple-cone conodont taxa from the Cobbs Arm Limestone (Middle Ordovician), New World Island, Newfoundland. Canadian Journal of Earth Sciences, **22**: 1171–1182.

Fåhræus, L.E., and Nowlan, G.S. 1978. Franconian (Late Cambrian) to early Champlainian (Middle Ordovician) conodonts from the Cow Head Group, western Newfoundland. Journal of Paleontology, **52**: 444–471.

Fåhræus, L.E., and Roy, K. 1993. Conodonts from the Cambro–Ordovician Cooks Brook and Middle Arm Point formations, Bay of Islands, western Newfoundland. Geologica et Palaeontologica, **27**: 1–53.

Ferri, F., Nelson, J., and Rees, C. 1995. Geology and mineralization of the Gataga River Area, northern Rocky Mountains (94I/7, 8, 9 and 10). *In* British Columbia Ministry of Energy, Mines and Petroleum Resources, Geological Fieldwork 1995. Mineral Resources Division, Geological Survey Branch, Victoria, pp. 277–295.

Ferri, F., Rees, C., and Nelson, J. 1996. Geology and mineralization of the Gataga Mountain Area, northern Rocky Mountains (94L/10, 11, 14 and 15). *In* British Columbia Ministry of Energy, Mines and Petroleum Resources, Geological Fieldwork 1995. Mineral Resources Division, Geological Survey Branch, Victoria, pp. 137–154.

Ferri, F., Rees, C., Nelson, J., and Legun, A. 1997. Geology of the Northern Kechika Trough (NTS 94L/14,15; 94M/3, 4, 5, 6, 12, 13; 104P/8, 9, 15, 16). *In* British Columbia Ministry of Energy, Mines and Petroleum Resources, Geological Fieldwork 1996. Mineral Resources Division, Geological Survey Branch, Victoria, pp. 125–144.

Fortey, R.A., Landing, E., and Skevington, D. 1982. Cambrian–Ordovician boundary sections in the Cow Head Group, western Newfoundland. *In* The Cambrian–Ordovician Boundary: Sections, Fossil Distributions and Correlations. *Edited by* M.G. Basset and W.T. Dean. National Museum of Wales, Geological Series 3, Cardiff, pp. 95–129.

Fortey, R.A., Harper, D.A.T., Ingham, J.K., Owen, A.W., and Rushton, A.W.A. 1995. A revision of Ordovician series and stages from the historical type area. Geological Magazine, **132**: 15–30.

Furnish, W.M. 1938. Conodonts from the Prairie du Chien (Lower Ordovician) beds of the Upper Mississippi Valley. Journal of Paleontology, **12**: 318–340.

Gabrielse, H. 1963. McDame Map Area, Cassiar District, British Columbia, Geological Survey of Canada, Memoir 319, pp. 1–138.

Gabrielse, H. 1981. Stratigraphy and structure of Road River and associated strata in Ware (west half) Map Area, Northern Rocky Mountains, British Columbia. Current Research, Part A, Geological Survey of Canada, Paper 81-1A, pp. 201–207.

Gabrielse, H., Dodds, C.J., and Mansy, J.L. 1977. Geology of Toodoggone (94E) and Ware (94F, west half) map-areas, British Columbia. Geological Survey of Canada, Open File 483.

Glass, D.J. 1990. Lexicon of Canadian Stratigraphy Volume 4. Western Canada, Including Eastern British Columbia, Alberta, Saskatchewan and Southern Manitoba. Canadian Society of Petroleum Geologists, Calgary.

Graves, R.W.J., and Ellison, S. 1941. Ordovician conodonts of the Marathon Basin, Texas. University of Missouri School of Mines and Metallurgy Bulletin, Technical Series, **14**: 1–26.

Greggs, R.G., and Bond, I.J. 1971. Conodonts from the March and Oxford formations in the Brockville Area,

Ontario. Canadian Journal of Earth Sciences, **8**: 1455–1471.

Hadding, A.R. 1913. Undre dicellograptusskiffern i Skåne jämte några darmed ekvivalenta bildningar. Lunds Universitets Årsskrift N. F. Avd. 2, **9**: 1–90.

Hamar, G. 1966. The Middle Ordovician of the Oslo Region, Norway. 22. Preliminary report on conodonts from the Oslo–Asker and Ringerike districts. Norsk Geologisk Tidsskrift, **46**: 27–83.

Harris, A.G., Dumoulin, J.A., Repetski, J.E., and Carter, C. 1995*a*. Correlation of Ordovician rocks of Northern Alaska. *In* Ordovician Odyssey: Short Papers for the Seventh International Symposium on the Ordovician System. *Edited by* J.D. Cooper, M.L. Droser, and S.C. Finney. Society of Economic Paleontologists and Mineralogists, Fullerton, Las Vegas, pp. 21–26.

Harris, A.G., Repetski, J.E., Stamm, N.R., and Weary, D.J. 1995*b*. Conodont age and CAI data for New Jersey. US Geological Survey, Open File Report 95-557, pp. 1–31.

Harris, R.W. 1962. New conodonts from the Joins (Ordovician) Formation of Oklahoma. Oklahoma Geology Notes, **62**: 119–211.

Harris, R.W. 1964. Subgenera of the conodont genus *Multioistodus* in Simpson–Burgen (Ordovician) of Oklahoma. Oklahoma Geology Notes, **24**: 108–118.

Harris, R.W., and Harris, B. 1965. Some West Spring Creek (Ordovician Arbuckle) conodonts from Oklahoma. Oklahoma Geology Notes, **25**: 34–47.

Hedberg, H.D. 1976. International Stratigraphic Guide: A Guide to Stratigraphical Classification, Terminology, and Procedure: International Subcommission on Stratigraphic Classification of IUGS Commission on Stratigraphy. John Wiley & Sons, New York.

Hein, F.J., and Nowlan, G.S. 1998. Regional sedimentology, conodont biostratigraphy and correlation of Middle Cambrian – Lower Ordovician (?) strata of the "Finnegan" and Deadwood formations, Alberta subsurface, Western Canada Sedimentary Basin. Bulletin of Canadian Petroleum Geology, **46**: 166–188.

Hills, L. 1981. Lexicon of Canadian Stratigraphy, Volume 2, Yukon–Mackenzie (southwestern Yukon Territory and western British Columbia). Canadian Society of Petroleum Geologist, Calgary, 240 pp.

Jackson, D.E. 1980. Geological Survey of Canada Paleontological Report, 2-DEJ-1980.

Jackson, D.E., and Lenz, A. 1962. Zonation of Ordovician and Silurian graptolites of northern Yukon, Canada. Bulletin of American Association of Petroleum Geologists, **46**: 30–45.

Ji, Z., and Barnes, C.R. 1994. Lower Ordovician conodonts of the St. George Group, Port au Port Peninsula, western Newfoundland, Canada. Palaeontographica Canadiana, **11**: 1–149.

Ji, Z., and Barnes, C.R. 1996. Uppermost Cambrian and Lower Ordovician conodont biostratigraphy of the Survey Peak Formation (Ibexian/Tremadoc), Wilcox Pass, Alberta, Canada. Journal of Paleontology, **70**: 871–890.

Johnston, D.I., and Barnes, C.R. 1999. Early and Middle Ordovician (Arenig) conodonts from St. Pauls Inlet and Martin Point, Cow Head Group, Western Newfoundland, Canada. Geologica et Palaeontologica, **33**: 21–70.

Jones, P.J. 1971. Lower Ordovician conodonts from the Bonaparte Gulf Basin and the Daly River Basin, north-western Australia. Australia Bureau of Mineral Resources, Geology and Geophysics Bulletin, **117**: 1–80.

Kennedy, D.J. 1980. A restudy of conodonts described by Branson and Mehl, 1933, from the Jefferson City Formation, Lower Ordovician, Missouri. Geologica et Palaeontologica, **14**: 45–76.

Kennedy, D.J. 1994. *Colaptoconus* (Conodonta), a replacement name for *Glyptoconus* Kennedy, 1981, non *Glyptoconus* von Moellendorff, 1894. Journal of Paleontology, **68**: 1417.

Khodalevich, A.N., and Tschernich, V.V. 1973. New subfamily Bellodellinae (Conodonts). Trudy Sverdlovskogo Gornogo Instituta, **93**: 42–47.

Landing, E. 1983. Highgate Gorge: Upper Cambrian and Lower Ordovician continental slope deposition and biostratigraphy, north-western Vermont. Journal of Paleontology, **57**: 1149–1187.

Landing, E. 1993. Cambrian–Ordovician boundary in the Taconic Allochthon, eastern New York, and its interregional correlation. Journal of Paleontology, **67**: 1–19.

Landing, E., and Barnes, C.R. 1981. Conodonts from the Cape Clay Formation (Lower Ordovician), southern Devon Island, Arctic Archipelago. Canadian Journal of Earth Sciences, **18**: 1609–1628.

Landing, E., Barnes, C.R., and Stevens, R.K. 1986. Tempo of earliest Ordovician graptolite faunal succession: conodont-based correlations from the Tremadocian of Quebec. Canadian Journal of Earth Sciences, **23**: 1928–1949.

Landing, E., Westrop, S.R., and Knox, L.A. 1996. Conodonts, stratigraphy and relative sea-level changes of the Tribes Hill Formations (Lower Ordovician, east-central New York). Journal of Paleontology, **70**: 656–680.

Lee, H. 1970. Conodonten aus der Choson-Gruppe (Unteres Ordovizium) von Korea. Neues Jahrbuch für Geologie und Paläontologie, Abhandlungen, **136**: 303–344.

Lee, H. 1975. Conodonts from the Dumugol Formation (Lower Ordovician), South Korea. Journal of the Geological Society of Korea, **11**: 75–93.

Lehnert, O. 1995. Ordovizische Conodonten aus der Präkordillere Westargentiniens: Ihre Bedeutung für Stratigraphie und Paläogeographie. Erlanger Geologische Abhandlungen, **125**: 1–193.

Lenz, A.C., and Jackson, D.E. 1986. Arenig and Llanvirn graptolite biostratigraphy, Canadian Cordillera. *In* Palaeoecology and Biostratigraphy of Graptolites. *Edited by* C.P. Hughes, R.B. Rickards, and A.J. Chapman. Geological Society of London, London, pp. 27–45.

Lindström, M. 1955. Conodonts from the lower-most Ordovician strata of south-central Sweden. Geologiska Föreningens i Stockholm Förhandlingar, **76**: 517–604.

Lindström, M. 1970. A suprageneric taxonomy of the conodonts. Lethaia, **3**: 427–445.

Lindström, M. 1971. Lower Ordovician conodonts of Europe. *In* Symposium on Conodont Biostratigraphy. *Edited by* W.C. Sweet and S.M. Bergström. Geological Society of America, Memoir 127, pp. 21–61.

Linnaeus, C. 1758. Systema naturae per regna tria naturae (10th edn). Laurentii Salvii, Stockholm.

Löfgren, A. 1978. Arenigian and Llanvirnian conodonts from Jämtland, northern Sweden. Fossils and Strata, **13**: 1–129.

Löfgren, A. 1985. Early Ordovician conodont biozones at Finngrundet, south Bothnian Bay, Sweden (Geology of the southern Bothnian Sea, Part 3). Bulletin of the Geological Institutions of the University of Uppsala, N. S., **10**: 115–128.

Löfgren, A. 1990. Non-platform elements of the Ordovician conodont genus *Polonodus*. Paläontologische Zeitschrift, **64**: 245–259.

Löfgren, A. 1993. Arenig conodont successions from central Sweden. Geologiska Föreningens i Stockholm Förhandlingar, **115**: 193–207.

Löfgren, A. 1994. Arenig (Lower Ordovician) conodonts and biozonation in the Eastern Siljan District, Central Sweden. Journal of Paleontology, **68**: 1350–1368.

Löfgren, A. 1995. The Middle Lanna/Volkhov Stage (Middle Arenig) of Sweden and its conodont fauna. Geological Magazine, **132**: 693–711.

Löfgren, A. 1996. Lower Ordovician conodonts, reworking and biostratigraphy of the Orreholmen quarry, Västergötland, south-central Sweden. Geologiska Föreningens i Stockholm Förhandlingar, **118**: 169–183.

Löfgren, A. 1997. Reinterpretation of the Lower Ordovician condont apparatus *Paroistodus*. Paleontology, **40**: 913–929.

Löfgren, A., Repetski, J.E., and Ethington, R.L. 1998. Some trans-Iapetus conodont faunal connections in the Tremadocian. Bollettino della Societa Paleontologica Italiana, **37**: 159–173.

MacIntyre, D.G. 1992. Geological setting and genesis of sedimentary exhalative barite and barite-sulfide deposits, Gataga District, northeastern British Columbia. Exploration and Mining Geology, **1**: 1–20.

McClay, K.R., Insley, M.W., Way, N.A., and Anderton, R. 1988. Tectonics and mineralization of the Kechika Trough, Gataga area, northeastern British Columbia. Current Research, Part E, Geological Survey of Canada, Paper 88-1E, pp. 1–12.

McCracken, A.D. 1989*a*. Preliminary report on Ordovician–Devonian conodont collections from carbonate and fine grained clastic facies of northern Yukon Territory and northwest District of Mackenzie, N.W.T. Current Research, Part G, Geological Survey of Canada Paper 89-1G, pp. 69–76.

McCracken, A.D. 1989*b*. *Protopanderodus* (Conodonta) from the Ordovician Road River Group, Northern Yukon Territory, and the evolution of the genus. Geological Survey of Canada, Bulletin 388, pp. 1–39.

McCracken, A.D. 1991. Middle Ordovician conodonts from the Cordilleran Road River Group, northern Yukon Territory, Canada. *In* Ordovician to Triassic Conodont Paleontology of the Canadian Cordillera. *Edited by* M.J. Orchard and A.D. McCracken. Geological Survey of Canada, Bulletin 417, pp. 41–64.

McCracken, A.D., and Lenz, A.C. 1987. Middle and Late Ordovician conodont faunas and biostratigraphy of graptolitic strata of the Road River Group, northern Yukon Territory. Canadian Journal of Earth Science, **24**: 643–653.

McTavish, R.A. 1973. Prioniodontacean conodonts from the Emanuel Formation (Lower Ordovician) of

Western Australia. Geologica et Palaeontologica, **7**: 27–51.

Mei, S. 1993. The biostratigraphy of the Middle and Late Cambrian conodonts and trilobites from Wanxin, Hebei. Journal of Stratigraphy, **17**: 11–24.

Mens, K., Heinsalu, H., Jegonjan, K., Kurvits, T., Puura, I., and Viira, V. 1996. Cambrian–Ordovician boundary beds in the Pakri Cape Section, NW Estonia. Proceedings of the Estonian Academy of Sciences, Geology, **45**: 9–21.

Miller, J.F. 1969. Conodont fauna from the Notch Peak Limestone (Cambro–Ordovician), House Range, Utah. Journal of Paleontology, **43**: 413–439.

Miller, J.F. 1980. Taxonomic revisions of some Upper Cambrian and Lower Ordovician conodonts with comments on their evolution. University of Kansas Paleontological Contributions, Paper 99, pp. 1–39.

Miller, J.F. 1984. Cambrian and earliest Ordovician conodont evolution, biofacies, and provincialism. Geological Society of America Special Paper, **196**: 43–68.

Miller, J.F. 1988. Conodonts as biostratigraphic tools for redefinition and correlation of the Cambro–Ordovician boundary. Geological Magazine, **125**: 349–362.

Miller, R.H., and Dockum, M.S. 1983. Ordovician conodonts from metamorphosed carbonates of the Salton Trough, California. Geology, **11**: 410–412.

Mound, M.C. 1965. A conodont fauna from the Joins Formation (Ordovician), Arbuckle Mountains, Oklahoma. Micropaleontology, **14**: 383–434.

Mound, M.C. 1968. Conodonts and biostratigraphy of the Iowa Arbuckle Group (Ordovician), Arbuckle Mountains, Oklahoma. Micropaleontology, **14**: 393–434.

Müller, K.J. 1959. Kambrische Conodonten. Zeitschrift für deutsche geologische Gesellschaftung, **111**: 431–485.

Müller, K.J. 1973. Late Cambrian and Early Ordovician conodonts from northern Iran. *In* Geological Survey of Iran Report Number 30, pp. 1–77.

Müller, K.J., and Hinz, I. 1991. Upper Cambrian conodonts from Sweden. Fossils and Strata, **28**: 1–153.

Nicoll, R.S. 1990. The genus *Cordylodus* and a latest Cambrian – earliest Ordovician conodont biostratigraphy. BMR Journal of Australian Geology and Geophysics, **11**: 529–558.

Nicoll, R.S. 1992. Analysis of conodont apparatus organisation and the genus *Jumudontus* (Conodonta), a coniform–pectiniform apparatus structure from the Early Ordovician. Journal of Australian Geology and Geophysics, **13**: 213–228.

Nicoll, R.S., Miller, J.F., Nowlan, G.S., Repetski, J.E., and Ethington, R.L. 1999. *Iapetonudus* (N. gen.) and *Iapetognathus* Landing, unusual earliest Ordovician multielement conodont taxa and their utility for biostratigraphy. Brigham Young University Geology Studies, **44**: 27–101.

Nogami, Y. 1967. Kambrische Conodonten von China, Teil 2. Conodonten aus den hoch oberkambrischen Yencho–Schichten. Memoirs of the College of Science, University of Kyoto, Geology and Mineralogy, Series B, **33**: 211–218.

Norford, B.S. 1969. Ordovician and Silurian stratigraphy of the southern Rocky Mountains. Geological Survey of Canada, Bulletin 176, pp. 1–90.

Norford, B.S. 1991. Ordovician and Silurian stratigraphy, paleogeography and depositional history in the Peace River Arch area, Alberta and British Columbia. *In* Geology of the Peace River Arch. *Edited by* S.C. O'Connell and J.S. Bell. Bulletin of Canadian Petroleum Geology, Special Volume 38A, pp. 45–54.

Nowlan, G.S. 1983. Geological Survey of Canada Paleontological Report, 001-GSN-1983.

Nowlan, G.S. 1985. Late Cambrian and Early Ordovician conodonts from the Franklinian miogeosyncline, Canadian Arctic Islands. Journal of Paleontology, **59**: 96–122.

Nowlan, G.S. 1995. Geological Survey of Canada Paleontological Report, 10-GSN-1995.

Orchard, M.J. 1995. Geological Survey of Canada Paleontological Report, MJO-1995-40.

Orndorff, R.C. 1988. Latest Cambrian and earliest Ordovician conodonts from the Conococheague and Stonehenge Limestones of Northwestern Virginia. U.S. Geological Survey Bulletin, **1837A**: A1–A18.

Pander, C.H. 1856. Monographie der fossilen Fische des Silurischen Systems der Russisch–Baltischen Gouvernements. St. Petersburg, **10**: 1–91.

Parsons, B.P., and Clark, D.L. 1999. Conodonts and the Cambrian–Ordovician boundary in Wisconsin. Geoscience Wisconsin, **17**: 1–10.

Pohler, S.M.L. 1994. Conodont biofacies of Lower to Lower Middle Ordovician megaconglomerates, Cow Head Group, western Newfoundland. Geological Survey of Canada, Bulletin 459, pp. 1–71.

Pohler, S.M.L., and Barnes, C.R. 1990. Conceptual models in conodont paleoecology. Courier Forschungsinstitut Senckenberg, **118**: 409–440.

Pohler, S.M.L., and Orchard, M.J. 1990. Ordovician conodont biostratigraphy, western Canadian Cordillera. Geological Survey of Canada, Paper 90-15, pp. 1–37.

Pohler, S.M.L., Barnes, C.R., and James, N.P. 1987. Reconstructing a lost faunal realm: conodonts from mega-conglomerates of the Ordovician Cow Head Group, western Newfoundland. *In* Conodonts: Investigative Techniques and Applications. *Edited by* R.L. Austin. Ellis Harwood Limited, Chichester, pp. 341–362.

Pyle, L.J., and Barnes, C.R. 2000. Upper Cambrian to Lower Silurian stratigraphic framework of platform-to-basin facies, northeastern British Columbia. Bulletin of Canadian Petroleum Geology, **48**: 123–149.

Pyle, L.J., and Barnes, C.R. 2001. Conodonts from the Kechika Formation and Road River Group (Lower to Upper Ordovician) of the Cassiar Terrane, Northern British Columbia. Canadian Journal of Earth Sciences, **38**: 1387–1401.

Rao, R.I., and Tortello, M.F. 1998. Tremadoc conodonts and trilobites from the Cardonal Formation, Incamayo Creek, Salta Province, Northwest Argentina. *In* Proceedings of the Sixth European Conodont Symposium (ECOS VI). *Edited by* H. Szaniawski. Palaeontologia Polonica **58**: 31–45.

Rejebian, V.A., Harris, A.G., and Huebner, J.S. 1987. Conodont color and textural alteration: An index to regional metamorphism, contact metamorphism and hydrothermal alteration. Bulletin of the Geological Society of America, **99**: 471–479.

Repetski, J.E. 1982. Conodonts from El Paso Group (Lower Ordovician) of westernmost Texas and southern New Mexico. New Mexico Bureau of Mines and Mineral Resources, Memoir 40, pp. 1–121.

Repetski, J.E., and Alberstadt, L. 1989. A Lower Ordovician sponge/algal facies in the Southern United States and its counterparts elsewhere in North America. Palaios, **4**: 225–247.

Repetski, J.E., and Ethington, R.L. 1983. *Rossodus manitouensis* (Conodonta), a new early Ordovician index fossil. Journal of Paleontology, **57**: 289–301.

Ross, R.J.J., Hintze, L.F., Ethington, R.L., Miller, J.F., Taylor, M.E., and Repetski, J.E. 1997. The Ibexian, lowermost Series in the North American Ordovician. *In* Early Paleozoic Biochronology of the Great Basin. *Edited by* M.E. Taylor. U.S. Geological Survey Professional Paper 1579, pp. 1–50.

Sergeeva, S.P. 1963. Conodonts from the Lower Ordovician of the Leningrad region. Paleontologicheskiy Zhurnal, Akademiya Nauk SSSR, **2**: 93–108.

Serpagli, E. 1967. I conodonti dell'Ordoviciano superiore (Ashgiliano) delle Alpi Carniche. Bolletino della Societá Paleontologica Italiana, **6**: 30–111.

Serpagli, E. 1974. Lower Ordovician conodonts from Precordilleran Argentina (Province of San Juan). Bolletino della Societa Paleontologica Italiana, **13**: 17–98.

Simpson, A.J. 1995. Silurian conodont biostratigraphy in Australia: A review and critique. *In* Contribution to the First Australian Conodont Symposium (AUSCOS 1). *Edited by* R. Mason and J.A. Talent. Courier Forschungsinstitut Senckenberg 182, pp. 325–345.

Smith, G.L., and Clark, D.L. 1996. Conodonts of the Lower Ordovician Prairie du Chien Group of Wisconsin and Minnesota. Micropaleontology, **42**: 363–373.

Smith, M.P. 1991. Early Ordovician conodonts of East and North Greenland. Meddelelser om Grønland, Geoscience, **26**: 1–81.

Stone, J. 1987. Review of investigative techniques used in the study of conodonts. *In* Conodonts: Investigative Techniques and Applications. *Edited by* R.L. Austin. Ellis Horwood Ltd., Chichester, pp. 17–34.

Stouge, S. 1984. Conodonts of the Middle Ordovician Table Head Formation, western Newfoundland. Fossils and Strata, **16**: 1–145.

Stouge, S., and Bagnoli, G. 1988. Early Ordovician conodonts from Cow Head Peninsula, western Newfoundland. Palaeontographia Italica, **75**: 89–179.

Stouge, S., and Bagnoli, G. 1990. Lower Ordovician (Volkhovian–Kundan) conodonts from Hägudden, northern Öland, Sweden. Palaeontographia Italica, **77**: 1–54.

Sweet, W.C. 1988. The Conodonta: morphology, taxonomy, paleoecology, and evolutionary history of a long-extinct animal phylum. Oxford Monographs on Geology and Geophysics, No. 10, Clarendon Press, Oxford, 212 pp.

Sweet, W.C., and Bergström, S.M. 1962. Conodonts from the Pratt Ferry Formation (Middle Ordovician) of Alabama. Journal of Paleontology, **36**: 1214–1252.

Sweet, W.C., and Schönlaub, H.P. 1975. Conodonts of the genus *Oulodus* Branson and Mehl, 1933. Geologica et Palaeontologica, **9**: 41–59.

Sweet, W.C., Turco, A.C., Warner, E.J., and Wilkie, L.C. 1959. The American Upper Ordovician Standard I, Eden conodonts from the Cincinnati region of Ohio and Kentucky. Journal of Paleontology, **33**: 1029–1068.

Szaniawski, H., and Bengtson, S. 1998. Late Cambrian euconodonts from Sweden. *In* Proceedings of the Sixth European Conodont Symposium (ECOS VI). *Edited by* H. Szaniawski. Palaeontologia Polonica 58, pp. 7–29.

Taylor, M.E., and Landing, E. 1982. Biostratigraphy of the Cambrian–Ordovician transition in the Bear River Range, Utah and Idaho, Western United States. *In* The Cambrian–Ordovician Boundary: Sections, Fossil Distributions, and Correlations. *Edited by* M.G. Bassett and W.T. Dean. National Museum of Wales, Geological Series 3, Cardiff, pp. 181–191.

Thompson, R.I. 1989. Stratigraphy, tectonic evolution and structural analysis of the Halfway River map area (94B), northern Rocky Mountains, British Columbia, Geological Survey of Canada, Memoir 42, pp. 1–119.

Tipnis, R.J. 1979. Geological Survey of Canada Paleontological Report, 8-RST-1979.

Tipnis, R.S., Chatterton, B.D.E., and Ludvigsen, R. 1978. Ordovician conodont biostratigraphy of the southern district of Mackenzie, Canada. *In* Western and Arctic Canadian Biostratigraphy. *Edited by* C.R. Stelck and B.D.E. Chatterton. Geological Association of Canada Special Paper 18, pp. 39–91.

Uyeno, T.T., and Barnes, C.R. 1983. Conodonts of the Jupiter and Chicotte formations (Lower Silurian), Anticosti Island, Quebec. Geological Survey of Canada Bulletin, **355**: 1–49.

van Wamel, W.A. 1974. Conodont biostratigraphy of the Upper Cambrian and Lower Ordovician of north-western Öland, south-eastern Sweden. Utrecht Micropaleontology Bulletin, **10**: 1–126.

Viira, V., and Männik, P. 1997. Conodonts. *In* Geology and Mineral Resources of Estonia. *Edited by* A. Raukas and A. Teedumäe. Estonian Academy of Sciences, Tallinn, pp. 241–246.

Viira, V., Sergeyeva, S., and Popov, L. 1987. Earliest representatives of the genus *Cordylodus* (Conodonta) from Cambro–Ordovician boundary beds of North Estonia and Leningrad Region. Proceedings of the Academy of Sciences of the Estonian SSR, Geology, **36**: 145–153.

Walcott, C.D. 1928. Pre-Devonian Paleozoic formations of the Cordilleran provinces of Canada. Smithsonian Miscellaneous Collections, **75**(5): 175–368.

Walliser, O.H. 1964. Conodonten des Silurs. *In* Abhandlungen des Hessischen Landesamtes für Bodenforschung, **41**: 1–106.

Wang, Z. 1986. Late Cambrian and Early Ordovician conodonts from North and Northeast China with comments on the Cambrian–Ordovician boundary. *In* Stratigraphy and Paleontology of Systemic Boundaries in China: Cambrian–Ordovician Boundary. *Edited by* J. Chen and Z. Wang. Nanjing Institute of Geology and Paleontology, Anhui Science and Technology Publishing House, Nanjing, pp. 195–258.

Webby, B.D. 1995. Towards an Ordovician time scale. *In* Ordovician Odyssey: Short Papers for the Seventh International Symposium on the Ordovician System. *Edited by* J.D. Cooper, M.L. Droser, and S.C. Finney. Society of Economic Paleontologists and Mineralogists, Las Vegas, pp. 5–10.

Webby, B.D. 1998. Steps toward a global standard for Ordovician stratigraphy. Newsletter of Stratigraphy, **36**: 1–33.

Williams, S.H., Nowlan, G.S., Barnes, C.R., and Batten, R.S.R. 1999. The Ledge section at Cow Head, western Newfoundland: new data and discussion of the graptolite, conodont and chitinozoan assemblages. Acta Universitatis Carolinae Geologica, **43**: 65–68.

Zhang, J. 1998. Conodonts from the Guniutan Formation (Llanvirnian) in Hubei and Hunan Provinces, south-central China. Stockholm Contributions in Geology, **46**: 1–161.

Appendix A

The lithologies of seven measured sections which yielded conodonts are described. The sample numbers of the conodont samples are placed in stratigraphic order, and their heights in metres above the base of the section are given. The sample numbers are not always in consecutive order because of methods in which ridge sections were measured and sampled in segments.

Section 13 (of Cecile and Norford 1979)

Section is located 4 km southwest of Mount Sheffield, trending east–west toward two glacial lakes to the west (94F/10; 57°42´N, 124°35´W). Camp was located on a grassy area along creek running east from the glacial lakes. Section was collected in three parts: A, B, and C.

Sample number	Description	Thickness (m) Sample	Total from Base
	Kechika Formation		
	Walked east along north side of the creek about 2 km and then north along tree line to collect the basal parts of the Kechika. Contact with underlying Cambrian units was not observed. Started to collect at lowest main outcrop on north side of main gully.		
Lloyd George Member			
Highly cleaved shale with rare limestone interbeds, generally boudinaged.			
96-13-C1	Lime mudstone, medium to dark grey, weathers grey brown, fine crystalline, slightly folded, compact, within highly cleaved shale.	0	0
Passed over compact, finely laminated, calcareous shale with very thin 0.3–1 cm beds of limestone, partly talus covered.			
96-13-C2	Lime mudstone–Wackestone, dark grey, 10–15 cm thick, internal fine lamination, fine crystalline, compact silty beds (calcisiltite).	10.5	10.5
96-13-C3	Lime mudstone–Wackestone, dark grey to black, weathers dark grey, finely laminated, fine crystalline, compact (calcisiltite).	31.5	42
96-13-C4	Lime mudstone–Wackestone, dark grey to black, weathers dark grey, compact thin to medium bed, fine crystalline (calcisiltite); less calcareous than early samples.	31.5	73.5
Passed one similar medium-bedded, compact black limestone bed, slightly recessive subunit in cliff north of creek, entered a more resistant unit of finely laminated, silty/argillaceous black limestone.		3.5	77
Prominent buff–orange-weathered fine dolostone bed, more structural flexure in creek bed.		1	78
Quentin Member		12	90
Base of "putty shale" unit; entered second cliff forming unit below top of cliff, light-grey–brown-weathered massive unit, phyllitic grey brown to silver shale with interbeds (sparse) of 5–10 cm compact limestone, often partly boudinaged.			
96-13-C5	Lime mudstone, dark grey, 5 cm bed, fine crystalline, compact, partly boudinaged.	13.5	103.5
96-13-C6	Lime mudstone–Wackestone, medium grey, weathers buff, 4 cm bed, nodular, dense, and compact.	22.5	126

Section 13 (of Cecile and Norford 1979) (*continued*).

Sample number	Description	Thickness (m) Sample	Total from Base
Interbeds of buff, putty grey shale.			
96-13-C7	Packstone, medium grey, 3 cm, finely laminated, dense, and compact.	162	288
96-13-C8	Lime mudstone–Packstone, medium grey, 10 cm interval, 2–4 cm beds, laminated, grading into wackestone–packstone in cores.	31.5	319.5
96-13-C9	As -C8; but more homogenous packstone, dense and lenticular.	18	337.5
Passed through interval of phyllitic putty grey shales, which abruptly changed into orange-weathered 1 cm bedded limestone, laminated, dense, forming upper waterfall.			
96-13-C10	Lime mudstone, weathers orange, two 1 cm beds.	64.5	402
96-13-C11	Wackestone–Packstone, light grey, weathers orange, nodular.	31.5	433.5
Interval of phyllitic shale and light grey limestone packages.			
96-13-C12	Wackestone–Packstone, medium grey, 4 cm interval.	27	460.5

Section A: Walked from camp down creek to tree line where Kechika was no longer prominent and underlay south grassy and bush-covered hill and the creek. Collected south side up creek and hill and noted slump breccias and folding in blocks of creek bed. Argillaceous, dolomitic to brown-weathered beds and light-grey-weathered beds 1–2 cm. Started collecting samples prefixed 96-13A where the creek takes a bend southward.

Quentin Member

Unit homogenous, undulatory, thin bedding of laminated calcilutites, siltites, and 5% grey-weathered, interstratified, argillaceous, silty, and laminated limestone comprising the beds collected.

96-13-A1	Wackestone, medium grey, weathers buff and light grey, dense 5 cm bed, finely laminated, silty.	0	0
96-13-A2	Wackestone, medium grey, weathers buff and light grey, 10–15 cm thick, medium crystalline.	27	27
96-13-A3	Wackestone, medium grey, 0–3 cm medium crystalline, among thin beds of calcisiltite, undulating beds.	21	48
96-13-A4	Wackestone, medium grey, weathers buff to light grey, 4 cm bed.	27	75
96-13-A5	Wackestone, light grey, weathers light grey, 2 cm bed, laminated.	24	99
96-13-A6	Packstone–Grainstone, medium grey, 3 cm bed, coarse crystalline.	21	120
96-13-A7	Wackestone–Packstone, medium grey, weathers light grey, 3 cm bed, medium crystalline.	34.5	154.5
96-13-A8	Wackestone, medium grey, weathers light grey, 2 cm bed, medium crystalline.	27	181.5
96-13-A9	Packstone, light to medium grey, weathers light grey, 2 cm bed, thin bedded, cleaved, and laminated.	27	208.5
Covered interval.		45	253.5

Walked up to exposure on north side of camp, collected -A10 1 m above base. It is a mid-slope bluff on line with waterfall below camp and grassy knoll from which covered interval was recorded, on line with convergence of two creeks, last area of outcrop before fault. Estimate 10 m missing from base.

Section 13 (of Cecile and Norford 1979) (*continued*).

Sample number	Description	Thickness (m) Sample	Total from Base
96-13-A10	Wackestone, medium grey, weathers light grey and buff, thin-bedded interval up to 4 cm thick beds, finely laminated.	11	264.5
96-13-A11	As -A10; light grey, three compact thin beds, 3 cm each.	21	285.5
96-13-A12	Packstone, medium grey, weathers buff, laminated.	20	305.5
96-13-A13	As -A10.	19.5	325
Covered interval to level of glacier south of main creek.			
Grey Peak Member			
Collected in east wall of creek bed trending N–S toward camp. Unit is made of alternating metre scale packages of resistant limestone, and recessive shale, in equal proportion, medium grey thin to medium beds, no longer laminated or silty.			
96-13-A14	Packstone, medium grey, weathers light grey and buff, 2 cm thick.	105	430
96-13-A15	Packstone, medium grey, weathers light grey, compact 4 cm interval.	21	451
96-13-A16	Grainstone, medium grey, weathers light grey, 1 m thick series of beds with shale partings, trilobites.	20	471
Top of creek outcrop, after this is the grassy valley in which we camped with no exposure, likely the fault zone.			
96-13-A17	Wackestone–Packstone, medium grey, weathers light grey, 4 cm bed, pseudonodular, resistant, buff, argillaceous, dolostone partings, cleaved, trilobites and intraclasts.	20	491
Covered interval from camp to the south side of creek fed by glacial lakes to the west.		185	676
Section B continues with the Grey Peak Member west of camp (likely repeated after the fault zone). Shale weathers buff to light grey with thin, competent interbeds of limestone up to 10 cm.			
96-13-B1	Packstone–Grainstone, weathers buff to light grey (intraclasts?).	3	679
Interval of shale and laminated calcilutite, thin intraclastic beds, and more calcareous packstone.			
96-13-B2	Packstone, medium grey, weathers buff to light grey, medium crystalline.	37	716
96-13-B3	Packstone–Grainstone, medium grey, weathers buff–grey, up to 30 cm thick, intraclastic–trilobitic.	20	736
96-13-B4	Packstone, medium grey, weathers light grey – buff, 10 cm bed.	20	756
96-13-B5	As -B4.	20	776
96-13-B6	As -B4; 30 cm bed.	34.5	810.5
96-13-B7	As -B4.	22.5	833
96-13-B8	As -B4.	30	863
Increasing shale upsection.			
96-13-B9	Lime mudstone, medium grey, fine crystalline, laminated.	30	893
96-13-B10	Packstone, medium grey, 2 cm thick among thicker bedded carbonate and platy, cleaved shale.	30	923
Back down to creek to start up waterfalls, to outcrop in main creek bed, which contains cleaved shale.		27	950

Section 13 (of Cecile and Norford 1979) (*continued*).

		Thickness (m)	
		Total from	
Sample number	Description	Sample	Base
Rubbly weathered limestone outcrops in creek; nodular, light grey limestone with buff argillaceous partings, collected the basal part in the creek bed, overlain by a recessive unit before the Haworth Member comprising the waterfalls begins.			
96-13-B11	Lime mudstone, medium grey, weathers light grey, fine crystalline, laminated, buff argillaceous partings.	21	971
96-13-B12	Packstone, medium grey, weathers light grey, 3 cm thick compact bed, rubbly.	69	1040
Haworth Member		19.5	1059.5
Base of waterfall; more resistant, massive unit of largely shale to argillaceous thin-bedded limestone and nodular limestone.			
96-13-B13	Lime mudstone–Wackestone, light grey, weathers buff, lenticular bed.	21	1080.5
At the top of next main waterfall, more calcareous grainstone, light grey beds and buff, argillaceous beds, rubbly, cleaved calcisiltite interbeds. This resistant unit continues until the 1st glacial lake.			
96-13-B14	Grainstone, medium grey, weathers light grey.	82.5	1163
96-13-B15	Lime mudstone–Wackestone, light grey, weathers grey–buff, 3 cm bed forming lip of falls, laminated.	25	1188
96-13-B16	Grainstone, light grey, weathers light grey, 3 cm thick, dolomitic.	42	1230
96-13-B17	Packstone, light grey, weathers buff and grey, bioclastic, rubbly.	45	1275
Mt. Sheffield Member		13.5	1288.5
Contact snow covered; unit comprises cleaved, lenticular limestone interbedded with buff, orange-weathered shale surrounding limy cores, massive bedded.			
Covered interval, by snow and talus and by moraine from glacier to south of creek.		62	1350.5
96-13-B18	Packstone, medium grey, weathers light grey, massive bedded, medium crystalline, cleaved limy cores appear nodular and lenticular, maybe once lenses or thin beds, laminated.	1	1351.5
Continued up creek to waterfalls and climbed moraines up to outcrop to south of creek rather than climb steep creekbed, toward cliff-forming part of unit.			
96-13-B19	Packstone, medium grey, weathers light grey, laminated, pyritic, cleaved.	30	1381.5
96-13-B20	Lime mudstone–Wackestone, dark grey, weathers light grey and buff, 2 cm up to 4 cm thick, concoidal fracture.	30	1411.5
96-13-B21	Lime mudstone, with packstone intraclasts, medium grey, weathers light grey.	33	1444.5
96-13-B22	Lime mudstone, medium grey, weathers light grey, cores within shale, fine crystalline.	40.5	1485
Continued down to the top of the first lake's delta and head waters. At top of waterfall, collected limestone interbedded with shale and lime mudstone.			
96-13-B23	Packstone–Grainstone, medium grey, weathers medium grey, 3 cm thick fossiliferous (orthid brachiopod), black bioclasts.	33	1518

Section 13 (of Cecile and Norford 1979) (*concluded*).

Sample number	Description	Thickness (m) Total from Sample	Base
Light grey and buff shale interval along creek beds; combination of lithologies in lenses, from lime mudstone to grainstone.			
96-13-B24	Lime mudstone–Wackestone, dark grey, weathers medium grey, lenticular, laminated.	37.5	1555.5
96-13-B25	Wackestone–Packstone, dark grey, weathers medium grey, 2 cm bed.	37.5	1593
In between the two lakes in top of grass-covered cliff. More calcareous in this interval with more lime mudstone interbeds.			
96-13-B26	Packstone–Grainstone, medium grey, weathers light and medium grey, nodular.	30	1623
Medium grey shale and talus weather orange.			
96-13-B27	Grainstone, medium grey, 10 cm interval of thin beds.	37.5	1660.5
Limy beds more prominent with many fossils in orange scree; more orange up section.			
96-13-B28	Grainstone, medium grey, weathers buff–orange–grey.	46	1706.5
96-13-B29	Packstone–Grainstone, medium grey, nodular 2 cm bed.	34.5	1741
	Skoki Formation, Sikanni Chief Member		
A change to orange and burrow-mottled unit with fewer fossils on weathered surface and beginning of jointing and platy breaking of massive beds (compared to thin-bedded unit below), fossiliferous and burrow mottled.			
96-13-B30	Grainstone, medium grey, weathers orange, massive, coarse crystalline, slightly dolomitic, burrow mottled.	33	1774
Through interval of Skoki talus and change to platy-weathered, orange- and medium-grey-weathered dolomitic limestone, fossiliferous, vertical jointing.			
96-13-B31	Wackestone–Packstone, medium grey, weathers orange, medium crystalline.	37.5	1811.5
96-13-B32	As -B31; massive without jointed surfaces.	58.5	1870
Subunit of massive-bedded, burrow-mottled limestone, weathers orange and light grey, dolomitic, and interbedded with hackly, rubbly, thin beds of medium grey wackestone, weathers ight grey, less than 0.5 m thick in intervals.			
96-13-B33	Wackestone, medium grey, weathers orange and light grey, massive, burrow mottled.	58.5	1928.5
Change to rubbly weathered, platy, more grey and pale orange.			
96-13-B34	Packstone, weathers grey and pale orange.	42	1970.5
Hackly limestone; overlain by grey dolostone talus with macluritid gastropods.		18	1988.5

Section 1 (of Cecile and Norford 1979)

Section is located 5 km north of Muskwa River (94F/16; 57°52′N, 124°05′W). Camped in a cirque south of the east–west trending ridge measured and collected. The first unit of Kechika collected was an outcrop on the east ridge. Walked down a steep edge to examine the Cambrian unit and base of the Kechika.

		Thickness (m)	
		Total from	
Sample number	Description	Sample	Base
	Cambrian unit		
	Quartzite, white to light grey, massive bedded, well-rounded and sorted quartz grains, cross-bedding prominent, sandy dolomite interbeds.		
	Quartzite and siltstone, grey, weathers brown.	8	8
	Kechika Formation		
	Comprises four distinct units, which have gradational contacts, within section.		
	Lloyd George Member	62	70
	Dolostone, silty with sandy laminae, light grey, weathers orange.		
	Quentin Member		
	Calcilutite, 70–80% of the lower part of the unit is light grey, weathers grey–yellow, thin bedded, infaunal indicators such as inarticulate brachiopods and Bathyurid trilobite. Other facies are dolomitic calcirudites with intraformational laminated clasts 4–8 cm up to 10 cm, flat and platy within trilobitic grainstones and calcirudites, some with lag surfaces of phosphate and borings, weathers orange, medium bedding. The upper part of the unit is more dolomitic and weathers orange, pale yellow – grey shale partings and interbeds throughout.		
96-1-1	Calcirudite, intraclasts, medium grey, weathers darker grey than cleaved calcilutite interbeds, 6 cm thick, medium crystalline, disseminated pyrite.	7.5	77.5
96-1-2	As -1; macrofossils are spar filled, upper bed 10 cm thick atop another 10 cm bed.	6	83.5
	Cleaved calcilutite interval.		
96-1-3	Grainstone, weathers light orange, less intraclasts, 3 cm bed, trilobite genal spines up to 10 cm.	5	88.5
	Intraclastic bed weathers darker grey, lutites and cross-laminated dolomitic siltstones, weather brown to orange, thin to medium bedded.	0.5	89
	Abundant traces on bottom of bedding plane.	4	93
	Cleaved calcareous shale interval and competent, silty lutites with traces and scours.	5	99
96-1-4	Calcirudite, intraclastic, light grey, weathers light grey and orange, 0–5 cm thick among shale, medium crystalline, shale interbedded with lutite.	1	100
96-1-5	Grainstone–Calcirudite, light grey, weathers light grey and orange, 5 cm thick bed among competent shales and lutites, medium crystalline, slightly dolomitic.	9	109
	Lutite, silty, light grey.	6	115
	Covered interval of orange, dolomitic siltstone talus on north ridge, abundant traces.	10	125

Section 1 (of Cecile and Norford 1979) (*continued*).

		Thickness (m)	
			Total from
Sample number	Description	Sample	Base
Grey Peak Member			
Abundant trace fossils in orange, silty dolostone alternating with limy intervals of shale, lutites, and calcirudites, which creates banding.			
96-1-6	Packstone–Grainstone, medium grey, weathers dark grey and orange.	7.5	132.5
96-1-7	Lime mudstone–Calcareous shale, medium grey, weathers buff, argillaceous partings, limestone interbeds cleaved within shale.	16.5	149
Shale interval, weathers light grey.			
96-1-8	Packstone, medium grey, weathers light grey, 1 cm bed, platy, burrow mottled, horizontal burrows weather yellow, compact.	10	159
Interval of light grey, cleaved thin-bedded shales–lutites and thin-bedded silty beds, weather buff and grey.			
96-1-9	Packstone–Grainstone, medium grey, weathers buff–grey, 6 cm compact bed.	9	168
Calcirudite interbedded with silty, argillaceous lime mudstone with abundant traces. Banded unit of grey and orange buff–brown beds alternating in packages.			
96-1-10	Calcirudite, intraclastic, medium grey, weathers medium grey, medium crystalline.	15	183
Grey interval of slabby–platy intraclastic beds; less traces.			
96-1-11	Wackestone, medium grey, weathers light grey – buff, 10 cm thick.	10	193
Intraclastic beds and hackly weathered beds, 1–2 m scale of siltstones, weather darker buff–orange, less traces.			
96-1-12	Wackestone, medium grey, weathers grey–buff, compact 2 cm bed, uniform, silty.	10.5	203.5
Calcirudite alternating with metre scale units of intraclastic beds weathers rough; above interval of dark orange, silty dolostone with some traces.			
96-1-13	Calcirudite, compact, light grey, weathers light grey, platy.	13	216.5
Covered interval.		17	233.5
Mt. Sheffield Member			
Largely grey-weathered limestone intervals and grey to buff–orange dolostone intervals.			
Grey and buff dolomitic beds.		3	236.5
96-1-14	Packstone–Wackestone, light grey, dolomitic, thin bed, platy, burrow mottled.	1	237.5
Buff and orange dolostone interval.			
96-1-15	Packstone, light grey, weathers light grey, thin bedded, platy.	9	246.5
Burrow-mottled, cross-laminated dolostone, weathers sandy brown.			
96-1-16	Lime mudstone, medium grey, weathers buff–grey, argillaceous partings.	10.5	257
96-1-17	Calcisiltite, medium grey, weathers light grey – pale yellow, 2–3 cm interval, burrow mottled, cross-laminated, platy.	9	266
96-1-18	As -17.	10.5	276.5
Grey interval, burrow mottled.		36	312.5

Section 1 (of Cecile and Norford 1979) (*concluded*).

Sample number	Description	Thickness (m) Sample	Total from Base
Chert nodules and interbedded chert, black, in upper 30 m of subunit; subtidal–intertidal, horizontal laminated silty dolostone, weathers sandy grey–brown, some beds comprise shell hash, medium to thick bedded (5–20 cm), fewer traces, some burrow mottling.		54	396.5
Increased burrow mottling, medium to thick beds, laminated, dolomitic siltstone, and grey dolostone, weathers smooth, creamy grey–buff, with lighter colored clasts (buff) and black chert interval (0.5 m).		16	412.5
Platy dolostone, weather darker grey and chert nodules, black.		2	414.5
96-1-19	Calcisiltite, medium grey, weathers light grey, finely laminated.	6	420.5
Buff – sandy beige dolostone overlain by light grey interval.			
96-1-20	Wackestone, medium grey, weathers light grey, compact, medium bed.	4.5	424.5
96-1-21	Wackestone, medium grey, weathers light grey, 2 cm bed, rubbly.	7.5	432
Dolostone, grey, up to a zone of mineralization (pyritic).		9	441
96-1-22	Packstone, medium grey, weathers light grey, 1 cm bed, medium crystalline, gastropods on bedding surface, pyritic.	0	441
Thin beds, weather grey.			
96-1-23	Wackestone, medium grey, weathers buff, 6 cm bed above thinner bedded ones.	10.5	451.5
To top of snow-covered ridge.		29.5	481
Skoki and Beaverfoot formations			
Although partially snow covered, the top of the ridge comprises dolostone, medium grey, weathers light grey – buff, massive bedded with some platy interbeds, laminated.			

Section 5 (of Cecile and Norford 1979)

Section is located 12 km southwest of where the Prophet River bifurcates (94F/8; 57°30′N, 124°26′W). Camped in cirque, walked down to gully of Kechika and measured and collected the Skoki in creek beds toward camp and up the ridge behind camp.

		Thickness (m)	
			Total from
Sample number	Description	Sample	Base
	Kechika Formation, Uppermost Unit		
The unit is largely light grey shale with few lenticular intervals of medium grey wackestone, weathers buff–orange and grey, thin bedded, disappears into the gully at tree line. Collected in cliff east of camp.			
96-5-K1	Wackestone, medium grey, lenses up to 2 cm thick, medium crystalline.	0	0
Contact with Skoki.		6	6
	Skoki Formation		
Comprises distinct units with gradational contacts, largely dolomitic limestone and shale, massive bedded. Moved from east cliff down to easternmost creek to begin measuring and collecting.			
Sikanni Chief Member			
Lowermost subunit of dolomitic shale, weathers orange, massive bedded, and dolostone–limestone lenses, 1 cm long, weather dark grey among massive-bedded shale, argillaceous partings within limestone.			
96-5-1	Wackestone, medium grey, weathers dark grey, 0–10 cm thick lenses, medium crystalline, dolomitic.	1	7
96-5-2	As -1; weathers orange.	20	27
96-5-3	Wackestone, medium grey, weathers light buff to grey, 1 cm lenses, medium crystalline, dense.	17	44
More lenticular limestone in this interval within dolomitic shale.			
96-5-4	Wackestone, medium grey, weathers dark grey, 7–10 mm lenses, medium crystalline.	24	68
Horizontal spar in interbeds.			
96-5-5	Lime mudstone, weathers orange and dark grey, 5 mm lenses.	16	84
96-5-6	Wackestone, medium grey, weathers buff, lenses up to 2 cm thick, medium crystalline, argillaceous partings.	30	114
Dolostone, weathers dark grey and orange–yellow, mottled, spar interbeds, massive bedded.		15	129
96-5-7	Wackestone, medium grey, mottled, medium crystalline.	15	144
96-5-8	Lime mudstone, medium grey, weathers buff, fine crystalline, argillaceous partings.	50	194
Subunit 2 is predominantly grey weathered and more calcareous, burrow mottled.			
96-5-9	Lime mudstone–Wackestone, medium grey, weathers buff and light grey, mottled.	15	219
Covered interval; fault in east bank, covered on west bank, overlain by massive dolomite.		20	239
96-5-10	Wackestone, medium grey, weathers light grey, massive bedded, medium crystalline.	7	246
Interval of dolostone, weathers light grey.			

Section 5 (of Cecile and Norford 1979) (*continued*).

		Thickness (m)	
		Total from	
Sample number	Description	Sample	Base
96-5-11	Wackestone–Packstone, light grey, weathers light grey, massive bedded, medium crystalline, argillaceous.	21.5	267.5
96-5-12	Packstone, medium grey, weathers light grey.	11.5	279
Carried on up the creek to the base of first major cliff/waterfall.			
96-5-13	Packstone, medium grey, weathers light grey and yellow, massive bedded.	14.5	293.5
96-5-14	As -13.	5	298.5
Dolostone, massive, weathers grey.		7	305.5
Covered interval.		7	312.5
Dolostone, 1 m thick beds, weather blue–grey.		25.5	338
Subunit 3 comprises massive dolostone, weathers orange and grey, more carbonate, medium beds have hackly fracture, weather medium grey.			
96-5-15	Wackestone, light grey, weathers light grey and hackly, thick bedded, medium crystalline.	1	339
Fossiliferous zone of dolomitic limestone, medium grey, thick to massive bedded, medium crystalline.			
96-5-15A	Packstone, dark grey, weathers light grey and hackly, massive bedded, some chert nodules, fossiliferous.	4.5	343.5
Fossiliferous zone of dolomitic limestone, medium grey, weathers light grey – yellow, and hackly, macluritid gastropods prominent.			
96-5-16	Packstone, medium grey, weathers light grey and yellow, hackly, macluritid gastropods.	4.5	348
96-5-16A	As -5-16.	5	353
Dolostone, medium grey, weathers light grey and rubbly, thick to massive bedded, north of camp toward ridge, 4.5 m thick.		20	373
Covered interval.		54	427
Dolostone, blue–grey, thick bedded, medium crystalline, base of first cliff outcrop.		15	442
Covered interval.		6	448
Keily Member		6	454
Burrow-mottled unit of dolostone, dark grey, weathers light grey, increasing yellow dolostone and less packstone upsection. Unit is rich in traces and bioclasts.			
Dolostone, burrow-mottled, fossiliferous.		42	496
96-5-17	Packstone, dark grey, weathers yellow and light grey, burrow mottled, massive, argillaceous partings, partly dolomitized, 1–5 cm chert nodules.	94.5	590.5
96-5-18	Packstone, dark grey, weathers yellow and light grey, burrow mottled.	12	602.5
96-5-19	Wackestone, dark grey, weathers light grey, burrow mottled, thick to massive beds, irregular shale partings, dolomitized in joints.	33	635.5
Dolostone, weathers light grey, rich in gastropods.		35.5	671
Gastropod subunit, doloarenite, weathers lighter grey, cross-laminated, trough cross-bedding, ooids and local brecciation.		40.5	711.5

Section 5 (of Cecile and Norford 1979) (*concluded*).

	Thickness (m)	
	Total from	
Sample number Description	Sample	Base
Top of doloarenite unit. Overlain by dolostone subunit, light grey, weathers orange–grey, platy.	15	726.5
Dolomitic mudstone, light grey, pale yellow – grey, massive beds, fine regular laminae and rip up clasts, shallow subtidal, forms middle cliff of the mountain.	46.5	773
Subunit of dolostone, dark grey and massive.	21.5	794.5
Redfern Member	108	902.5
Banded unit comprising massive dolostone, weathers dark grey and alternates with light grey – yellow beds.		
Next cliff of dolostone, weathers dark grey.	60	962.5
Doloarenite, dark grey, weathers black, laminated with cross-bedding, overlies dolostone, weathers light grey.	30	992.5
Dolostone, weathers yellow – light grey, laminated.	87	1079.5
Upper light grey to buff unit of dolostone.	40.5	1119.5

Grey Peak Section

The stratigraphic section at Grey Peak (94F/14; 57°48´N, 125°13´W) was sampled in eight segments, beginning at the Kechika – Road River contact and progressing upward through the Ospika Formation, eastward along the ridge to the cliff-forming platformal carbonate debris of the Middle Recessive Member. The overlying orange-weathering Kwadacha Formation was walked out over a spur to the east. The Kechika was then collected in seven sections, working westward toward Grey Peak for Sections 2 through 5 and then upwards from the Cambrian units for Sections 6 through 8.

		Thickness (m)	
Sample number	Description	Total from Sample	Base
	Cambrian Strata, Unit 1		
Sandstone, mostly conglomerate with large clasts (up to 1 m) in thick to massive beds, coarse to granular, *Skolithos* present in one-half of them, cross-bedding in a few. Some beds have conglomeratic bases with rounded clasts up to 5 cm. Overlain by thick to massive dolostone.			
GP-94-6-1	Limestone/Sandstone (4 m), lowest 10 cm of massive bed of sandy, oolitic limestone, some larger clasts visible up to 2 cm across.	2	2
GP-94-6-2	Oolite, top 10 cm exposed, medium grey, medium sand-sized ooids, no clasts here, scattered quartz grains.	2	4
Covered interval.		40	44
	Kechika Formation		
The Kechika can be divided into units based on subtle changes in weathering colour of shale (70%) interbedded with thin-bedded limestone. Unit contacts are gradational.			
Lloyd George Member			
Platy limestone within a largely massive-bedded sequence of calcirudite.			
GP-94-6-3	Grainstone, medium grey, all beds are massive and oolitic, some prominently laminated, scattered quartz grains, attenuating with massive grey beds of conglomerate with intraformational limestone clasts of bedded grey limestone up to 1 m across, which were sampled.	2	46
GP-94-6-4	Lime mudstone–Wackestone, medium grey, most likely matrix of conglomerate.	20	66
GP-94-6-5	Oolite, black to grey, middle of slightly recessive thin-bedded interval where beds are finely oolitic and pyritic.	22	88
GP-94-6-6	Conglomerate, dark grey, fine intraclasts, clasts 2–4 cm, some dolomitic veining.	24	112
Quentin Member			
Change to brown-weathering shale with thin lenticular, rather widely spaced pods of dolomitic limestone.			
GP-94-6-7	Lime mudstone–Wackestone, dark grey, weathers buff, 0–15 cm bed, slightly dolomitic, ?scattered ooids.	30	142
GP-94-6-8	Lime mudstone–Wackestone, medium grey, compact 4 cm bed with slight lamination of limited lateral aspect (10 cm thick, extends over 1 m), few limestone beds or large lenses in this interval, sparse trilobite fragments.	30	172

Grey Peak Section (*continued*).

Sample number	Description	Thickness (m) Sample	Total from Base
GP-94-6-9	Lime mudstone–Wackestone, medium grey, lens 0–4 cm thick over 25 cm interval, within black shale.	36	208
GP-94-6-10	As -6-9.	60	268
GP-94-7-8	Lime mudstone, argillaceous, compact bed, more calcareous than calcareous shale in this interval.	130	398
GP-94-7-7	Wackestone, medium grey – brown, dolomitic, 3 cm bed, most calcareous of the overlying beds.	45	443
Subunit of thin-bedded shale, medium grey to grey–brown with thin 1–3 cm platy to lenticular limestone beds, often laminated and argillaceous/dolomitic.			
GP-94-7-6	Lime mudstone–Wackestone, medium grey, 15 cm compact bed breaking into 3–4 laminated beds.	19	462
GP-94-7-5	As -7-6; platy breakage.	24	486
GP-94-7-4	As -7-6.	19	505
GP-94-7-3	Lime mudstone–Wackestone, medium grey, beds less argillaceous, 3 compact, 1 cm beds.	20	525
GP-94-7-2	Lime mudstone, medium grey, weathers yellow–brown–grey, striped because of thin-bedded to laminated bedding which is then cleaved to accentuate striped pattern, argillaceous.	20	545
GP-94-7-1	As -7-2.	15	560
GP-94-7-0	Wackestone, medium grey, compact 3 cm bed in shale.	20	580
Subunit of shale weathers light grey and slightly buff locally, similar color when fresh with thin 2 cm to thick lime mudstone beds at widely separated intervals, 1–3 m.			
GP-94-5-1	Lime mudstone, medium grey, argillaceous, 2 cm bed.	20	600
GP-94-5-2	Lime mudstone, medium grey, finely laminated, compact 2 cm bed.	20	620
GP-94-5-3	As -5-2; 3 cm bed, more dolomitic in this interval.	20	640
GP-94-5-4	Lime mudstone, medium grey, finely laminated, not as argillaceous as below, compact 8 cm bed.	20	660
Limestone beds rare in this interval.			
GP-94-5-5	As -5-2; 1 cm bed, argillaceous top and bottom.	24	684
GP-94-5-6	Lime mudstone, medium grey – brown, 3 cm bed.	20	704
Grey Peak Member			
Forms the core of Grey Peak, becoming more calcareous. Beds are dominantly thin-bedded (1–2 cm) lime mudstone to wackestone separated by thin grey shale partings with dolomitic stringers, the alternation giving banded appearance to unit.			
GP-94-5-7	Wackestone, medium brown – grey, 5 cm compact bed, some fine laminations.	20	724
GP-94-5-8	As -5-7.	24	748
GP-94-5-9	Wackestone, medium brown – grey, slightly irregular bedded, compact 4 cm bed breaking into 2 by 2 cm blocks, with some trilobite fragments and rare pyrite, more shale towards the top of this unit.	20	768

Grey Peak Section (*continued*).

Sample number	Description	Thickness (m) Total from Sample	Base
Subunit forming platy talus of Grey Peak's east crest, limestone now weather brown but are not dolomitic and are predominant as in the previous unit except the limestone is brown as are the shale partings.			
GP-94-5-10	Wackestone, medium brown – grey.	27	795
Bedded limestone interval.			
GP-94-5-11	Wackestone, medium brown – grey, 3 cm bed.	20	815
GP-94-5-12	Lime mudstone–Wackestone, medium grey, finely laminated, compact 2–5 cm bed.	24	839
GP-94-5-13	Lime mudstone, medium grey, finely laminated 0–5 cm bed, pinches out laterally.	20	859
Subunit begins as shale, weathers dominantly grey to silvery and light grey with widely spaced limestone beds in 1–3 m intervals.			
GP-94-5-14	Lime mudstone, medium grey, 3 cm compact bed, finely laminated.	23	882
GP-94-5-15	As -5-14; abundant trace fossils in shaly–limestone talus, weathers brown.	22	904
GP-94-5-16	Lime mudstone, medium grey, 2 cm bed, lots of trace fossils in following interval.	29	933
Haworth Member			
This part of the section contains almost no limestone, only small pods within largely grey- to grey–brown-weathering medium grey shale, seen as a darker interval below samples numbered 4-.			
GP-94-5-17	Wackestone, light to medium grey, irregular 4 cm bed.	20	953
GP-94-5-18	As -5-17.	37	990
Unit consists of thin-bedded grey shale with interbedded nodular limestone which are rare but prominent 10–20 cm beds.			
GP-94-4-1	Conglomerate, medium grey – brown, clasts up to 2 cm, coarse packstone matrix, 10 cm compact bed.	21	1011
GP-94-4-2	Packstone–Grainstone, light grey, 15 cm compact bed, trilobite bioclasts.	20	1031
GP-94-4-3	As -4-2.	21	1052
GP-94-4-4	Grainstone, light grey, 30 cm fine pelmatozoan bed with 25 cm of grainstone at the base and 5 cm of mottled wackestone on top.	21	1073
GP-94-4-5	Packstone–Grainstone, light brown – grey, 15 cm bed with some 2 mm pelmatozoan ossicles, bed has dolomitized burrow-mottled patches.	13	1086
Bioclastic beds are becoming rarer towards the top of this unit, occurring only every 5–6 m instead of every 2–5 m.			
GP-94-4-6	Grainstone, medium grey – brown, 15 cm compact bed with slightly mottled surface, trilobite bioclasts.	13	1099
GP-94-3-1	Wackestone, medium brown – grey, thin 10 cm bed below which is a unit of fissile dark grey shale with very few carbonate beds.	20	1119
14 m of a subunit of grey shales with 30% limestone beds, some of which are dolomitic and some are trilobite packstone. The next interval is of silver–grey-weathering shale with thin, widely spaced limestone beds, only 1–2 cm thick and continues up to the Road River contact.			

Grey Peak Section (*continued*).

Sample number	Description	Thickness (m) Total from Sample	Base
GP-94-3-2	Packstone, medium grey, 5 cm layer, trilobite bioclasts, part of a 1 m interval of rich trilobite packstone–grainstone with some pelmatozoan ossicles.	14	1133
Interval of phyllitic shale, weathers white–silvery to pale yellow.			
GP-94-3-3	Lime mudstone, medium brown – grey, 2 cm compact bed.	21	1154
GP-94-3-4	Lime mudstone, medium grey, 3 cm compact bed.	25	1179
GP-94-3-5	Lime mudstone, medium grey, weathers grey–brown, especially the bioclastic beds, thin 0–2 cm beds in 10 cm interval of lenticular beds.	20	1199
GP-94-3-6	Lime mudstone, medium grey – brown, 2 cm bed.	20	1219
Interval of 70% white-weathering dark grey shale.			
GP-94-3-7	Lime mudstone, medium brown – grey, 2 cm bed.	23	1242
Subunit of pseudonodular dolomitic limestone and shale with rare 10–15 cm beds of trilobite packstone to grainstone.			
GP-94-3-8	Packstone, compact bed, trilobite bioclasts and rare pelmatozoan ossicles and some calcite veins.	20	1262
Interval has been dominantly shale with a change from silver weathering to grey–brown weathering with thin limestone beds every 50 cm.			
GP-94-3-9	Lime mudstone, medium brown – grey, 3 cm bed.	23	1285
Fissile shale, grey to medium brown with thin dolomitic limestone beds. Limestone beds are typically 1–4 cm up to 10 cm towards contact with Road River and separated by 10–30 cm of shale, therefore shale is 70% of section. Shale has two cleavages, which creates pencil shale talus and which may also cause the pseudonodular form of most limestones.			
GP-94-2-1	Lime mudstone, medium grey – brown, weathers grey–brown, thin bedded, dolomitic.	5	1290
Change to brown weathering.			
GP-94-2-2	Lime mudstone, grey–brown, 2 cm bed, dolomitic.	23	1313
GP-94-2-3	Lime mudstone, medium grey – brown, 2 cm bed, 10 cm shale partings.	21	1334
GP-94-2-4	As -2-3; 2–4 cm bed.	20	1354
GP-94-2-5	Lime mudstone, medium grey – brown, 5 cm thick bed alternating with 10 cm of grey shale.	19	1373
Limestone rare, calcareous 15 cm vein in this interval.			
GP-94-2-6	Lime mudstone – Wackestone, medium brown – grey.	20	1393
Interval of 80% pseudonodular limestone beds 5–10 cm thick interbedded with shale, dark grey, weathers brown–grey, two thinner dark grey intervals of predominantly dark grey shale.			
GP-94-2-7	Lime mudstone – Wackestone, medium brown – grey, top bed of Kechika Group.	15	1408

Grey Peak Section (*continued*).

Sample number	Description	Thickness (m) Sample	Thickness (m) Total from Base
	Road River Group		
	Ospika Formation		
Conformably overlies Kechika and comprises several distinct units with gradational contacts.			
Cloudmaker Member			
Comprises dark grey, cleaved black shale and siltstone, weathers light brown and medium to dark grey. The cleavage makes the lime mudstone beds rubbly. Largely pencil shale, weathers dark brown and cleaved with interbedded thin lime mudstone beds and nodular lime mudstone and lime siltstone, and the upper part of the unit comprises dark grey to light grey – brown shale.			
GP-94-1-1	Lime mudstone, dark grey, weathers dark grey, 10 cm thick bed, fine crystalline, silty, cleaved, thin bed, among pencil shales.	5	1413
Weakly calcareous siltstone and shale, 10 cm calcite veined silty limestone marker bed.		5	1418
GP-94-1-2	Lime siltstone, dark grey, weathers dark grey, 10 cm bed, fine crystalline, cleaved.	8	1426
GP-94-1-3	Lime mudstone, medium grey, weathers grey–brown, fine crystalline, pseudonodular.	1	1427
Nodular lithology present at the Kechika – Road River contact may be a repeated lithology moved by faulting.			
Marker bed of calcite, banded, weathers brown–grey to white and contrasts surrounding dark pencil shale. About 15 cm thick on ridge, partially talus covered.			
GP-94-1-4	Lime siltstone, medium grey, weathers brown to medium grey, thin bed, weakly calcareous, compact/dense but because of cleavage it breaks quite easily.	8	1435
Veins of calcite form a prominent knob which could follow a fault, brecciated.		3	1438
GP-94-1-5	Lime mudstone, medium grey, weathers grey–brown, compact nodular bed, fine crystalline, finely laminated, slightly recrystallized.	0	1438
GP-94-1-6	Lime mudstone, medium grey, weathers brown–grey, nodular, fine crystalline, dense, almost concoidal fracture, recrystallized.	7.5	1445.5
Predominance of pseudonodular limestone beds and dark shale, white calcite vein.		10	1455.5
GP-94-1-7	As -1-6; widely spaced laminations.	2.5	1458
GP-94-1-8	Lime mudstone – Wackestone, medium grey, weathers brown–grey, compact nodular bed, medium crystalline, uniform.	3	1461
Interval of siltstones and shale, with no strongly calcareous beds.			
GP-94-1-9	Lime mudstone, medium grey, weathers brown–grey, pseudonodular, fine crystalline.	26	1487
Change of color of shale to light brown – grey.			
GP-94-1-10	As -1-9.	9	1496

Grey Peak Section (*continued*).

Sample number	Description	Thickness (m) Sample	Total from Base
Finlay Limestone Member			
Color change to light brown – grey shale from dark grey shale, over a covered interval. Unit comprises a "sandwich" of lime mudstone, weathered dark grey, light grey, and dark grey, marking the first thick, prominent limestone, laminated, and forms a resistant cap on the ridge.		34.5	1530.5
GP-94-1-11	Lime mudstone, dark grey, weathers medium–dark grey and light brown, finely and evenly laminated, fine crystalline.	10	1540.5
A change in weathering color to light grey.			
GP-94-1-12	Lime mudstone, dark grey, weathers light grey, fine crystalline, slightly pyritic and laminated.	8.5	1549
Change back to dark grey weathering.			
GP-94-1-13	As -1-11; 2 m below top of unit.	9	1558
Chesterfield Member			
A thick unit with bands of resistant limestone and platy limestone, shale, and siltstones. Basal part is shale and platy limestone (Subunit A); overlain by tripart "sandwich" with dark middle between two lighter, buff platy limestone beds (B); overlain by a light band of shale and platy limestone and dark band (C); dolomitic conglomerate forming the second cap along the ridge (D); dark subunit of shale and platy, dolomitic limestone, weathers orange (E); top of unit is breccia, weathers yellow–grey.			
GP-94-1-14	Lime mudstone, dark grey, weathers buff, 1 m thick bed, fine crystalline, platy.	11.5	1569.5
Shale, thin-bedded limestone (*Subunit A*).		4.5	1574
Tripart limestone, lower part is limestone, dark grey, weathers yellow–grey, flaggy (1 m) (*Subunit B*).		1	1575
Middle is dark grey, weathers medium grey (2.5 m thick).		1.5	1576.5
GP-94-1-15	Lime mudstone, dark grey, weathers buff, from middle of tripart bed.	1	1577.5
Top is limestone, dark grey, weathers pale buff, thin, platy (2 m).		2	1579.5
Interval of darker weathered, platy limestone and shale (*Subunit C*).			
GP-94-1-16	Wackestone, dark grey, weathers dark brown – grey, medium crystalline, bits of spar, compact, some calcite veining.	13	1592.5
GP-94-1-17	Lime mudstone, medium grey, weathers brown buff, 1.5–2.0 m thick, platy, minor calcite veining.	10	1602.5
GP-94-1-18	Lime mudstone, dark grey fresh, weathers buff, thin, compact bed, fine crystalline.	10	1612.5
GP-94-1-19	Wackestone, dark grey, weathers medium grey, thin compact bed, medium crystalline.	20	1632.5
Thick polymictic conglomerate, dolomitized with chert clasts. The top of the conglomerate has slumped into a series of flow rolls (*Subunit D*).		10	1642.5
Dolomitic limestone and shale, orange, platy (*Subunit E*).			

Grey Peak Section (*continued*).

Sample number	Description	Thickness (m) Total from Sample	Base
GP-94-1-20	Packstone, medium grey, weathers yellow – light grey, layer on top of orange-weathered, laminated dolomite, grain-supported, medium crystalline, compact.	20	1662.5
GP-94-1-21	Grainstone–Calcirudite, medium grey, weathers yellow–brown, compact 20 cm bed, coarse crystalline, bioclastic, compact lower half and planar laminated upper half (graded from -20).	0.5	1663
GP-94-1-22	Calcirudite, polymictic, up to 2 cm long clasts, weathers orange, 20 cm thick bed, black lime granules and chert pebbles.	6	1669
Breccia unit, weathers yellow–grey (*Subunit F*).			
GP-94-1-23	Breccia, grainstone and micritic clasts round to platy, up to 30 cm long and 6 cm wide, lime matrix, weathers yellow, at least 6 m thick bed, cannot determine the base exactly.	9	1678
	To top of the breccia.	11	1689
Covered interval of breccia talus; Subunit F comprises four distinct bands of breccias with wavy bedding, indicating slumping and (or) scour and fill.		10	1699
GP-94-1-24	Lime mudstone, dark grey, weathers light grey and orange along joints, fine crystalline.	1	1700
Interval of flow breccias, oligomictic, clasts weather dark grey, matrix weathers light grey.			
GP-94-1-25	Lime mudstone, dark grey, weathers medium grey, laminated, uniform.	10	1710
GP-94-1-26	Lime mudstone, dark grey, weathers light grey, fine crystalline, irregular, dolomitic with horizontal burrows and bioturbation.	7	1717
GP-94-1-27	Conglomerate, oligomictic, clasts are round, 1–2 cm in diameter, dark grey, weathers light brown and grey.	12	1729
GP-94-1-28	Conglomerate, oligomictic, fine-grained lime mudstone clasts up to 4 cm long, well rounded, dark grey, weathers light grey.	10	1739
GP-94-1-29	Conglomerate, fine-grained limestone clasts, dark grey, weathers light grey.	10	1749
Finbow Shale Member			
Interbedded dolostone–shale, weathers brown–orange.			
Covered interval of shale.		10	1759
Bed with chert at the top and bottom.		0.5	1759.5
Shale, dark grey, weathers light grey – brown.		10	1769.5
Dolostone conglomerate with chert nodules.		2	1771.5
Shale and thin-bedded dolostone.		7	1778.5
Dolostone with chert clasts, 4 m thick, 20 cm oolite bed 1 m from the top of the bed and unit, dolomitized, weathers blue–grey, coarse crystalline. Top 10 cm of the bed and unit consists of blue–black chert, which also occurs throughout in the conglomerate.		4	1781.5
Subunit of mostly shale, weathers brown, dark grey, and orange–red, silty wackestone beds near the base of unit. A few blocks of lime mudstone are present, dark grey, bioturbated similar to -1-26, produce float in the gulley.			

Grey Peak Section (*concluded*).

		Thickness (m)	
Sample number	Description	Sample	Total from Base
GP-94-1-30	Dolostone, light grey, weathers orange, thin bed.	2.5	1784
GP-94-1-31	Dolowackestone, light grey, weathers light grey – brown, medium grained – silty laminated, weakly calcareous.	33	1817
Kwadacha Formation		50	1867
Shale and siltstone, weather red, orange, grey, and brown, extend to the synclinal axis.			

Section 4 (of Cecile and Norford 1979)

The ridge section, located 6 km north of the Akie River (94F/7; 57°26´N, 124°35´W), trends northeast to southwest with the Kechika exposed in a col to the northeast and the Road River making up the ridge peak to the southwest. Camp was established on a flat, grassy col within the Road River Group. Measurements and collections proceeded from the Kechika to the north up to the ridge to the south of camp.

Sample number	Description	Thickness (m) Sample	Thickness (m) Total from Base
	Kechika Formation		
Began collecting in the saddle north of camp, as the Kechika is largely recessive shale and falls off into the next valley to the north. Comprises light grey, cleaved phyllitic shale and thin-bedded limestone.			
Interval of light grey shale.		30	30
96-4-K1	Lime mudstone, medium grey, weathers medium grey, 0.5 cm thick, fine crystalline, lenses from 20 cm interval of light grey shale.	0	30
Light grey shale, interbedded with >2 m outcrop of thin centimetre limestone beds and lenses.			
96-4-K2	Packstone, medium grey, weathers buff–brown, 4 cm thick lens, coarse crystalline, trilobite bioclasts.	108	138
Outcrop of light grey – buff shale and thin lenses of limestone.		1	139
96-4-K3	Wackestone–Packstone, medium grey, two continuous beds, one irregular 2 cm thick and 1 cm bed overlying it, medium–coarse crystalline.	14	153
Packstone, weathers buff, interbedded with increased number of thin limestone beds.			
96-4-K4	Packstone, medium grey, weathers buff, 2 cm bed and 1 cm bed above, coarse crystalline.	25.5	178.5
	Road River Group, Ospika Formation		
Comprises distinct units with gradational contacts; largely shale with limestone beds. The Cloudmaker Member is cut out by an unconformity.			
Finlay Limestone Member			
Platy limestone, weathers yellow, thin bedded (1–2 cm), argillaceous limestone weathers buff.			
96-4-R1	Wackestone, medium grey, weathers yellow, 1–2 cm bed, medium crystalline.	27	205.5
Interval of shale.			
96-4-1	Lime mudstone, medium grey, weathers grey–yellow, 10 cm bed, fine crystalline, dense and resistant.	75	280.5
Dolomitic limestone interbeds, 1 cm thick.			
96-4-2	Lime mudstone, dark grey, weathers medium grey, 10 cm bed, fine crystalline.	60	340.5
Prominent, resistant cliff of limestone, coarser grained intraclasts in some beds.			
96-4-3	Wackestone, medium grey, weathers orange–grey, 10 cm bed, medium crystalline, graded and laminated with chert nodules and some chert layers.	13.5	354

Section 4 (of Cecile and Norford 1979) (*concluded*).

Sample number	Description	Thickness (m) Total from Sample	Base
Within grassy knob along strike with main cliff of Road River limestone, 3 m wide, medium to thick bedded, next prominent knob is dolomitic and veined.			
96-4-4	Wackestone, medium grey, weathers light grey, medium crystalline.	13.5	367.5
Chesterfield Member		12	379.5
Recessive unit of dolomitic shale, dark grey to green with dolostone interbeds. Knob of veined dolostone (2 m), white.			
Dolostone, veined (2 m), in float, within dark grey shale weathers green–buff to rust.		24	403.5
Dolostone (1 m).		10.5	422.5
Finbow Shale Member		3	425.5
Change to shale, weathers dark grey – orange. Dolostone, weathers buff and dark grey, 40 cm thick, veined. This point is about 30 m from camp on a grassy saddle before the cliff forming rocks making up the ridge begin.			
To another dolostone in float, 2 m wide.		24	449.5
Ware Member		42	491.5
Black shale interval, some weather orange among thick-bedded dolostone and rarer quartzite, weathers blue–grey. The interval is partly grass covered, but mostly dolostone and quartzite.			
96-4-5	Dolosiltstone, brown–grey, weathers buff, two thin 1.5 cm beds, weakly calcareous, medium crystalline.	24	515.5
Cross-bedded quartzite beds 50 cm thick.		10	525.5
Interval of thick-bedded quartzite ending with massive-bedded ones; covered interval of snow.		108	633.5
	Pesika Formation		
Unit is mostly dolomitic lime mudstone and dark shale.			
96-4-6	Grainstone, medium grey, weathers buff, dolomitic, coarse crystalline, overlies a dark grey graptolitic shale.	192	825.5
96-4-7	Lime mudstone, medium grey, 5–7 cm bed, fine crystalline, weakly alcareous, compact.	9.5	835
Talus covered.		30	865
	Kwadacha Formation	23.5	888.5
Cliff-forming silty dolostone, weathers orange, abundant horizontal, fan-shaped traces in the talus.			

Road River Core

Core from Inmet Mining was sampled at their core shack (94F/7; 57°23´N, 124°52´W), located east of a Repeater Station (57°24´N, 124°53´W), 5 km north of the Akie River and 37 km northeast of Finbow. The core is from hole A-95-19, "Cardiac Creek." Beds dipping at about 40° in core; measurements below are directly of core and not corrected for dip.

Sample number	Description	Thickness (m) Total from Sample	Base
	Road River Group, ?Kwadacha Formation		
Basal black graphitic, silty, rarely calcareous shale and siltstones, fine grained and foliated. Upper units of core contain varying amounts of siltstone and black shale and slate with rare calcareous siltstone beds, thin bedded throughout.			
Shale, black, fine grained, foliated, graptolitic. Fault zone at 18 m. Lime mudstone is medium grey.		65.0	65.0
RR-96-1	Lime mudstone, dark grey, 35 cm interval, fine crystalline.	0.0	65.0
RR-96-2	As -1.	22.7	87.7
RR-96-3	Lime mudstone, medium grey, silty, laminae (1 mm).	5.3	93.0
RR-96-4	As -3.	23.0	116.0
RR-96-5	As -3; quartz and calcite veins.	7.4	123.4
RR-96-6	As -3; slate partings.	28.8	152.2
RR-96-7	As -3; slate partings.	11.8	164.0
RR-96-8	As -3; slate partings.	12.4	176.4
RR-96-9	As -3; slate partings.	9.2	185.6
RR-96-10	As -3.	15.3	200.9
Interval of increased thin-bedded limestone, calcareous siltstone, black shale.			
RR-96-11	Calcareous siltstone, medium grey, medium crystalline, 1 mm laminae.	18.1	219.0
Interval of black shale, minor limestone, and lime siltstone, light to medium grey, rubbly and sheared, brecciated veins of quartz and calcite.			
RR-96-12	Calcareous siltstone, medium grey, silty, graphitic, veined.	55.5	274.5
Unit is largely siltstone (70%), weakly calcareous, light to medium grey, 1 cm beds interbedded with black shale/slate, sheared.			
RR-96-13	Siltstone, medium grey, laminated, platy, dense, black slate interbeds (40%), slightly graphitic.	50.1	324.6
RR-96-14	As -13; largely black slate (80%) in this interval.	14.4	339.0
RR-96-15	Calcareous siltstone, light grey, thinly bedded with minor slate interbeds.	16.0	355.0
RR-96-16	As -13; mainly slate, only minor siltstone.	47.0	402.0
Interval is mainly siltstone with 15% black shale and argillite.			
RR-96-17	Calcareous siltstone, medium grey, thin bedded, pyrite laminae.	18.0	420.0
RR-96-18	As -17; increased shale interbeds.	20.7	440.7
Interval of black shale (80%).			
RR-96-19	Siltstone, light grey, thin bedded, laminated, dense.	24.3	465.0
RR-96-20	As -19; medium grey.	47.0	512.0
RR-96-21	As -19; siltstone beds up to 20 cm thick, less shale in this interval (15%).	39.5	551.5
RR-96-22	As -19; 1–2 cm beds, 25% black shale.	50.5	602.0

Gataga Mountain Section

The section is located 12 miles south of Terminus Mountain Camp, on the east side of Gataga Mountain (94L/10; 58°38´N, 126°55´W). Camp was established in a cirque east of the mountain. The section was measured in segments, beginning at the boundary of Kechika and Road River in a saddle marked by a 20 m interval of frost-heaved volcanics. The section comprises ridges of the mountain below the overthrust Cambrian groups making up the main peak of Gataga Mountain. The Kechika was measured and sampled along the north–south ridge east of camp in two segments (GBK-95-A and -B). The fault-repeated sequence of Kwadacha Formation and Ospika Formation was measured toward Gataga Mountain from the volcanics. One lens of Ospika Formation was collected in a creek bed (GB-95-1). Two samples were taken from the top of section in the large bluff of light-grey-weathering Proterozoic(?) carbonates and volcanics.

Sample number	Description	Thickness (m) Sample	Thickness (m) Total from Base
	Kechika Formation, Unit 1		
Consists of cleaved and folded white–grey phyllitic shale with rare interbeds of platy lime mudstone, lenticular wackestone to packstone, and some distal turbidite beds 1–2 cm thick. The section is likely tectonically thickened.			
GBK-95-B1	Wackestone–Packstone, medium grey, pitted weathering, compact 4 cm bed, medium crystalline.	0	0
Interval of fissile, cleaved, grey shale with thin compact 0–1 cm, platy limestone beds. Becomes more shaly upsection, 80–90%.			
GBK-95-B2	Lime mudstone, medium grey, four beds in 20 cm interval, fine crystalline.	68	68
GBK-95-B3	As -B2; 0–3 cm bed, lenticular and folded.	12	80
GBK-95-B4	As -B2; argillaceous.	40	120
GBK-95-B5	Lime mudstone, medium grey, 0–2 cm, argillaceous, folded and lenticular.	40	160
GBK-95-B6	Wackestone, medium grey, 5 cm bed, medium crystalline, folded.	32	192
GBK-95-B7	As -B5.	48	240
GBK-95-B8	As -B5.	20	260
First outcrop along ridge that swings north toward Section A.			
GBK-95-B9	Lime mudstone, medium grey, three thin beds, fine crystalline.	20	280
GBK-95-B10	As -B9; compact set of three beds.	20	300
GBK-95-B11	As -B9.	10	310
GBK-95-B12	As -B9; finely laminated.	30	340
Fissile, grey shale interval.			
GBK-95-B13	As -B9; three beds, each 1 cm thick, laminated.	30	370
Unit now comprises brown–grey limestone and cleaved grey shale.			
GBK-95-B14	Wackestone, medium grey – brown, 20 cm composite bed.	38	408
GBK-95-B15	As -B14; 10 cm bed.	22	430
	Unit 2		
Consists mainly of cleaved grey shale with rare limestone beds, boudinaged.			
GBK-95-A1	Wackestone–Packstone, medium brown – grey, 1 cm bed, lenticular.	31	461
Interval of black and grey shale, minor calcareous distal turbidites.			
GBK-95-A2	Lime mudstone, medium grey, compact 2 cm bed, fine crystalline.	20	481

Gataga Mountain Section (*concluded*).

Sample number	Description	Thickness (m) Total from Sample	Base
GBK-95-A3	As -A2.	20	501
GBK-95-A4	Lime mudstone, medium grey – brown, 0–2 cm bed, fine crystalline, argillaceous, lenticular.	20	521
GBK-95-A5	As -A4; compact bed.	20	541
Interval of finely laminated, alternating shale and limestone.			
GBK-95-A6	Lime mudstone, medium to dark grey, rubbly, 2 cm bed, fine crystalline, argillaceous.	11	552
GBK-95-A7	Lime mudstone, medium grey, lenticular, argillaceous stringers separate lime mudstone.	21	573
GBK-95-A8	As -A7; 2–3 cm bed.	24	597
GBK-95-A9	Lime mudstone, medium grey, five 1–2 cm beds, fine crystalline, thin shale partings.	24	621
Scree covered interval of fissile dark grey shale, likely basal Road River.		80	701
Saddle contains interval of frost-heaved, intermediate to basic volcanics, weather green.		20	721
Road River Group			
Consists of poor exposures along the ridge of more resistant units of finely laminated dolomitic siltstone of the Kwadacha Formation, which weather orange and recessive units of the Ospika Formation comprising black, cleaved pencil shale. The units are fault repeated.			
Black shale, cleaved.		27.5	748.5
Siltstone, weathers orange, laminated, dolomitic.		22.5	771
Black shale.		21	792
Siltstone.		3	795
Black shale.		63	858
Talus covered to thrust fault, black shale noted along strike.		120	978
One sample was taken in a black shale unit within a creek bed west of the cirque.			
GB-95-1	Wackestone, medium grey, weathers medium grey, compact concretion (remobilized carbonate forming striations or a prismatic-like alignment of carbonate), medium crystalline, some calcite veining.		
Proterozoic(?) Unit			
Measured up west gully through fault-repeated section up to overthrust strata (247.5 m). The vertical cliff weathers light grey, is massive and highly fractured. Samples were collected above base of outcrop of cliff.			
GB-95-1C	Lime mudstone, medium grey, weathers light grey, compact, fine crystalline, white marbleized blebs, minor mottling.	1	1
GB-95-2C	Lime mudstone, medium grey, weathers light grey, 1 m thick, laminated, horizontal tectonic lineations.	2	3

Appendix B

Appendix B contains the numerical distribution of conodont species from productive sections and samples. Genera and species are listed alphabetically in the left-hand column. Sample numbers, including the measured distance above the base of each section, are listed in ascending stratigraphic order from left to right. An asterisk in front of a taxon indicates species not illustrated (as listed in Tables 15–17).

The tables include:

Tables 1–4	Section 13
Tables 5 and 6	Section 1
Tables 7 and 8	Section 5
Tables 9–11	Grey Peak Section
Table 12	Section 4
Table 13	Road River Core
Table 14	Gataga Mountain Section
Table 15	Summary table of total species abundance of the Early Tremadocian
Table 16	Summary table of total species abundance of the Late Tremadocian
Table 17	Summary table of total species abundance of the Early Arenigian

Table 1. Conodont distribution table, Section 13, Kechika Formation, Lloyd George and Quentin members.

Section 13, Kechika Fm.	Lloyd George Member	Quentin Member										
Distance above base (m)	10.5	0	27	48	75	120	154.5	181.5	264.5	305.5	325	
Species/Sample no.	96-13-C2	96-13-A1	96-13-A2	96-13-A3	96-13-A4	96-13-A6	96-13-A7	96-13-A8	96-13-A10	96-13-A12	96-13-A13	Total
Acodus oneotensis					16	154	160	4				334
Acodus quentinensis n. sp.		8		4	11	280	37					340
Cordylodus angulatus										1		1
C. caboti		29		14	20	284	31			13		391
C. caseyi		1	1		30	33	1					66
C. delicatus n. sp.				2		34	4		2	1		43
C. intermedius						27	2			4		33
C. lindstromi		73	4	61	199	906	526	2	14	21	1	1807
C. primitivus		2				1						3
C. proavus		2	1		5	18				1		27
Drepanoistodus pervetus						20	26		6			52
E. notchpeakensis	4	35										39
Iapetognathus aengensis		2	1		2	3	11					19
I. fluctivagus		1					4					5
I. sprakersi						12	1					13
I. sp.		1										1
Granatodontus cf. *G. ani*							6					6
Polycostatus falsioneotensis									16			16
Semiacontiodus nogamii			1	2	11	38	23		1			76
Teridontus gracillimus				2			14		1			17
T. nakamurai		22	1	2	54	185	18		6			288
T. obesus		4				20	28					52
Utahconus utahensis		18	6	13	4	55	65		3	3		167
Coniform indet. E					4	5	4					13
Coniform indet. H							54	1				55
Total	4	198	15	100	356	2075	1015	7	49	44	1	3864

Table 2. Conodont distribution table, Section 13, Kechika Formation, Grey Peak Member.

Section 13, Kechika Fm.	Grey Peak Member											
Distance above base (m)	430	451	471	491	679	716	736	756	776	810.5	833	
Species/Sample no.	96-13-A14	96-13-A15	96-13-A16	96-13-A17	96-13-B1	96-13-B2	96-13-B3	96-13-B4	96-13-B5	96-13-B6	96-13-B7	Total
Acanthodus lineatus	17	64	150	144	94	152	199	388	982			2190
A. uncinatus					3	2	31		6			42
Colaptoconus sp.						5	10		24			39
Cordylodus angulatus	4	4	17	16	4	6	15	13	27	15		121
C. caseyi		1	5		2	1			3			12
C. delicatus n. sp.			2									2
C. intermedius	4	11	40		1	2			3			61
C. lindstromi	7	12	2						6			27
?*Diaphorodus* n. sp.										14		14
Drepanoistodus nowlani											12	12
D. pervetus		9	66	60	26	111	50	32	98	26		478
Glyptodontus sp.					1			4				5
Laurentoscandodus cf. *L. triangularis*					14	14	8	22	25	4		87
Macerodus? n. sp. B									5			5
Macerodus? n. sp. A							3					3
?*Parapaltodus* sp.											24	24
Paltodus deltifer							4			6		10
Paroistodus numarcuatus										6	18	24
Polycostatus falsioneotensis			2									2
Polycostaus cf. *sulcatus*							5	4	1	5		15
R. kwadachaensis n. sp.						12	6		6			24
Rossodus manitouensis		6	61	187	90	112	284	237	348	120		1445
R. muskwaensis n. sp.					2	4	5	6	10			27
R. sheffieldensis n. sp.							19	4	58			81
Rossodus subtilis n. sp.									9	16		25
R. tenuis	1	27	54	26	5	29			37	11		190
Striatodontus prolificus							2	1	25	6		34
Striatodontus strigatus n. sp.					32	45	2		21			100
Stultodontus costatus									9			9
Tricostatus cf. *T. glyptus*								49	15	18		82
Tricostatus terilinguis n. sp.					12		4	6				22
Utahconus longipinnatus	4	11	37	59	2	2			3			118
U. utahensis	3	26	94						4	19		146
Variabiloconus bassleri	4	7	26	45	50	83	59	74	235	102		685
Ventricodus spurius n. gen.			76	60								136
Coniform indet. F					11	6						17
Coniform indet. D						5						5
Coniform indet. A									5			5
Total	44	178	632	597	349	591	706	840	1960	368	54	6319

Table 3. Conodont distribution table, Section 13, Kechika Formation, Grey Peak, Haworth, and Mt. Sheffield members.

Section 13, Kechika Fm.	Grey Peak Member				Haworth Member						
Distance above base (m)	863	923	971	1040	1163	1188	1230	1275	1444.5	1485	
Species/Sample no.	96-13-B8	96-13-B10	96-13-B11	96-13-B12	96-13-B14	96-13-B15	96-13-B16	96-13-B17	96-13-B21	96-13-B22	Total
Acanthodus lineatus	167										167
Acodus deltatus	1802						46			5	1853
A. kechikaensis n. sp.									1		1
Colaptoconus bolites							65				65
C. greypeakensis n. sp.					9						9
C. floweri								47			47
C. quadraplicatus					24						24
Cornuodus longibasis							72				72
Drepanoistodus concavus					92		65	37			194
D. nowlani	70										70
Eucharodus parallelus								4			4
?*Paltodus jemtlandicus*		61	11	7	211	1	632				923
Paracordylodus gracilis							21			1	22
Parapanderodus striatus							8	3			11
Paroistodus proteus		1	3	20	8	3	68	28	2		133
Graciliconus concinnus n. gen. n. sp.		4	2	38		1	7				52
Kallidontus princeps n. gen. n. sp.			2	4							6
Scolopodus subrex	21		2	7	275		41		9		355
Semiacontiodus aff. *S. cornuformis*					7		16				23
Tropodus australis									2		2
T. comptus	18										18
Total	2078	66	20	76	626	5	1041	119	14	6	4051

Table 4. Conodont distribution table, Section 13, Kechika Formation, Mt. Sheffield Member, and Skoki Formation.

Section 13	Kechika Fm., Mt. Sheffield Member							Skoki Formation					
Distance above base (m)	1518.0	1555.5	1593.0	1623.0	1660.5	1706.5	1741.0	1774	1811.5	1870	1928.5	1970.5	
Species/Sample no.	96-13-B23	96-13-B24	96-13-B25	96-13-B26	96-13-B27	96-13-B28	96-13-B29	96-13-B30	96-13-B31	96-13-B32	96-13-B33	96-13-B34	Total
Acodus deltatus		6	9										15
A. kechikaensis n. sp.	262	30	401	141	32								866
A. longibasis			64	47									111
A. neodeltatus n. sp.				140	257								397
Bergstroemognathus extensus										6			6
Cornuodus longibasis		3		13							2	2	20
Diaphorodus russoi			92	8									100
Drepanodus arcuatus								2					2
Drepanoistodus amoenus				20	16								36
D. forceps						2		2	4				8
Drepanoistodus n. sp. A			17										17
Drepanoistodus n. sp. B											1		1
Drepanoistodus n. sp. C												5	5
Fahraeusodus marathonensis			6					9	25	2	20		62
Juanognathus jaanusoni											5	22	27
Juanognathus variabilis											44		44
Kallidontus serratus n. gen. n. sp.	2	10	89	35	5		1						142
K. nodosus n. gen. n. sp.				18									18
K. princeps n. gen. n. sp.	3	3		4	8		2						20
Oelandodus elongatus		1	78	10	16		3	9					117
Oepikodus communis						16	51	52	53	66	6	101	345
O. evae								5			51	17	73
Oistodus bransoni								3					3
Oistodus lanceolatus				1	2		1	6	5				15
?*Paltodus jemtlandicus*	32	7											39
Paltodus subaequalis											4	5	9
Paracordylodus gracilis			2	8	90								100
Parapanderodus striatus				2			2	1	4		1	13	23
Paroistodus parallelus			4			1	1	6		3	6	18	39
P. proteus	49	26	196	86	34	1							392
Prioniodus elegans				1	3								4
Protopanderodus elongatus										2		16	18
P. gradatus											7	43	50
P. leonardi									8	5	1	13	27
Scolopodus krummi	78	21	32	34	6				3	2	6	14	196
Stolodus stola		2	29	11			2						44
Tropodus australis	24	15		12	12		2	18	18	19	9	50	179
T. sweeti								13	10	18	33	169	243
Platform indet.											4		4
Coniform indet. G			19										19
Coniform indet. K											4		4
Total	450	124	1038	591	481	20	65	126	130	123	204	488	3840

Table 5. Conodont distribution table, Section 1, Kechika Formation, Quentin and Grey Peak members.

Section 1, Kechika Fm.	Quentin Member					Grey Peak Member							
Distance above base (m)	77.5	83.5	88.5	100	109	132.5	149	159	168	183	193	203.5	
Species/Sample no.	96-1-1	96-1-2	96-1-3	96-1-4	96-1-5	96-1-6	96-1-7	96-1-8	96-1-9	96-1-10	96-1-11	96-1-12	Total
Acanthodus lineatus	295	249	288	680	502	218	64	174	612	149	1	13	3245
A. uncinatus	21	13			6	5	2	1	3				51
Cordylodus angulatus	433	186	136	221	66	63	3	26	57	30		4	1225
C. caboti		15		3									18
C. caseyi		6		5	2	2		1	1				17
C. intermedius		2		2	2	2		2					10
C. lindstromi		5		1	2	0	6	4					18
Drepanoistodus minutus n. sp.				18	21					13			52
D. pervetus	510	211	120	246	110	137	5	50	126	121		4	1640
D. expansus				18	12				29	13		7	79
Laurentoscandodus sinuosus n. sp.	66	31	32	8	2	9			19	1			168
Loxodus latibasis	1			24	2				3	1			31
Polycostatus cf. *sulcatus*		3			3	1							7
Rossodus manitouensis	125	87	176	14	109	7	3	391	1008	252		12	2184
R. muskwaensis n. sp.	68	7	24		32	4		17	176	19			347
R. tenuis	2	34	56	54	115	3	2	165	419	408		13	1271
Striatodontus strigatus n. sp.	3	5		3	2	6	9		18	11		1	58
?*Stultodontus* sp.						1			8				9
Tricostatus infundibulum n. sp.			1	30		2			2				35
Utahconus longipinnatus	140	45	144	81	69	65	16	21	48	11			640
U. utahensis	4	27	184	216	270	71	7	11	82	183	1	12	1068
Variabiloconus bassleri	59	59	80	53	102	52	13	72	96	22			609
Ventricodus spurius n. gen.			32	30	31	30	4	29	10	74		4	244
Westergaardodina sp.			1										1
Coniform indet. B						2							2
Coniform indet. C						2			6	2			10
Total	1727	985	1274	1707	1460	682	134	964	2723	1310	2	70	13038

Table 6. Conodont distribution table, Section 1, Kechika Formation, Mt. Sheffield Member.

Section 1	Grey Peak Member	Mt. Sheffield Member									
Distance above base (m)	216.5	237.5	246.5	257	266	276.5	420.5	424.5	432	441	
Species/Sample no.	96-1-13	96-1-14	96-1-15	96-1-16	96-1-17	96-1-18	96-1-19	96-1-20	96-1-21	96-1-22	Total
Acanthodus lineatus	1	67	15	59	5						147
A. uncinatus		13									13
Acodus deltatus										2	2
Colaptoconus bolites										8	8
C. floweri							2	5			7
C. greypeakensis n. sp.							4	6		19	29
Colaptoconus sp.			2	2							4
Cordylodus angulatus		9	2								11
Cornuodus longibasis										2	2
Drepanodus arcuatus							14	3		2	19
Drepanoistodus concavus							4	9	3	104	120
D. pervetus	3	8	29	20							60
D. expansus		2									2
Graciliconus concinnus n. gen. n. sp.							7	4	1		12
Laurentoscandodus sinuosus n. sp.		1	14								15
?*Paltodus jemtlandicus*							9	6	4	6	25
Paroistodus proteus								1	1	8	10
Rossodus manitouensis	6	10	12	10	1	1					40
R. muskwaensis n. sp.		4									4
R. tenuis	4	23		3	2						32
Scolopodus subrex							41	3	7	14	65
Semiacontiodus aff. *S. cornuformis*										20	20
Striatodontus strigatus n. sp.		1	3								4
Tropodus comptus										3	3
Utahconus longipinnatus			1								1
U. utahensis	2	9									11
Variabiloconus bassleri		2	7								9
Total	16	149	85	94	8	1	81	37	16	188	675

Table 7. Conodont distribution table, Section 5, Upper Kechika Formation and Skoki Formation.

Section 5	Upper Kechika Fm.	Skoki Fm., Sikanni Chief Member											
Distance above base (m)	0	7	27	44	68	84	114	144	194	246	267.5	279	
Species/Sample no.	96-5-K1	96-5-1	96-5-2	96-5-3	96-5-4	96-5-5	96-5-6	96-5-7	96-5-8	96-5-10	96-5-11	96-5-12	Total
Acodus kechikaensis n. sp.							27						27
Bergstroemognathus extensus							2	1		2	2	1	8
Cornuodus longibasis	2				1		5					5	13
Drepanodus arcuatus							2			3		1	6
Drepanoistodus basiovalis												3	3
Drepanoistodus n. sp. C											1		1
Eucharodus parallellus											3	15	18
Fahraeusodus marathonensis					8								8
Juanognathus variabilis									1			18	19
Kallidontus cf. *nodosus*	3						2						5
Oelandodus elongatus	6												6
Oepikodus communis	227	2		2	33	3	9	20		1	4		301
O. evae								6				24	30
O. intermedius	5	2								42		110	159
Oistodus lanceolatus							15	1			2		18
Oistodus sp.	3												3
Parapanderodus striatus							7			1	3		11
Paroistodus parallelus	8				1		9				2		20
Periodon flabellum										6			6
P. selenopsis											3		3
Prioniodus adami							9						9
Prioniodus n. sp.							11						11
Protopanderodus gradatus							34					7	41
Protopanderodus rectus					1							7	8
Protoprioniodus cowheadensis												7	7
Scolopodus amplus n. sp.	1						6	5	1	6			19
Scolopodus krummi	3												3
Stolodus stola	2												2
Tropodus australis					6			1		7	2	5	21
T. sweeti				2	7	3	5			5	5	8	35
New genus B n. sp.							6						6
Total	260	4	0	4	57	6	149	34	2	73	27	211	827

Table 8. Conodont distribution table, Section 5, Skoki Formation, Sikanni Chief and Keily members.

Section 5	Skoki Formation, Sikanni Chief Member						Keily Member			
Distance above base (m)	293.5	298.5	339	343.5	348	353	590.5	602.5	635.5	
Species/Sample no.	96-5-13	96-5-14	96-5-15	96-5-15A	96-5-16	96-5-16A	96-5-17	96-5-18	96-5-19	Total
Anodontus longus					4		4	2	1	11
Ansella jemtlandica								4		4
Ansella longicuspica							6			6
Bergstroemognathus extensus	1			14						15
Cornuodus longibasis	5		4		20	3			5	37
Drepanodus arcuatus	2		6	6		7		2		23
Drepanoistodus amoenus				6	5	19				30
Drepanoistodus forceps					8	6				14
Drepanoistodus n. sp. D						3				3
Drepanoistodus n. sp. E					11					11
Juanognathus jaanussoni			4	8	30					42
Juanognathus variabilis		1	100	86	155					342
Jumudontus gananda			1	2						3
Microzarkodina n. sp.				14						14
?*Multioistodus compressus*								1		1
Oepikodus communis	10		20	162	295					487
O. evae				4						4
O. intermedius				42	29					71
Oepikodus n. sp.					4					4
Oistodus lanceolatus				14	8	3				25
O. multicorrugatus		2	14	4	17	15	8	8	1	69
Parapanderodus striatus	13		4		4	11	3	4	6	45
Paroistodus originalis							3	3		6
P. parallelus	2		4	26	10	10				52
Paroistodus n. sp.					4					4
Periodon aculeatus	2		8	24			26	1	23	84
P. flabellum					19	47	13			79
P. selenopsis				4	8					12
Planusodus gradus n. gen. n. sp.		1	1	38	35					75
Prioniodus n. sp.				8						8
Protopanderodus elongatus	5									5
P. leonardii			4	36	46	21				107
P. rectus	1	3			?2					4
Protoprioniodus aranda						18				18
P. nyinti	8		8	24	10	75				125
Protoprioniodus n. sp.				4						4
Pteracontiodus cryptodens								2		2
Tripodus laevis						3		1		4
Tropodus australis	25									25
T. sweeti	39									39
New genus C n. sp.						4				4
Coniform indet. J						15				15
Coniform indet. L					6					6
Coniform indet. M							4			4
Coniform indet. N								6		6
Coniform indet. O							3	6		9
Total	113	7	178	526	728	260	70	40	36	1958

Table 9. Conodont distribution table, Grey Peak Section, Kechika Formation, Lloyd George, Quentin, Grey Peak, and Haworth members.

Grey Peak Section, Kechika Fm.	Lloyd George Member		Quentin Member		Grey Peak Member				Haworth Member		
Distance above base (m)	66	112	525	640	768	933	953	990	1011	1031	
Species/Sample no.	GP-94-6-4	GP-94-6-6	GP-94-7-3	GP-94-5-3	GP-94-5-9	GP-94-5-16	GP-94-5-17	GP-94-5-18	GP-94-4-1	GP-94-4-2	Total
Cordylodus deflexus		2									2
C. hastatus		13									13
C. primitivus		59									59
C. proavus		34									34
Colaptoconus bolites									2		2
C. floweri									106		106
Colaptoconus greypeakensis n. sp.								9	98		107
Cornuodus longibasis										4	4
Drepanoistodus concavus									10	11	21
Eoconodontus notchpeakensis		87									87
Eucharodus parallelus									8	12	20
Graciliconus concinnus n. gen. n. sp.							22		33	23	78
Kallidontus serattus n. gen. n. sp.									4		4
K. princeps n. gen. n. sp.						1	4		17		22
?*Paltodus deltifer*			4								4
Paroistodus proteus							16	6	3	16	41
Phakelodus tenuis	3										3
Proconodontus muelleri		7									7
?*Protopanderodus inconstans*										8	8
Scolopodus subrex								20	56	24	100
Teridontus nakamurai		20									20
Tropodus comptus										2	2
?*Utahconus utahensis*				7							7
Total	3	222	4	7	0	1	42	35	337	100	751

Table 10. Conodont distribution table, Grey Peak Section, Kechika Formation, Haworth Member.

Grey Peak Section	Kechika Fm., Haworth Member								
Distance above base (m)	1052	1073	1086	1099	1119	1133	1262	1313	
Species/Sample no.	GP-94-4-3	GP-94-4-4	GP-94-4-5	GP-94-4-6	GP-94-3-1	GP-94-3-2	GP-94-3-8	GP-94-2-2	Total
Acodus deltatus	2	9							11
Colaptoconus bolites		1	2	2					5
C. floweri					5				5
C. greypeakensis n. sp.	2	14			5				21
Cornuodus longibasis	8	40	6	14					68
Drepanodus arcuatus			2	10					12
Drepanoistodus concavus	24	37	19	43					123
Eucharodus parallelus	60	20		8					88
Graciloconus concinnus n. gen. n. sp.	85	5							90
Granatodontus cf. *G. ani*	2								2
Kallidontus serratus n. gen. n. sp.	3	1		8					12
K. princeps n. gen. n. sp.	29	1		39					69
?*Paltodus jemtlandicus*	29	14							43
Parapanderodus striatus		7							7
Paroistodus proteus	8	13		2	2				25
?*Protopanderodus inconstans*				7		11			18
Scolopodus subrex	28	40	16	27			1		112
Stolodus stola								2	2
?*Stultodontus* sp.	2	5	6						13
New genus A n. sp.					9	26			35
Coniform indet. I							2		2
Total	282	207	51	160	21	37	1	2	763

Table 11. Conodont distribution table, Grey Peak Section, Road River Group, Ospika Formation, Cloudmaker, Chesterfield, and Finbow Shale members.

Grey Peak Section, Ospika Fm.	Cloudmaker Member	Chesterfield Member									Finbow Shale Member	
Distance above base (m)	1461	1592.5	1602.5	1632.5	1662.5	1678	1700	1729	1739	1749	1784	
Species/Sample no.	GP-94-1-8	GP-94-1-16	GP-94-1-17	GP-94-1-19	GP-94-1-20	GP-94-1-23	GP-94-1-24	GP-94-1-27	GP-94-1-28	GP-94-1-29	GP-94-1-30	Total
Acodus deltatus	1											1
Cornuodus longibasis		1			1							2
Drepanodus arcuatus					11							11
Drepanoistodus suberectus					12							12
D. bellburnensis					6							6
D. forceps					20							20
Erraticodon balticus						2						2
Multioistodus subdentatus							10			1		11
M. compressus?							1					1
Oistodus lanceolatus						1						1
Parapanderodus striatus					9	1						10
Paroistodus originalis		31		49	71	7						158
P. parallelus		27		5								32
Periodon aculeatus		16	1	413	23	20		3			1	477
P. flabellum		38		57		5						100
Prioniodus elegans	1											1
Protopanderodus elongatus				3	1	1						5
P. gradatus				1	24	1						26
P. leonardii		3				1						4
P. rectus										1		1
P. robustus				1	30							31
P. varicostatus					2							2
Protoprioniodus aranda		10		19	2	2						33
P. nyinti		2		2								4
Spinodus spinatus					2	1						3
?*Walliserodus nakholmensis*				6	1				3			10
Total	2	128	1	550	220	43	11	3	3	2	1	964

Table 12. Conodont distribution table, Section 4, Kechika Formation, Haworth Member, Ospika Formation, Finlay Limestone Member, Pesika Formation.

Section 4	Kechika Fm., Haworth Member			Ospika Fm., Finlay Limestone Member				Pesika Fm.	
Distance above base (m)	138	153	178.5	205.5	340.5	354	367.5	825.5	
Species/Sample no.	96-4-K2	96-4-K3	96-4-K4	96-4-R1	96-4-2	96-4-3	96-4-4	96-4-6	Total
Acodus deltatus	2								2
A. kechikaensis n. sp.		171	134						305
A. longibasis		121	74						195
A. warensis n. sp.		14	28						42
Ansella jemtlandica						4			4
Diaphorodus russoi	7								7
Distomodus staurognathoides								9	9
Drepanodus arcuatus							4		4
Drepanoistodus forceps			5						5
Drepanoistodus cf. *D. pitjanti*						13			13
Dzikodus tableheadensis						21	2		23
Kallidontus cf. *nodosus* n. gen. n. sp.	4	8	7						19
Oepikodus communis				2					2
Oelandodus elongatus	2	1	4						7
Oulodus sp. A								9	9
Panderodus unicostatus								22	22
Paroistodus horridus					7	61	4		72
Paroistodus originalis					7	46			53
P. parallelus				1					1
P. proteus	2	25	59						86
Periodon aculeatus					65	234	85		384
Prioniodus elegans			9						9
Protopanderodus calceatus						46			46
P. cooperi						32	7		39
P. varicostatus						5	1		6
Protoprioniodus sp.				1					1
Pseudooneotodus beckmanni								2	2
Pteracontiodus cryptodens					1				1
Scolopodus krummi	2	30	42						74
Triangulodus akiensis n. sp.						64			64
Walliserodus costatus						2	2		4
W. ethingtoni					3	29	3		35
Spinodus spinatus					1				1
Total	19	370	362	4	84	557	108	42	1546

Table 13. Conodont distribution table, Road River Core, Kwadacha Formation.

Road River Core, Kwadacha Fm.	
Distance above base (m)	116
Species/Sample no.	RR-96-4
Panderodus unicostatus	8
Oulodus sp. B	5
Total	13

Table 14. Conodont distribution table, Gataga Mountain Section, Ospika Formation.

Gataga Mountain Section, Ospika Fm.	
Distance above base (m)	c. 40–60
Species/Sample no.	GB-95-1
Phragmodus undatus	4
Total	4

Table 15. Summary table of total species abundance of the Early Tremadocian, *Iapetognathus* Zone to *Rossodus manitouensis* Zone.

Species	Total number of elements	% of total fauna (species)	% of total fauna (genus)
**Acanthodus lineatus*	5749	24.27	24.72
Acanthodus uncinatus	106	0.45	
Acodus oneotensis	334	1.41	2.85
Acodus quentinensis n. sp.	340	1.44	
Colaptoconus sp.	43	0.18	0.18
Cordylodus angulatus	1358	5.73	16.44
Cordylodus caboti	409	1.73	
Cordylodus caseyi	95	0.40	
Cordylodus delicatus n. sp.	45	0.19	
Cordylodus intermedius	104	0.44	
Cordylodus lindstromi	1852	7.82	
Cordylodus primitivus	3	0.01	
Cordylodus proavus	27	0.11	
?*Diaphorodus* n. sp.	14	0.06	0.06
Drepanoistodus expansus	81	0.34	9.98
Drepanoistodus minutus n. sp.	52	0.22	
Drepanoistodus pervetus	2230	9.41	
Eoconodontus notchpeakensis	35	0.15	0.15
Iapetognathus aengensis	19	0.08	0.16
Iapetognathus fluctivagus	5	0.02	
Iapetognathus sprakersi	13	0.05	
Iapetognathus sp.	1	0.00	
**Glyptodontus* sp.	5	0.02	0.02
Granatodontus cf. *G. ani*	8	0.03	0.03
Laurentoscandodus sinuosus n. sp.	183	0.77	1.14
L. cf. *L. triangularis*	87	0.37	
Loxodus latibasis	31	0.13	0.13
Macerodus? n. sp. A	5	0.02	0.03
Macerodus? n. sp. B	3	0.01	
**Paltodus deltifer*	10	0.04	0.06
*?*Paltodus deltifer*	4	0.02	
Polycostatus falsioneotensis	18	0.08	0.17
Polycostatus cf. *sulcatus*	22	0.09	
Rossodus kwadachaensis n. sp.	24	0.10	23.94

Table 15 (*concluded*).

Species	Total number of elements	% of total fauna (species)	% of total fauna (genus)
Rossodus manitouensis	3669	15.49	
Rossodus muskwaensis n. sp.	378	1.60	
Rossodus sheffieldensis n. sp.	81	0.34	
Rossodus subtilis n. sp.	25	0.11	
Rossodus tenuis	1493	6.30	
Semiacontiodus nogamii	76	0.32	0.32
Striatodontus prolificus	34	0.14	0.83
Striatodontus strigatus n. sp.	162	0.68	
Stultodontus costatus	9	0.04	0.04
Teridontus gracillimus	17	0.07	1.51
Teridontus nakamurai	288	1.22	
Teridontus obesus	52	0.22	
Tricostatus cf. *T. glyptus*	82	0.35	0.59
Tricostatus infundibulum n. sp.	35	0.15	
Tricostatus terilinguis n. sp.	22	0.09	
Utahconus longipinnatus	758	3.20	9.11
Utahconus utahensis	1392	5.88	
?*Utahconus utahensis*	7	0.03	
Variabiloconus bassleri	1303	5.50	5.48
Ventricodus spurius n. gen.	380	1.60	1.60
Westergaardodina sp.	1	0.00	0.00
Coniform indet. A	5	0.02	0.02
Coniform indet. B	2	0.01	0.01
Coniform indet. C	10	0.04	0.04
Coniform indet. D	5	0.02	0.02
Coniform indet. E	13	0.05	0.05
Coniform indet. F	17	0.07	0.07
Coniform indet. H	55	0.23	0.23
Total	23686	100	100

Table 16. Summary table of total species abundance of the Late Tremadocian, Low Diversity Interval to *Acodus kechikaensis* Zone.

Species	Total number of elements	% of total fauna (species)	% of total fauna (genus)
Acodus deltatus	1884	19.83	40.29
Acodus kechikaensis n. sp.	1199	12.62	
Acodus longibasis	306	3.22	
Acodus neodeltatus n. sp.	397	4.18	
Acodus warenis n. sp.	42	0.44	
Colaptoconus bolites	80	0.84	4.39
Colaptoconus floweri	147	1.55	
Colaptoconus greypeakensis n. sp.	166	1.75	
**Colaptoconus quadraplicatus*	24	0.25	
**Cornuodus longibasis*	218	2.29	2.29
Diaphordus russoi	107	1.13	1.13
**Drepanodus arcuatus*	77	0.81	0.81
Drepanoistodus amoenus	66	0.69	6.56
Drepanoistodus concavus	458	4.82	
Drepanoistodus nowlani	82	0.86	
Drepanoistodus n. sp. A	17	0.18	
Eucharodus parallelus	130	1.37	1.37
Graciliconus concinnus n. gen. n. sp.	232	2.44	2.44
Kallidontus nodosus n. gen. n. sp.	18	0.19	3.34
Kallidontus cf. *K. nodosus* n. gen. n. sp.	24	0.25	
Kallidontus princeps n. gen. n. sp.	117	1.23	
Kallidontus serratus n. gen. n. sp.	158	1.66	
Oelandodus elongatus	130	1.37	1.37
Oistodus lanceolatus	59	0.62	0.65
Oistodus n. sp.	3	0.03	
?*Paltodus jemtlandicus*	1030	10.84	10.84
Paracordylodus gracilis	122	1.28	1.28
*?*Parapaltodus* sp.	24	0.25	0.25
Parapanderodus striatus	107	1.13	1.13
**Paroistodus numarcuatus*	24	0.25	7.48
Paroistodus proteus	687	7.23	
Prioniodus elegans	14	0.15	0.15
*?*Protopanderodus inconstans*	26	0.27	0.27
**Protoprioniodus* sp.	1	0.01	0.01
Scolopodus krummi	273	2.87	9.53
Scolopodus subrex	632	6.65	
Semiacontiodus aff. *S. cornuformis*	43	0.45	0.45
Stolodus stola	48	0.51	0.51
?*Stultodontus* sp.	22	0.23	0.23
Tropodus australis	227	2.39	2.63
Tropodus comptus	23	0.24	0.24
Coniform indet. G	19	0.20	0.20
Coniform indet. I	2	0.02	0.02
New genus A	35	0.37	0.37
Total	9500	100	100

Table 17. Summary table of total species abundance of the Early Arenigian, *Oepikodus communis* Zone to *Tripodus laevis* Zone.

Species (genus)	Total number of elements	% of total fauna (species)	% of total fauna (genus)
Anodontus longus	11	0.22	0.22
Ansella jemtlandica	8	0.16	0.27
Ansella longicuspica	6	0.12	
Bergstroemognathus extensus	29	0.57	0.57
Drepanoistodus basiovalis	3	0.06	2.00
Drepanoistodus bellburnensis	6	0.12	
Drepanoistodus forceps	47	0.92	
Drepanoistodus cf. *D. pitjanti*	13	0.26	
Drepanoistodus suberectus	12	0.24	
Drepanoistodus n. sp. B	1	0.02	
Drepanoistodus n. sp. C	6	0.12	
Drepanoistodus n. sp. D	3	0.06	
Drepanoistodus n. sp. E	11	0.22	
Erraticodon balticus	2	0.04	0.04
Fahraeusodus marathonensis	70	1.37	1.37
Juanognathus jaanussoni	69	1.35	9.30
Juanognathus variabilis	405	7.95	
Jumudontus gananda	3	0.06	0.06
Microzarkodina n. sp.	14	0.27	0.27
Multioistodus compressus?	2	0.04	0.26
Multioistodus subdentatus	11	0.22	
Oepikodus communis	1135	22.28	28.97
Oepikodus evae	107	2.10	
Oepikodus intermedius	230	4.51	
Oepikodus n. sp.	4	0.08	
Oistodus bransoni	3	0.06	1.41
Oistodus multicorrugatus	69	1.35	
**Paltodus subaequalis*	9	0.18	0.18
Paroistodus horridus	72	1.41	8.58
Paroistodus originalis	217	4.26	
Paroistodus parallelus	144	2.83	
Paroistodus n. sp.	4	0.08	
Periodon aculeatus	945	18.55	22.47
Periodon flabellum	185	3.63	
Periodon selenopsis	15	0.29	
Planusodus gradus n. gen. n. sp.	75	1.47	1.47
Prioniodus adami	9	0.18	0.55
Prioniodus n. sp.	19	0.37	
Protopanderodus calceatus	46	0.90	8.09
Protopanderodus cooperi	39	0.77	
**Protopanderodus elongatus*	28	0.55	

Table 17 (*concluded*).

Species (genus)	Total number of elements	% of total fauna (species)	% of total fauna (genus)
Protopanderodus gradatus	117	2.30	
Protopanderodus leonardi	138	2.71	
Protopanderodus rectus	13	0.26	
Protopanderosus robustus	31	0.61	
Protoprioniodus aranda	51	1.00	3.75
Protoprioniodus cowheadensis	7	0.14	
Protoprioniodus nyinti	129	2.53	
Protoprioniodus n. sp.	4	0.08	
Pteracontiodus cryptodens	3	0.06	0.06
Scolopodus amplus n. sp.	19	0.37	0.37
Spinodus spinatus	4	0.08	0.08
Triangulodus akiensis n. sp.	64	1.26	1.26
Tripodus laevis	4	0.08	0.08
Tropodus sweeti	317	6.22	6.22
Walliserodus costatus	4	0.08	0.96
Walliserodus ethingtoni	35	0.69	
?*Walliserodus nakholmensis*	10	0.20	
Coniform indet. J	15	0.29	0.29
Coniform indet. K	4	0.08	0.08
Coniform indet. L	6	0.12	0.12
Coniform indet. M	4	0.08	0.08
Coniform indet. N	6	0.12	0.12
Coniform indet. O	9	0.18	0.18
New genus B	6	0.12	0.12
New genus C	4	0.08	0.08
Platform indet.	4	0.08	0.08
Total	5095	100	100

Plates

For each figured specimen, the element morphotype, view, sample number, repository number, and magnification are listed. The sample number abbreviates the locality and the year collected and correlates with data listed in Appendix A. Numbers may have a letter included for ridge sections measured and sampled in segments (e.g., 96-13-A6 is sample 6 from Section 13 near Mt. Sheffield, collected in 1996 from segment A of the section). Figures 4–8 show the stratigraphic distribution of all species, and Tables 1 through 17, Appendix A, list the total number of specimens from each section.

All figured specimens are deposited in the National Type Collections of the Geological Survey of Canada (GSC), Ottawa. All specimens, unless otherwise indicated, are hypotypes, and those held in open nomenclature are figured specimens.

Plates are arranged stratigraphically in which specimens figured on Plates 1–18 are from the Kechika Formation, Plates 19–22 are from the Road River Group, and Plates 23–29 are from the Skoki Formation; within each of these groups of plates, species are arranged in alphabetical order.

Plate 1

Figs. 1, 2. *Acanthodus uncinatus* Furnish

1. a element, inner lateral view, 96-13-B3, GSC 119279, ×70
2. e element, inner lateral view, 96-13-B3, GSC 119280, ×70

Figs. 3–10. *Acodus deltatus* Lindström

3. P element, inner lateral view, 96-13-B8, GSC 119232, ×86
4. Sd element, inner lateral view, 96-13-B8, GSC 119233, ×86

5–7. Sb elements, inner lateral views, 96-13-B8, GSC 119234, 119235, 119236, ×86

8. M element, inner lateral view, 96-13-B8, GSC 119237, ×86

9, 10. Sc elements, inner lateral views, 96-13-B8, GSC 119238, 119239, ×86

Figs. 11–17. *Acodus kechikaensis* n. sp.

11. Sb element, antero-lateral view, 96-13-B25, *paratype*, GSC 119250, ×67
12. Sa element, posterior view, 96-13-B25, *paratype*, GSC 119251, ×67
13. Sc element, inner lateral view, 96-13-B25, *paratype*, GSC 119252, ×67
14. Sd element, postero-lateral view, 96-13-B25, *paratype*, GSC 119253, ×67

15, 16. P element, inner and outer lateral views, 96-13-B25, *holotype*, GSC 119254, ×67

17. M element, inner lateral view, 96-13-B25, *paratype*, GSC 119255, ×67

Figs. 18–24. *Acodus longibasis* McTavish

18. M element, inner lateral view, 96-13-B26, GSC 119239, ×90
19. Sb element, inner lateral view, 96-13-B26, GSC 119240, ×90
20. Pb element, inner lateral view, 96-13-B26, GSC 119243, ×90
21. Sd element, inner lateral view, 96-13-B26, GSC 119242, ×90
22. Sa element, posterior view, 96-13-B26, GSC 119241, ×90
23. Pa element, inner lateral view, 96-13-B26, GSC 119244, ×65
24. Sc element, inner lateral view, 96-13-B26, GSC 119245, ×65

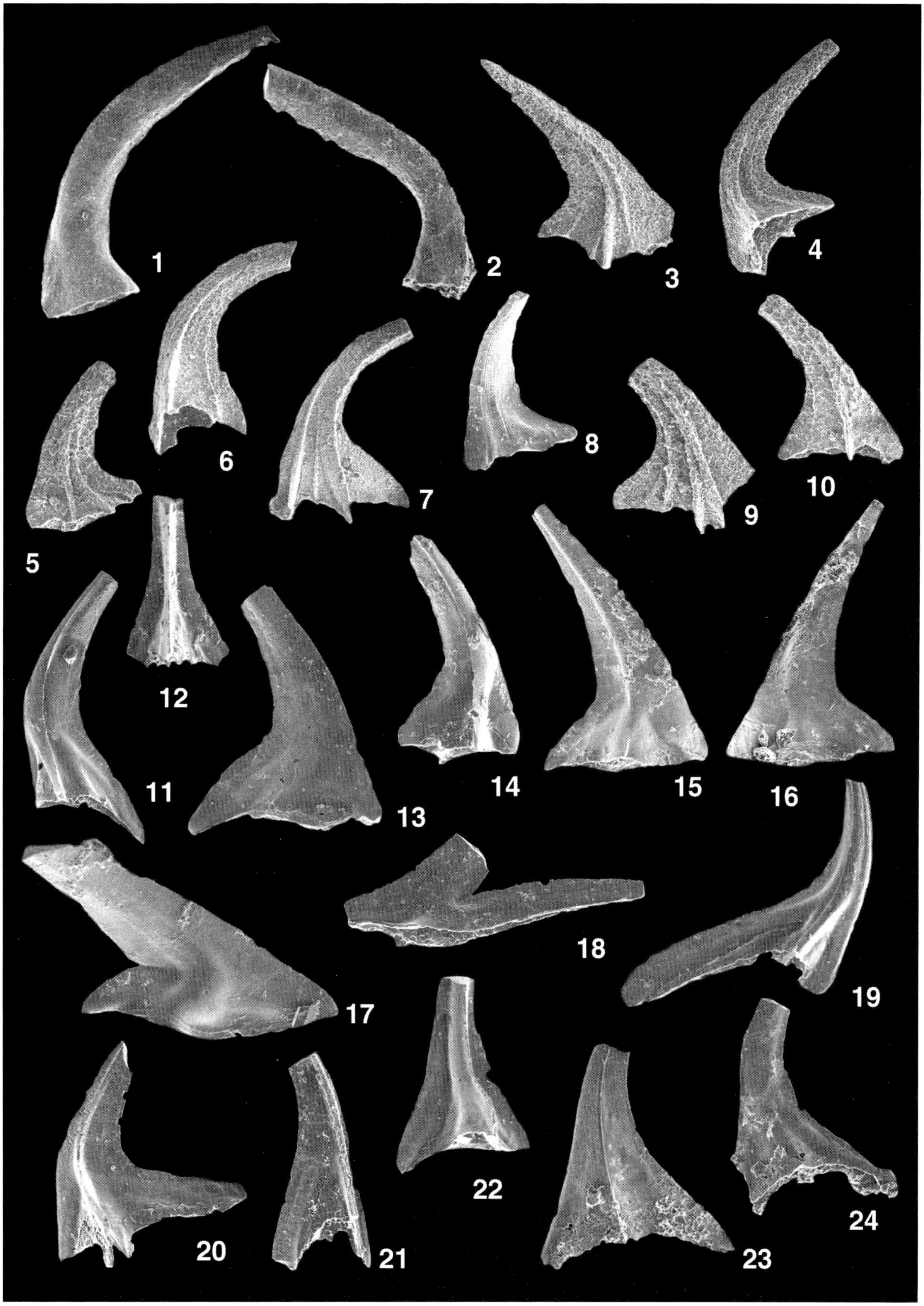
1
2
3
4
5
6
7
8
9
10
11
12
13
14
15
16
17
18
19
20
21
22
23
24

Plate 2

Figs. 1–7. *Acodus neodeltatus* n. sp.

1. Sb element, inner lateral view, 96-13-B27, *paratype*, GSC 119263, ×130
2. Sc element, inner lateral view, 96-13-B27, *paratype*, GSC 119264, ×130

3, 4. P element, inner and outer lateral views, 96-13-B27, *holotype*, GSC 119265, ×130

5, 6. Sd element, inner and outer lateral views, 96-13-B27, *paratype*, GSC 119266, ×130

7. M element, inner lateral view, 96-13-B27, *paratype*, GSC 119267, ×130

Figs. 8–12. *Acodus oneotensis* Furnish

8. Sb element, inner lateral view, 96-13-A6, GSC 119222, ×140
9. Sc element, inner lateral view, 96-13-A6, GSC 119223, ×140

10, 11. Sa element, anterior and posterior views, 96-13-A6, GSC 119224, ×140

12. M element, inner lateral view, 96-13-A6, GSC 119225, ×140

Figs. 13–19. *Acodus quentinensis* n. sp.

13, 14. Sb element, postero-lateral and antero-lateral views, 96-13-A6, *holotype*, GSC 119227, ×100

15. Sc element, inner lateral view, 96-13-A6, *paratype*, GSC 119228, ×84

16, 17. Sa element, anterior and posterior views, 96-13-A6, *paratype*, GSC 119229, ×100

18. P element, inner lateral view, 96-13-A6, *paratype*, GSC 119230, ×84
19. M element, inner lateral view, 96-13-A6, *paratype*, GSC 119231, ×84

Figs. 20–27. *Acodus warensis* n. sp.

20, 21. Sb element, inner and outer lateral views, 96-4-K4, *holotype*, GSC 119256, ×75

22. Sa element, posterior view, 96-4-K4, *paratype*, GSC 119257, ×75
23. M element, inner lateral view, 96-4-K4, *paratype*, GSC 119262, ×70
24. Sc element, inner lateral view, 96-4-K4, *paratype*, GSC 119258, ×75
25. Sd element, inner lateral view, 96-4-K4, *paratype*, GSC 119259, ×75
26. Pa element, inner lateral view, 96-4-K4, *paratype*, GSC 119260, ×75
27. Pb element, inner lateral view, 96-4-K4, *paratype*, GSC 119261, ×75

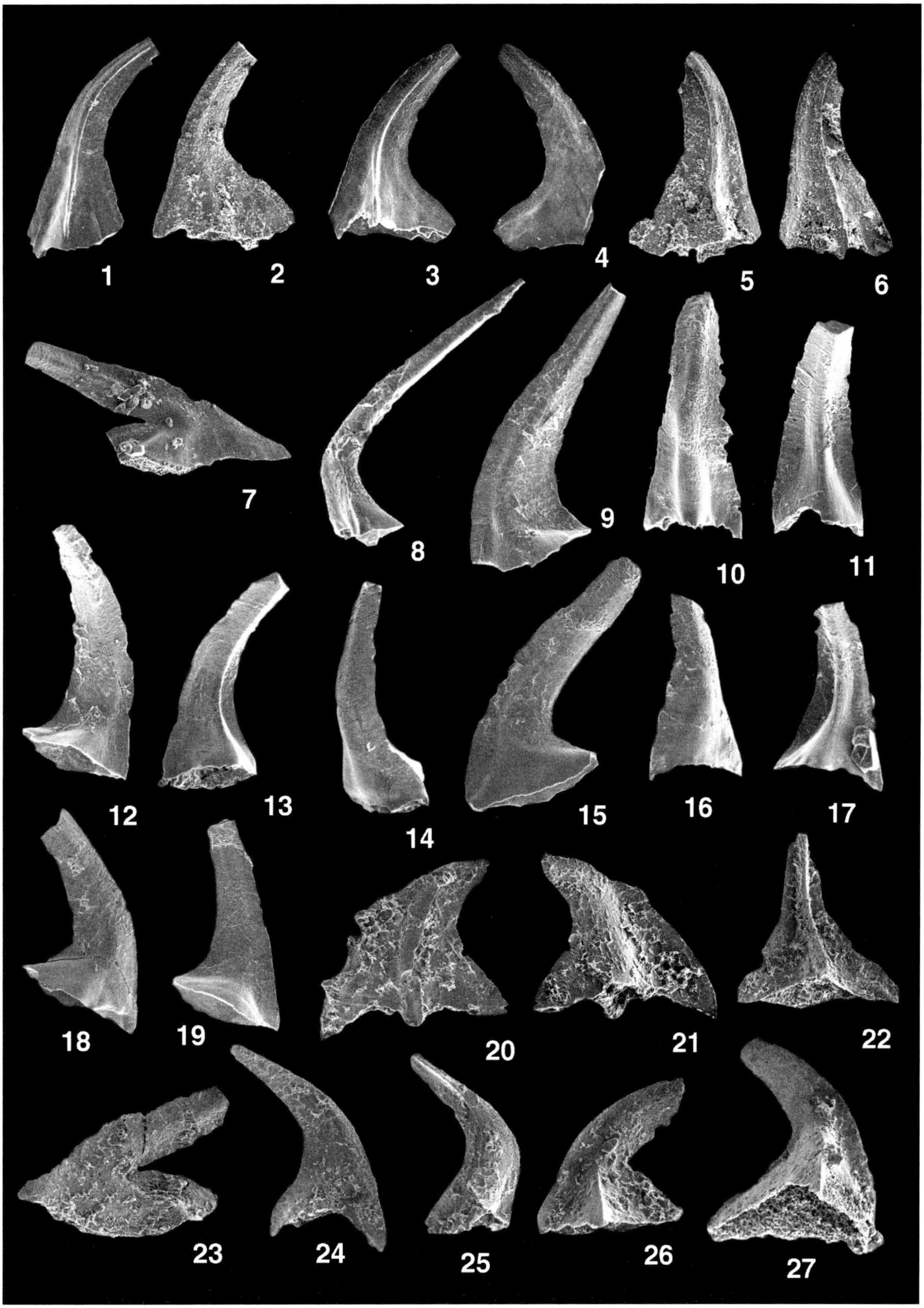
1
2
3
4
5
6
7
8
9
10
11
12
13
14
15
16
17
18
19
20
21
22
23
24
25
26
27

Plate 3

Figs. 1–4. *Colaptoconus bolites* (Repetski)

1. e element, inner lateral view, 96-13-B16, GSC 119299, ×100
2. a element, inner lateral view, 96-13-B16, GSC 119300, ×100
3. c element, posterior view, 96-13-B16, GSC 119301, ×100
4. a element, posterior view, 96-13-B16, GSC 119302, ×100

Figs. 5–8. *Colaptoconus floweri* (Repetski)

5. a element, postero-lateral view, 96-13-B17, GSC 119295, ×68
6. c element, posterior view, 96-13-B17, GSC 119296, ×68
7. b element, postero-lateral view, 96-13-B17, GSC 119297, ×55
8. e element, inner lateral view, 96-13-B17, GSC 119298, ×68

Figs. 9–14. *Colaptoconus greypeakensis* n. sp.

9. a element, inner lateral view, GP-94-4-1, *paratype*, GSC 119303, ×50

10, 11. c element, posterior and anterior views, GP-94-4-1, *holotype*, GSC 119304, ×50

12. a element, inner lateral view, GP-94-4-1, *paratype*, GSC 119305, ×50
13. b element, inner lateral view, GP-94-4-1, *paratype*, GSC 119306, ×50
14. e element, inner lateral view, GP-94-4-1, *paratype*, GSC 119307, ×50

Figs. 15, 16. *Colaptoconus* sp.

15. c element, posterior view, 96-13-B5, GSC 119183, ×83
16. c element, posterior view, 96-13-B5, GSC 119184, ×110

Figs. 17–20. *Cordylodus caboti* Bagnoli, Barnes, and Stevens

17. a element, inner lateral view, 96-13-A6, GSC 119068, ×105
18. e element, inner lateral view, 96-13-A6, GSC 119070, ×105
19. c element, outer lateral view, 96-13-A6, GSC 119069, ×105
20. a element, outer lateral view, 96-13-A6, GSC 119071, ×105

Fig. 21. *Cordylodus deflexus* s. f. Bagnoli, Barnes, and Stevens

21. outer lateral view, GP-94-6-6, GSC 119043, ×64

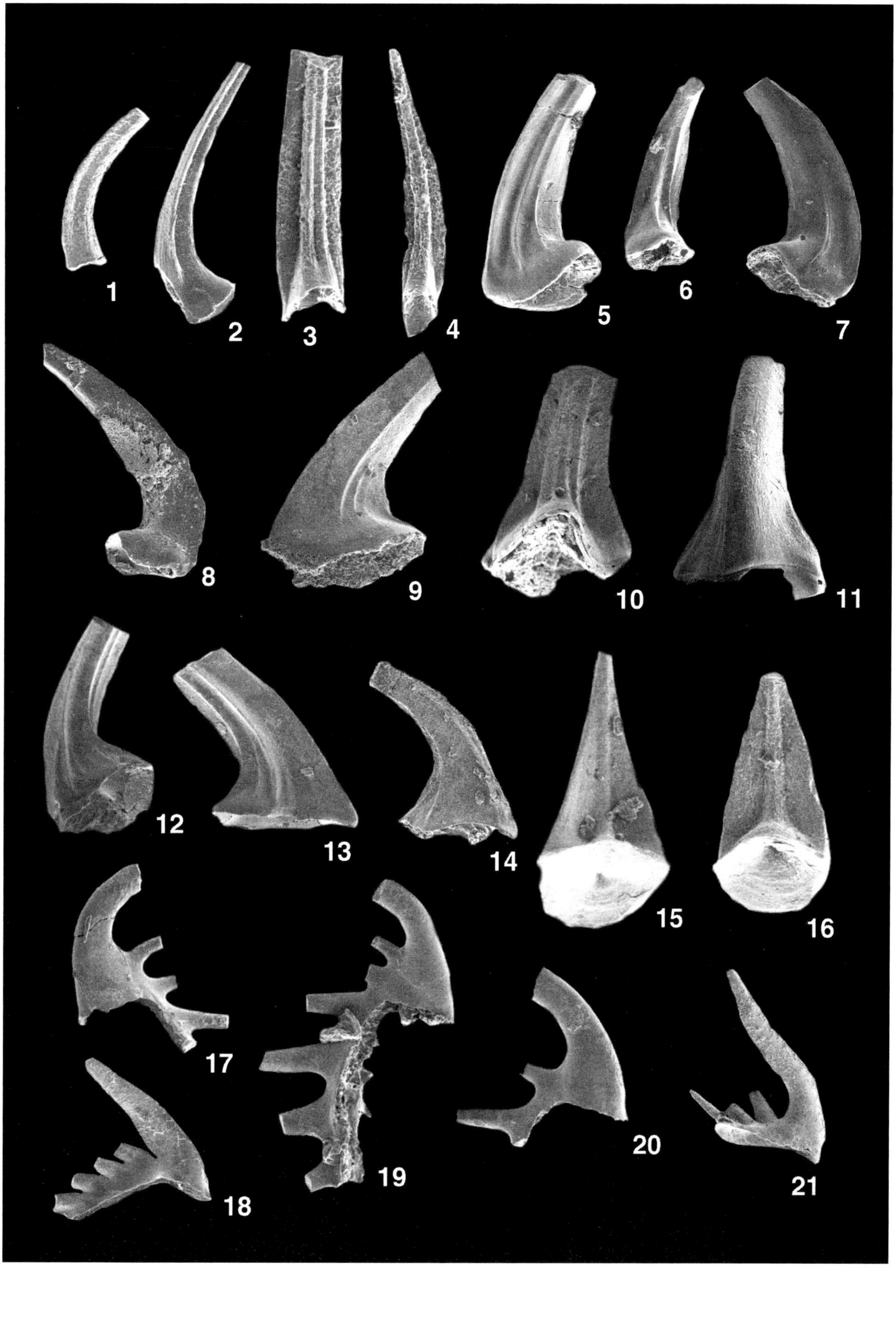
1
2
3
4
5
6
7
8
9
10
11
12
13
14
15
16
17
18
19
20
21

Plate 4

Figs. 1–3. *Cordylodus angulatus* Pander

1. a element, inner lateral view, 96-1-1, GSC 119078, ×58
2. c element, inner lateral view, 96-1-1, GSC 119079, ×58
3. e element, inner lateral view, 96-1-1, GSC 119080, ×58

Figs. 4–7. *Cordylodus caseyi* Druce and Jones

4. a element, inner lateral view, 96-13-A6, GSC 119074, ×58
5. e element, inner lateral view, 96-13-A6, GSC 119076, ×58
6. c element, inner lateral view, 96-13-A6, GSC 119075, ×58
7. e element, inner lateral view, 96-13-A6, GSC 119077, ×58

Figs. 8–11. *Cordylodus delicatus* n. sp.

8, 9. a element, outer and inner lateral views, 96-13-A6, *holotype*, GSC 119050, ×84

10. e element, inner lateral view, 96-13-A6, *paratype*, GSC 119051, ×84
11. c element, inner lateral view, 96-13-A6, *paratype*, GSC 119052, ×84

Figs. 12–14. *Cordylodus hastatus* Barnes

12. e element, inner lateral view, GP-94-6-6, GSC 119053, ×41
13. c element, inner lateral view, GP-94-6-6, GSC 119054, ×58
14. a element, outer lateral view, GP-94-6-6, GSC 119055, ×58

Figs. 15–17. *Cordylodus intermedius* Furnish

15. e element, inner lateral view, 96-13-A6, GSC 119063, ×100
16. c element, outer lateral view, 96-13-A6, GSC 119064, ×100
17. a element, inner lateral view, 96-13-A6, GSC 119065, ×100

Figs. 18–21. *Cordylodus lindstromi* Druce and Jones

18. a element, inner lateral view, 96-13-A6, GSC 119059, ×100
19. c element, outer lateral view, 96-13-A6, GSC 119060, ×100

20, 21. e elements, inner lateral views, 96-13-A6, GSC 119061, GSC 119062, ×100

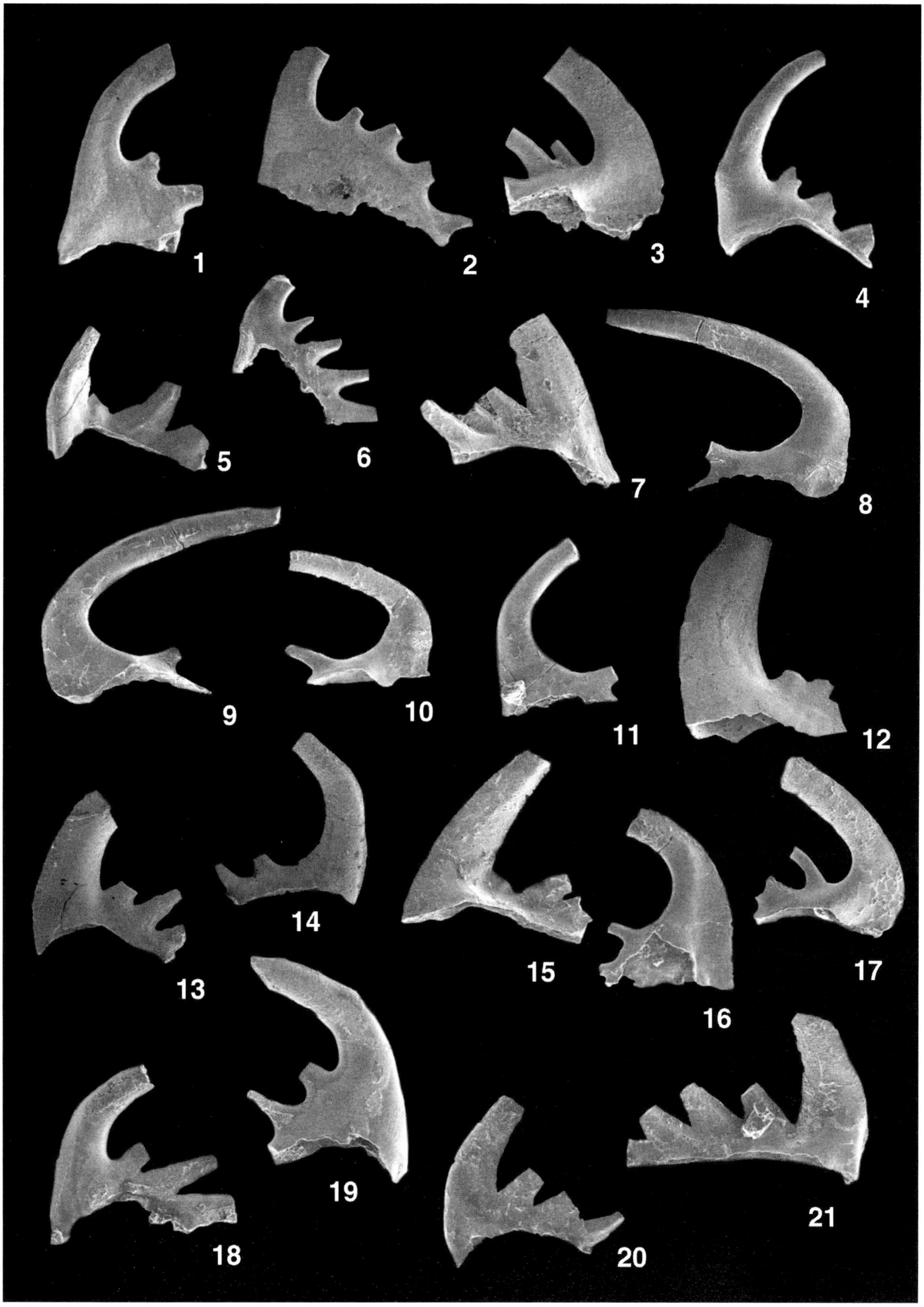
1
2
3
4
5
6
7
8
9
10
11
12
13
14
15
16
17
18
19
20
21

Plate 5

Figs. 1–3. *Cordylodus primitivus* Bagnoli, Barnes, and Stevens

1. a element, inner lateral view, GP-94-6-6, GSC 119047, ×62
2. c element, inner lateral view, GP-94-6-6, GSC 119048, ×62
3. e element, outer lateral view, GP-94-6-6, GSC 119049, ×62

Figs. 4–6. *Cordylodus proavus* Müller

4. c element, inner lateral view, GP-94-6-6, GSC 119056, ×60
5. a element, inner lateral view, GP-94-6-6, GSC 119057, ×60
6. e element, inner lateral view, GP-94-6-6, GSC 119058, ×60

Figs. 7–11. *Diaphorodus russoi* (Serpagli)

7. P element, inner lateral view, 96-13-B25, GSC 119273, ×150
8. Sc element, inner lateral view, 96-13-B25, GSC 119274, ×150
9. Sa element, posterior view, 96-13-B25, GSC 119275, ×150
10. M element, inner lateral view, 96-13-B25, GSC 119276, ×150
11. Sb element, inner lateral view, 96-13-B25, GSC 119277, ×150

Figs. 12–18. *?Diaphorodus* n. sp.

12, 13. Sa element, posterior and anterior views, 96-13-B6, GSC 119212, ×170

14. Sd element, inner lateral view, 96-13-B6, GSC 119212a, ×170
15. P element, inner lateral view, 96-13-B6, GSC 119213, ×130

16, 17. Sb element, antero-lateral and inner lateral views, 96-13-B6, GSC 119211, ×170

18. M element, inner lateral view, 96-13-B6, GSC 119214, ×130

Figs. 19–21. *Drepanoistodus amoenus* Lindström

19. e element, inner lateral view, 96-13-B27, GSC 119395, ×50
20. a element, antero-lateral view, 96-13-B27, GSC 119396, ×50
21. a element, inner lateral view, 96-13-B27, GSC 119397, ×64

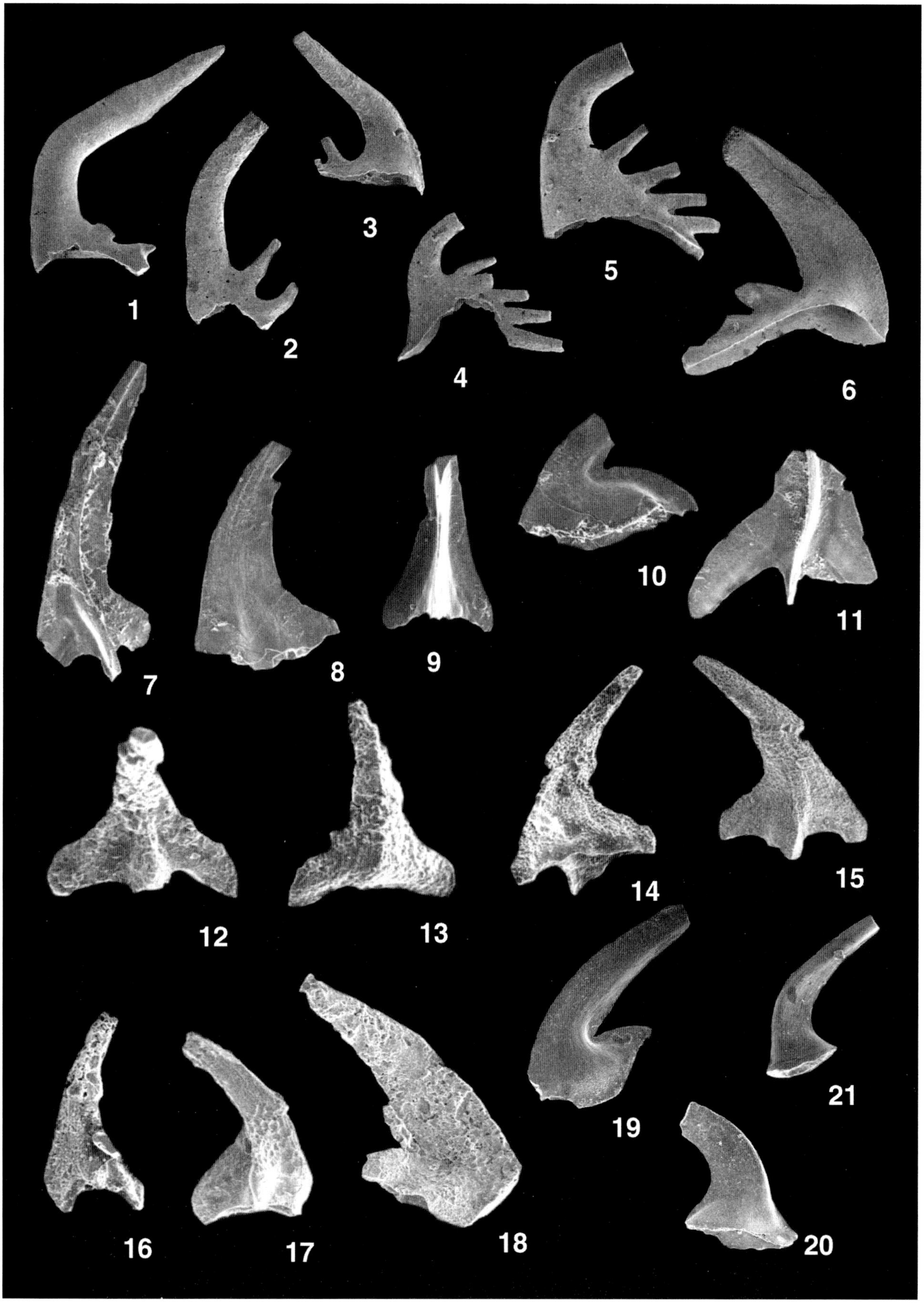
1
2
3
4
5
6
7
8
9
10
11
12
13
14
15
16
17
18
19
20
21

Plate 6

Figs. 1–4. *Drepanoistodus concavus* (Branson and Mehl)

1. a element, inner lateral view, 96-13-B14, GSC 119391, ×55
2. c element, inner lateral view, 96-13-B14, GSC 119392, ×55
3. e element, inner lateral view, 96-13-B14, GSC 119393, ×55
4. a element, inner lateral view, 96-13-B14, GSC 119394, ×55

Figs. 5–8. *Drepanoistodus expansus* Chen and Gong

5. e element, inner lateral view, 96-1-4, GSC 119381, ×115
6. c element, inner lateral view, 96-1-4, GSC 119382, ×115
7. a element, lateral view, 96-1-4, GSC 119383, ×115
8. a element, inner lateral view, 96-1-4, GSC 119384, ×115

Figs. 9–12. *Drepanoistodus minutus* n. sp.

9. e element, inner lateral view, 96-1-4, *paratype*, GSC 119388, ×130
10. a element, inner lateral view, 96-1-4, *paratype*, GSC 119389, ×130

11, 12. c element, inner and outer lateral views, 96-1-4, *holotype*, GSC 119390, ×130

Figs. 13–15. *Drepanoistodus nowlani* Ji and Barnes

13. a element, inner lateral view, 96-13-B8, GSC 119385, ×88
14. c element, inner lateral view, 96-13-B8, GSC 119386, ×70
15. e element, inner lateral view, 96-13-B8, GSC 119387, ×88

Figs. 16–18. *Drepanoistodus pervetus* Nowlan

16. e element, inner lateral view, 96-13-A7, GSC 119378, ×140
17. c element, inner lateral view, 96-13-A7, GSC 119379, ×140
18. a element, inner lateral view, 96-13-A7, GSC 119380, ×140

Figs. 19–21. *Drepanoistodus* n. sp. A

19. e element, inner lateral view, 96-13-B25, GSC 119398, ×50
20. c element, inner lateral view, 96-13-B25, GSC 119399, ×85
21. a element, inner lateral view, 96-13-B25, GSC 119400, ×85

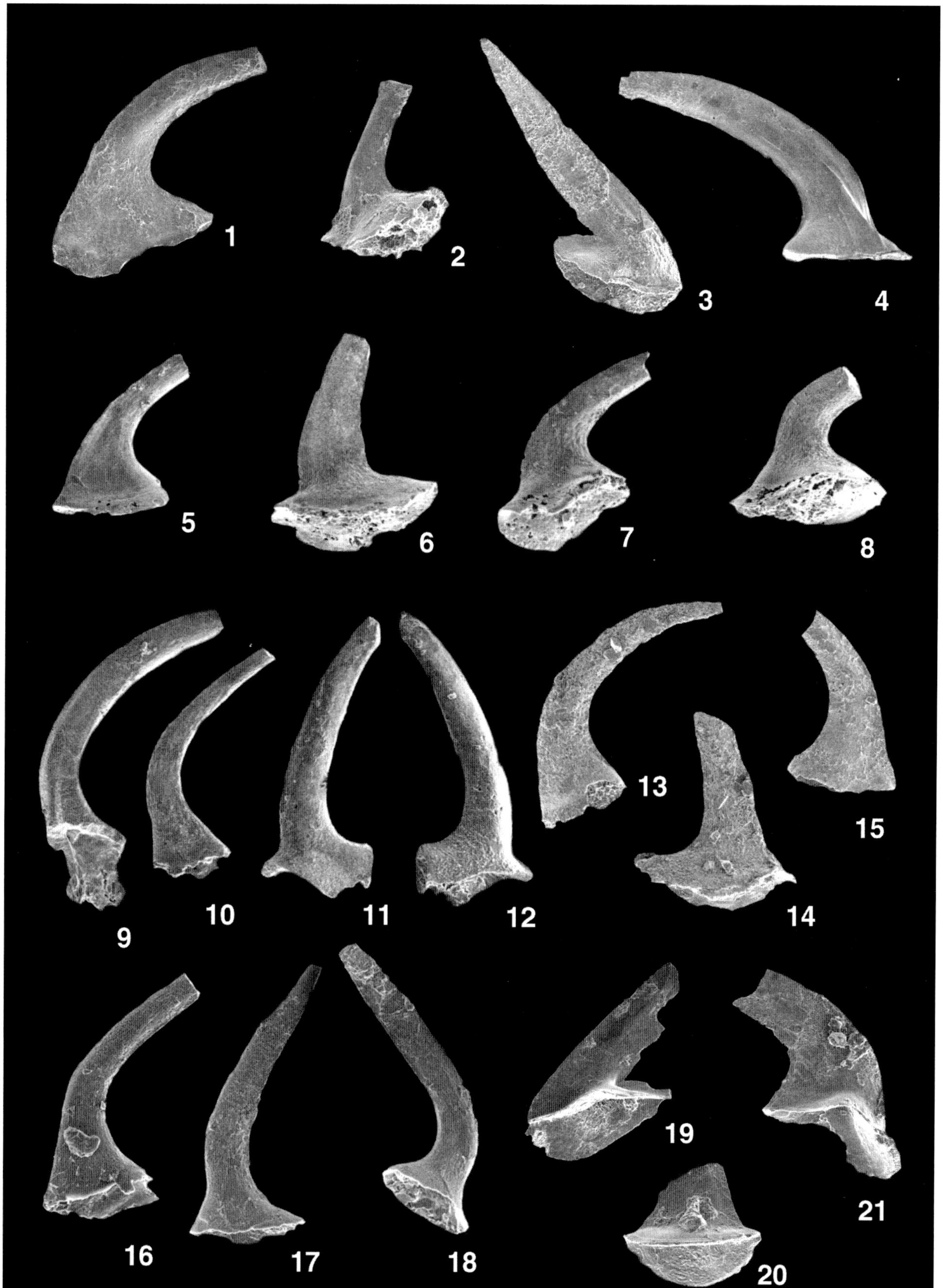
1
2
3
4
5
6
7
8
9
10
11
12
13
14
15
16
17
18
19
20
21

Plate 7

Figs. 1–3. *Eoconodontus notchpeakensis* (Miller)

1. a element, inner lateral view, GP-94-6-6, GSC 119040, ×96
2. a element, inner lateral view, GP-94-6-6, GSC 119041, ×96
3. e element, inner lateral view, GP-94-6-6, GSC 119042, ×96

Figs. 4, 5. *Iapetognathus aengensis* Lindström

4. Sc element, aboral view, 96-13-A6, GSC 119067, ×133
5. Sb element, posterior view, 96-13-A4, GSC 119084, ×180

Figs. 6–8. *Iapetognathus fluctivagus* Nicoll et al.

6. Sc element, postero-lateral view, 96-13-A7, GSC 119081, ×180
7. Sb element, oral view, 96-13-A1, GSC 119082, ×180
8. Sb element, posterior view, 96-13-A7, GSC 119083, ×180

Figs. 9, 10. *Iapetognathus sprakersi* Landing

9. Sc element, posterior view, 96-13-A6, GSC 119085, ×140
10. Sb element, inner lateral view, 96-13-A6, GSC 119086, ×140

Fig. 11. *Iapetognathus* sp.

11. posterior view, 96-13-A1, GSC 119066, ×74

Fig. 12. *Granatodontus* cf. *G. ani* (Wang)

12. outer lateral view, 96-13-A7, GSC 119046, ×148

Figs. 13–20. *Graciloconus concinnus* n. gen. n. sp.

13. Sc element, inner lateral view, GP-94-4-3, *paratype*, GSC 119324, ×110
14. Sa element, inner lateral view, GP-94-4-3, *paratype*, GSC 119325, ×110
15. Sb element, inner lateral view, GP-94-4-3, *paratype*, GSC 119326, ×110
16. P element, inner lateral view, GP-94-4-3, *paratype*, GSC 119327, ×110
17. ?Sd element, inner lateral view, GP-94-4-3, *holotype*, GSC 119328, ×110
18. P element, inner lateral view, GP-94-4-3, *paratype*, GSC 119331, ×110

19, 20. M elements, inner lateral views, GP-94-4-3, *paratypes*, GSC 119329, 119330, ×110

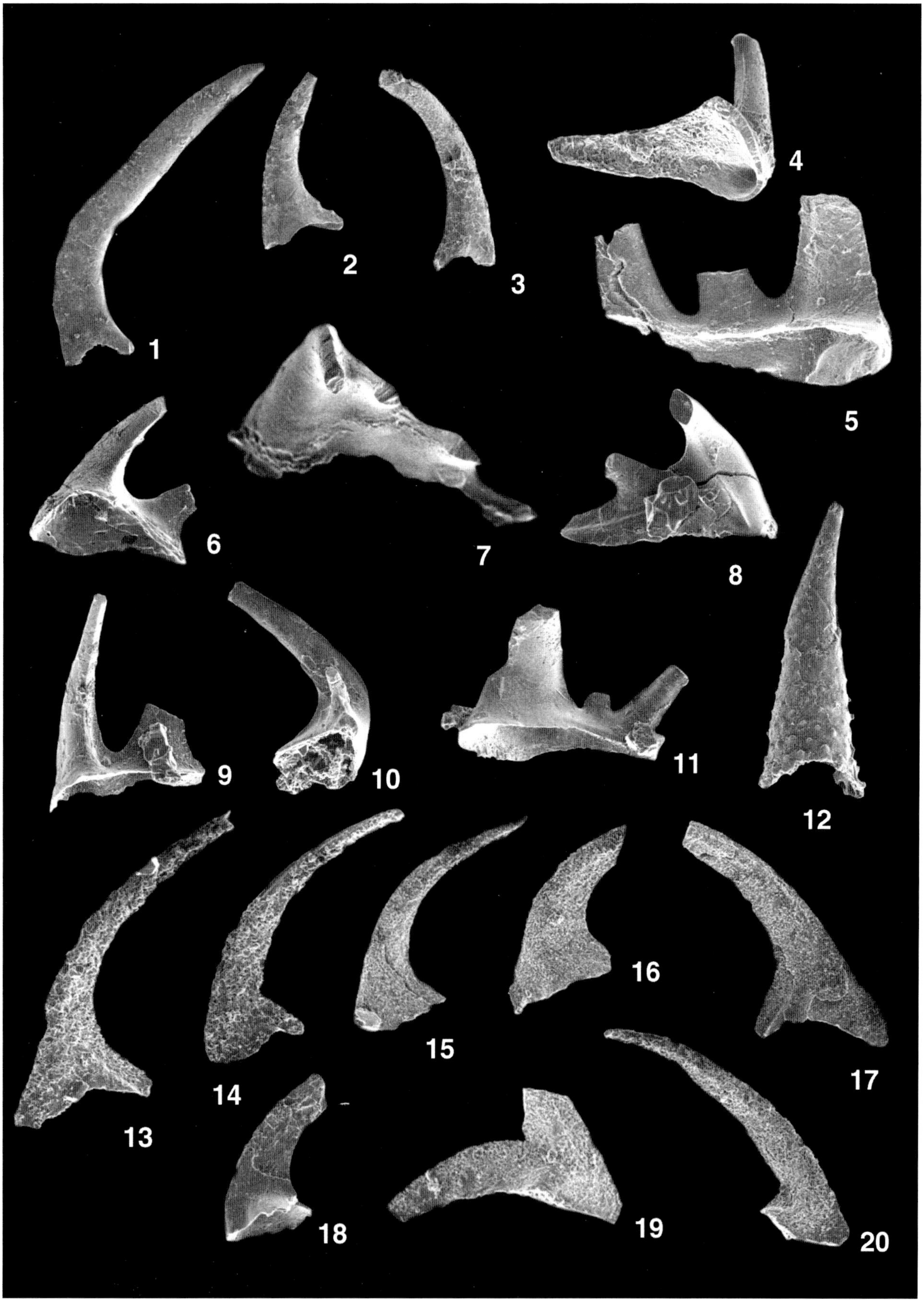
1
2
3
4
5
6
7
8
9
10
11
12
13
14
15
16
17
18
19
20

Plate 8

Figs. 1–15. *Kallidontus nodosus* n. gen. n. sp.

1. Pb element, oral view, 96-13-B26, paratype, GSC 119355, ×54
2. Pb element, oral view, 96-13-B26, *paratype*, GSC 119368, ×83
3. Pb element, oral view, 96-13-B26, *paratype*, GSC 119367, ×64

4, 5. Pb element, posterior and anterior views, 96-13-B26, *holotype*, GSC 119354, ×42

6. Pb element, oral view, 96-13-B26, *paratype*, GSC 119356, ×110
7. Pb element, aboral view, 96-13-B26, *paratype*, GSC 119372, ×100
8. Pb element, lateral view, 96-13-B26, *paratype*, GSC 119366, ×62
9. Pb element, posterior view, 96-13-B26, *paratype*, GSC 119370, ×150
10. Sa element, anterior view, 96-13-B26, *paratype*, GSC 119373, ×88
11. S element, lateral view, 96-13-B26, *paratype*, GSC 119352, ×62
12. Pb element, posterior view, 96-13-B26, *paratype*, GSC 119371, ×130
13. S element, posterior view, 96-13-B26, *paratype*, GSC 119353, ×88
14. Pb element, posterior view, 96-13-B26, *paratype*, GSC 119354, ×42
15. Pa element, oral view, 96-13-B26, GSC 119369, *paratype*, ×150

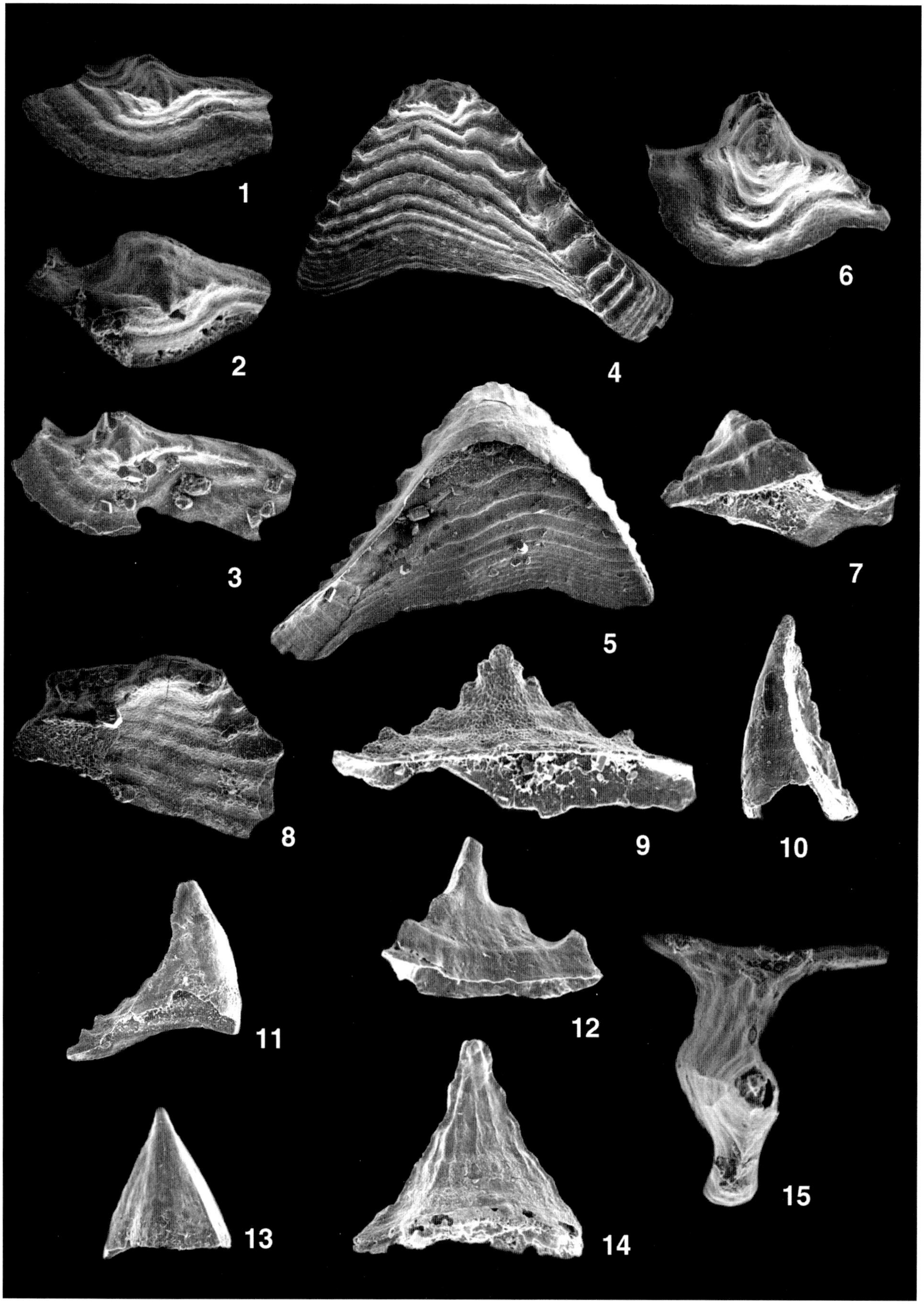
1
2
3
4
5
6
7
8
9
10
11
12
13
14
15

Plate 9

Figs. 1–7. *Kallidontus princeps* n. gen. n. sp.

1, 2. P element, posterior and anterior views, GP-94-4-6, *holotype*, GSC 119337, ×100

3, 4. S element, outer and inner lateral views, GP-94-4-6, *paratype*, GSC 119335, ×100

5. Sa element, posterior view, GP-94-4-6, *paratype*, GSC 119338, ×100

6. S element, inner lateral view, GP-94-4-6, *paratype*, GSC 119336, ×100

7. S element, outer lateral view, GP-94-4-6, *paratype*, GSC 119339, ×100

Figs. 8–15. *Kallidontus serratus* n. gen. n. sp. (early form)

8, 9. Pb element, posterior and anterior views, 96-13-B24, *holotype*, GSC 119343, ×80

10. S element, lateral view, 96-13-B24, *paratype*, GSC 119340, ×88

11. S element, lateral view, 96-13-B24, *paratype*, GSC 119341, ×88

12. S element, posterior view, 96-13-B24, *paratype*, GSC 119345, ×88

13. Pa element, lateral view, 96-13-B24, *paratype*, GSC 119346, ×85

14. Pa element, lateral view, 96-13-B24, *paratype*, GSC 119342, ×88

15. Pa element, oral view, 96-13-B24, *paratype*, GSC 119344, ×170

Figs. 16, 17. *Kallidontus serratus* n. gen. n. sp. (late form)

16, 17. Pa elements, anterior views, 96-13-B25, *paratypes*, GSC 119357, ×100

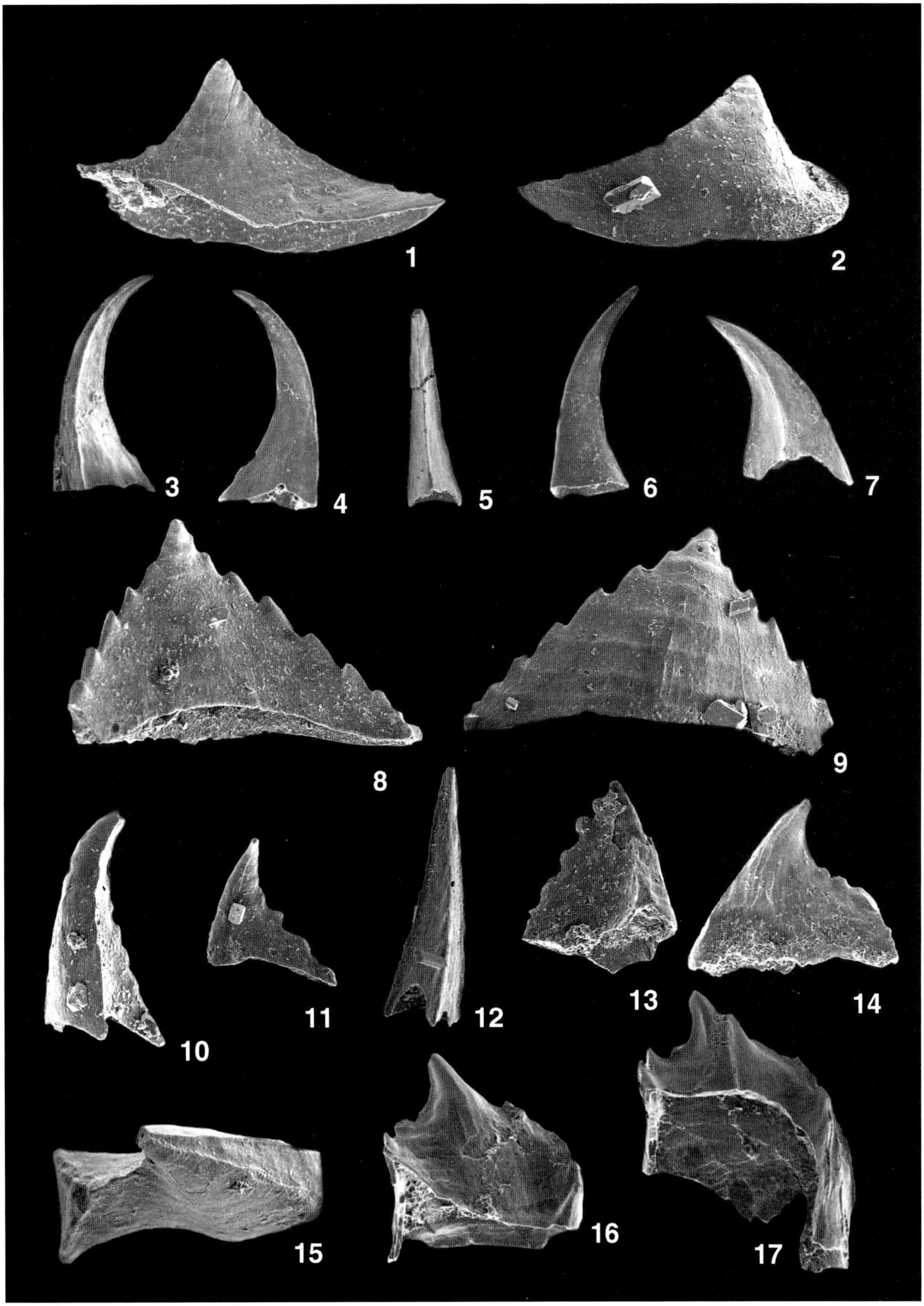
1
2
3
4
5
6
7
8
9
10
11
12
13
14
15
16
17

Plate 10

Figs. 1–14. *Kallidontus serratus* n. gen. n. sp. (late form)

1, 2. S elements, inner lateral views, 96-13-B25, *paratypes*, GSC 119361, ×50

3–5. S elements, posterior views, 96-13-B25, *paratypes*, GSC 119362, 119363, 119364, ×50

6. S element, inner lateral view, 96-13-B25, *paratype*, GSC 119365, ×50

7. S element, lateral view, 96-13-B25, *paratype*, GSC 119347, ×64

8. Pa element, oral view, 96-13-B25, *paratype*, GSC 119348, ×75

9. Pb element, posterior view, 96-13-B25, *paratype*, GSC 119349, ×90

10. S element, lateral view, 96-13-B25, *paratype*, GSC 119350, ×64

11. S element, lateral view, 96-13-B25, *paratype*, GSC 119351, ×130

12. Pa element, anterior view, 96-13-B25, *paratype*, GSC 119358, ×70

13. Pa element, lateral view, 96-13-B25, *paratype*, GSC 119359, ×145

14. Pa element, antero-lateral view, 96-13-B25, *paratype*, GSC 119360, ×145

Figs. 15–20. *Kallidontus* cf. *K. nodosus* n. gen. n. sp.

15. ?Sa element, posterior view, 96-4-K4, *paratype*, GSC 119376, ×70

16. S element, lateral view, 96-4-K4, *paratype*, GSC 119377, ×70

17. S element, lateral view, 96-4-K4, *paratype*, GSC 119375, ×70

18. P element, 96-5-K1, *paratype*, GSC 119332, ×55

19. S element, 96-5-K1, *paratype*, GSC 119333, ×55

20. P element, 96-5-K1, *paratype*, GSC 119334, ×55

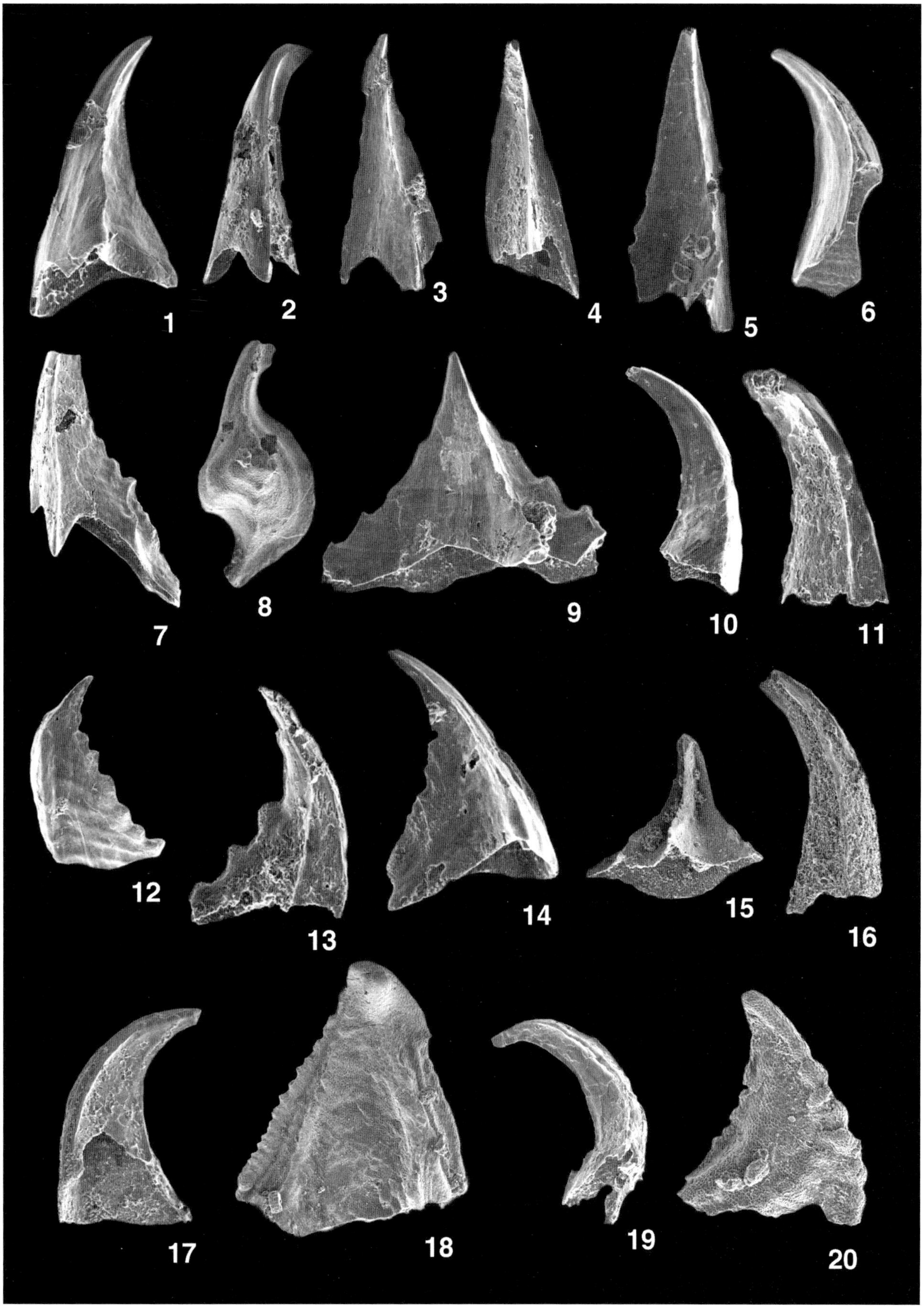
1
2
3
4
5
6
7
8
9
10
11
12
13
14
15
16
17
18
19
20

Plate 11

Figs. 1–5. *Laurentoscandodus sinuosus* n. sp.

1. e element, inner lateral view, 96-1-1, *paratype*, GSC 119100, ×53

2, 3. a element, inner lateral and antero-lateral views, 96-1-1, *holotype*, GSC 119101, ×53

4. c element, inner lateral view, 96-1-1, *paratype*, GSC 119102, ×53

5. b element, inner lateral view, 96-1-1, *paratype*, GSC 119103, ×53

Figs. 6–9. *Laurentoscandodus* cf. *L. triangularis* (Furnish)

6. e element, postero-lateral view, 96-13-B5, GSC 119087, ×80

7. a element, inner lateral view, 96-13-B5, GSC 119088, ×80

8. c element, inner lateral view, 96-13-B5, GSC 119089, ×65

9. b element, inner lateral view, 96-13-B5, GSC 119090, ×65

Figs. 10, 11. *Loxodus latibasis* Ji and Barnes

10. dextral element, inner lateral view, 96-1-4, GSC 119072, ×100

11. sinistral element, inner lateral view, 96-1-4, GSC 119073, ×100

Figs. 12–15. *Macerodus*? n. sp. B

12. a element, inner lateral view, 96-13-B5, GSC 119161, ×85

13, 14. e element, outer and inner lateral views, 96-13-B5, GSC 119162, ×85

15. e element, outer lateral view, 96-13-B5, GSC 119163, ×85

Figs. 16–18. *Macerodus*? n. sp. A

16. a element, inner lateral view, 96-13-B3, GSC 119159, ×80

17, 18. e element, inner and outer lateral views, 96-13-B3, GSC 119160, ×80

Fig. 19. *Oistodus* sp.

19. inner lateral view, 96-5-K1, GSC 119278, ×60

Figs. 20–23. *Oelandodus elongatus* van Wamel

20. P element, inner lateral view, 96-13-B25, GSC 119320, ×100

21. Sa element, posterior view, 96-13-B25, GSC 119321, ×100

22. M element, inner lateral view, 96-13-B25, GSC 119322, ×100

23. Sb element, inner lateral view, 96-13-B25, GSC 119323, ×100

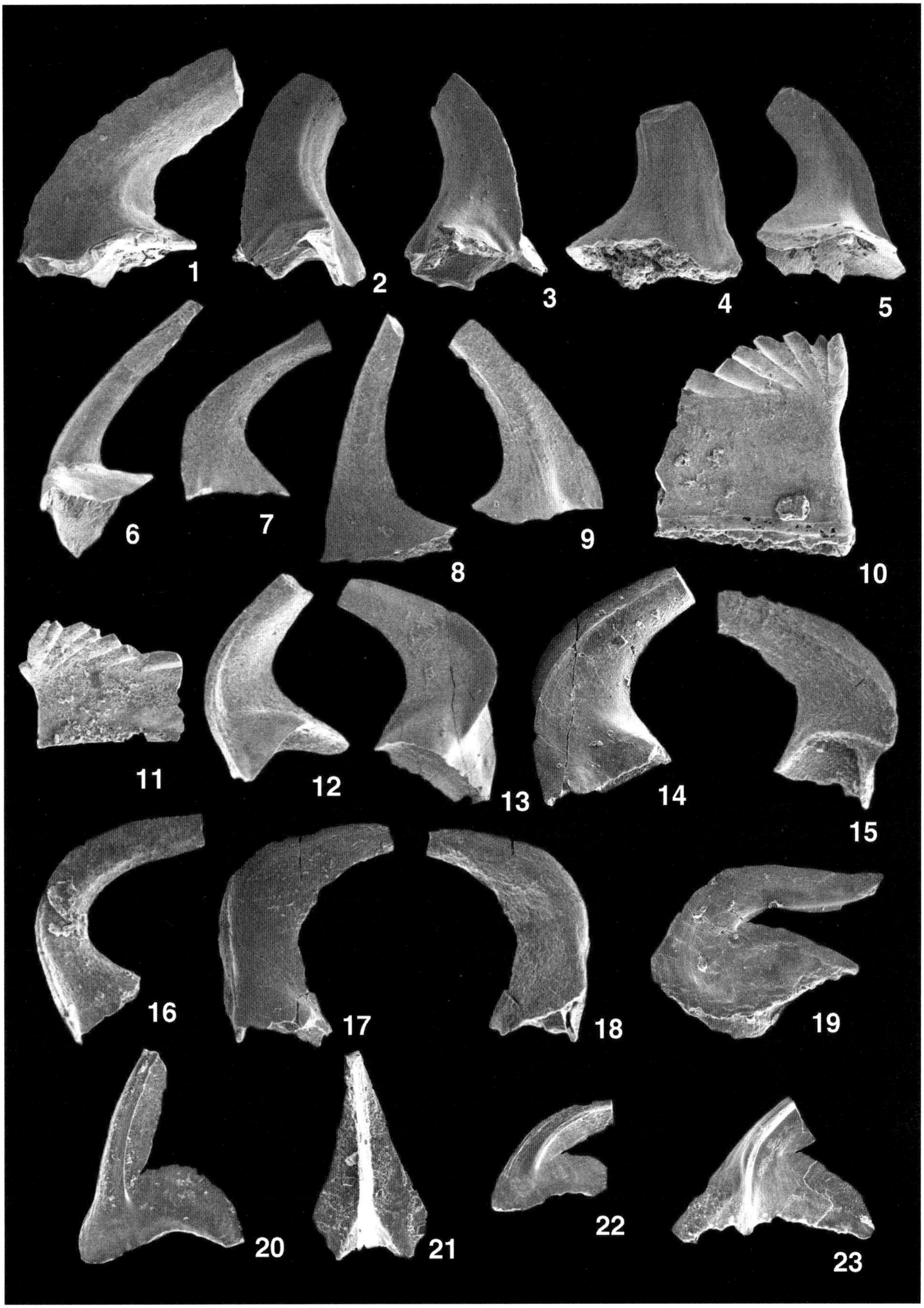
1
2
3
4
5
6
7
8
9
10
11
12
13
14
15
16
17
18
19
20
21
22
23

Plate 12

Figs. 1–5. ?*Paltodus jemtlandicus* Löfgren

1–3. S elements, 96-13-B16, inner lateral views, GSC 119290, 119291, 119292, ×88

4. M element, inner lateral view, 96-13-B16, GSC 119293, ×88

5. P element, inner lateral view, 96-13-B16, GSC 119294, ×110

Figs. 6–9. *Paracordylodus gracilis* Lindström

6. P element, inner lateral view, 96-13-B27, GSC 119268, ×105

7, 8. S elements, inner lateral views, 96-13-B27, GSC 119269, 119270, ×105

9. M element, inner lateral view, 96-13-B27, GSC 119271, ×105

Figs. 10–12. *Paroistodus proteus* Lindström

10. S element, inner lateral view, 96-13-B25, GSC 119403, ×75

11. M element, inner lateral view, 96-13-B25, GSC 119401, ×75

12. P element, inner lateral view, 96-13-B25, GSC 119402, ×75

Figs. 13–15. *Polycostatus falsioneotensis* Ji and Barnes

13. e element, inner lateral view, 96-13-A10, GSC 119191, ×88

14. c element, posterior view, 96-13-A10, GSC 119192, ×88

15. a element, posterior view, 96-13-A10, GSC 119193, ×94

Figs. 16–19. *Polycostatus* cf. *P. sulcatus* (Furnish)

16. a element, inner lateral view, 96-13-B3, GSC 119194, ×53

17. b element, inner lateral view, 96-13-B3, GSC 119195, ×53

18. c element, posterior view, 96-13-B3, GSC 119196, ×43

19. e element, inner lateral view, 96-13-B3, GSC 119197, ×53

Figs. 20, 21. *Proconodontus muelleri* Miller

20. a element, inner lateral view, GP-94-6-6, GSC 119044, ×64

21. e element, inner lateral view, GP-94-6-6, GSC 119045, ×64

Figs. 22–25. *Prioniodus elegans* Pander

22. M element, inner lateral view, 96-4-K4, GSC 119309, ×68

23. Pb element, inner lateral view, 96-4-K4, GSC 119310, ×68

24. Sd element, inner lateral view, 96-4-K4, GSC 119311, ×68

25. Sb element, inner lateral view, 96-4-K4, GSC 119312, ×68

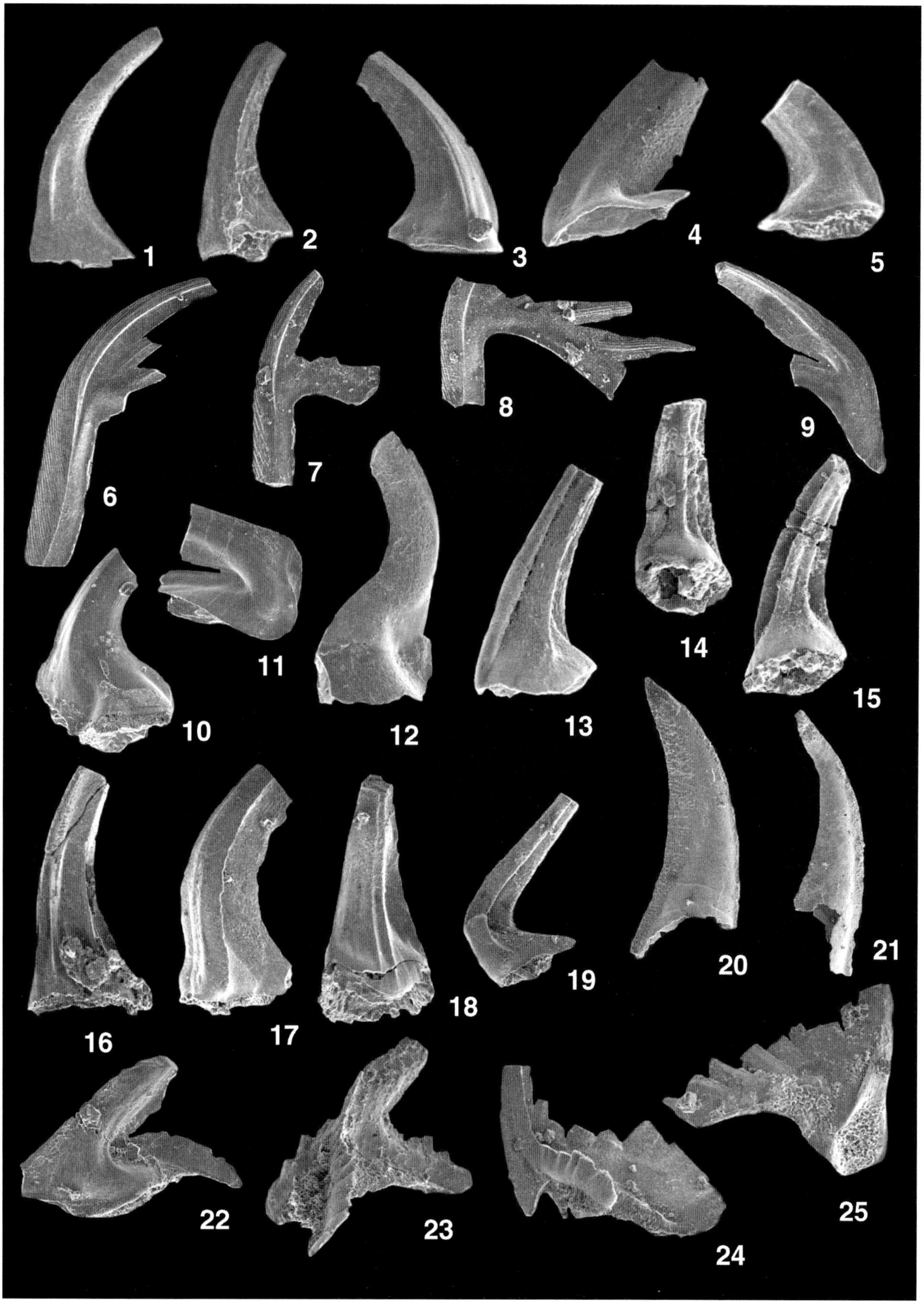
1
2
3
4
5
6
7
8
9
10
11
12
13
14
15
16
17
18
19
20
21
22
23
24
25

Plate 13

Figs. 1–6. *Rossodus manitouensis* Repetski and Ethington

1, 2, 6. e elements, inner lateral views, 96-1-9, GSC 119118, 119119, 119120, ×8

3. b element, postero-lateral view, 96-1-9, GSC 119121, ×84

4. c element, posterior view, 96-1-9, GSC 119122, ×93

5. a element, postero-lateral view, 96-1-9, GSC 119123, ×84

Figs. 7–15. *Rossodus muskwaensis* n. sp.

7, 13. e elements, inner lateral views, 96-1-9, *paratypes*, GSC 119124, 119125, ×80

8, 12, 14. a elements, postero-lateral views, 96-1-9, *paratypes*, GSC 119126, 119127, 19128, ×80

9, 15. b element, postero- and antero-lateral views, 96-1-9, *holotype*, GSC 119129, ×80

10, 11. c element, posterior and anterior views, 96-1-9, *paratype*, GSC 119130, ×80

Figs. 16–20. *Rossodus kwadachaensis* n. sp.

16. b element, inner lateral view, 96-13-B2, *paratype*, GSC 119137, ×100

17, 18. c element, posterior and anterior views, 96-13-B2, *holotype*, GSC 119138, ×100

19. e element, inner lateral view, 96-13-B2, *paratype*, GSC 119136, ×100

20. a element, postero-lateral view, 96-13-B2, *paratype*, GSC 119139, ×100

Figs. 21–26. *Rossodus tenuis* (Miller)

21, 22. e elements, inner lateral views, 96-1-9, GSC 119112, 119113, ×105

23. b element, inner lateral view, 96-1-9, GSC 119115, ×105

24. c element, posterior view, 96-1-9, GSC 119116, ×125

25. a element, inner lateral view, 96-1-9, GSC 119117, ×105

26. e element, inner lateral view, 96-1-9, GSC 119114, ×100

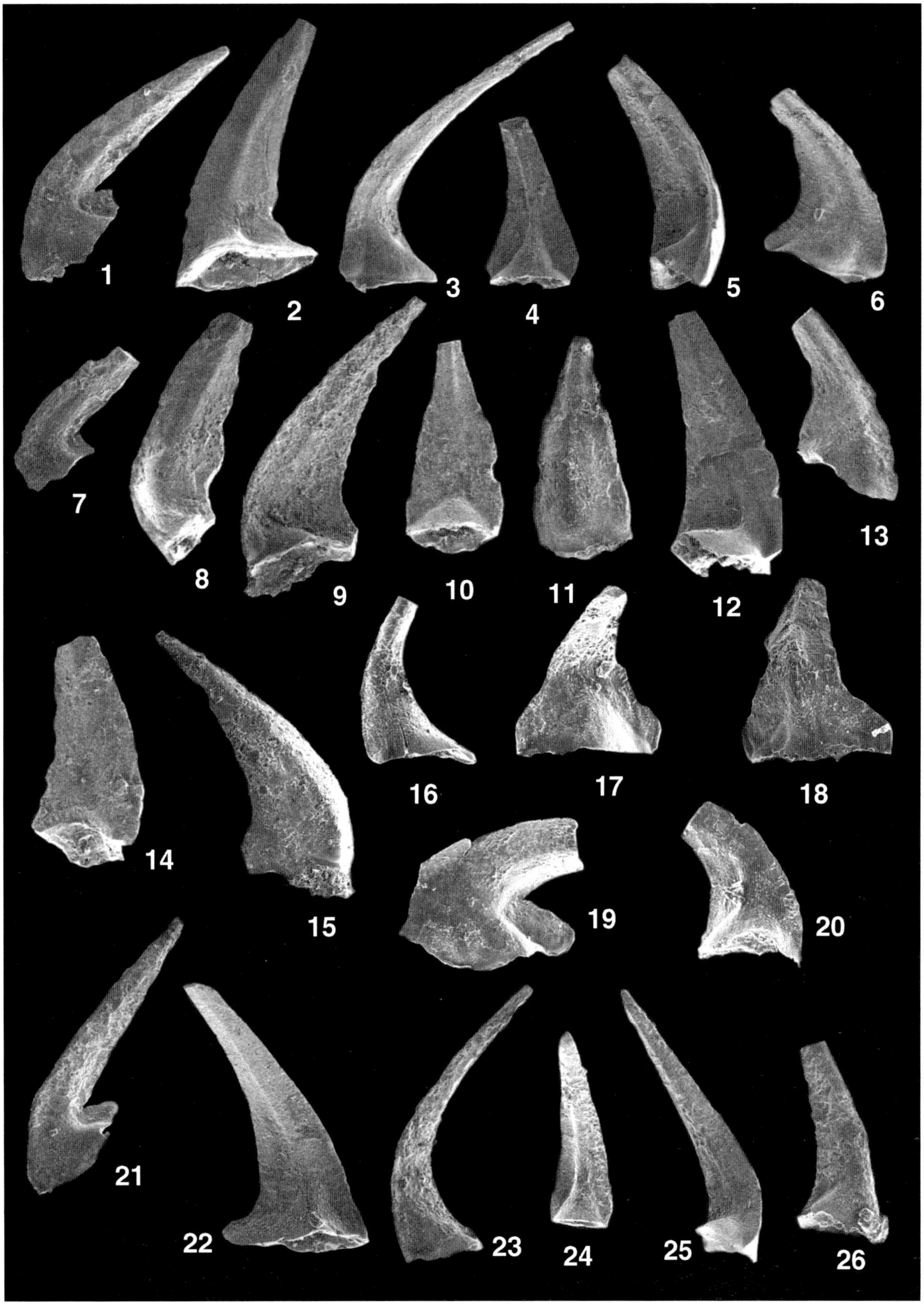
1
2
3
4
5
6
7
8
9
10
11
12
13
14
15
16
17
18
19
20
21
22
23
24
25
26

Plate 14

Figs. 1–6. *Rossodus sheffieldensis* n. sp.

1, 2. a element, postero-lateral and anterior views, 96-13-B5, *holotype*, GSC 119131, ×78

3. c element, posterior view, 96-13-B5, *paratype*, GSC 119134, ×115

4. a element, posterior view, 96-13-B5, *paratype*, GSC 119132, ×53

5. b element, inner lateral view, 96-13-B5, *paratype*, GSC 119133, ×78

6. e element, inner lateral view, 96-13-B5, *paratype*, GSC 119135, ×72

Figs. 7–11. *Rossodus subtilis* n. sp.

7. e element, inner lateral view, 96-13-B5, *paratype*, GSC 119143, ×160

8, 9. c element, anterior and posterior views, 96-13-B5, *holotype*, GSC 119144, ×160

10. b element, inner lateral view, 96-13-B5, *paratype*, GSC 119146, ×160

11. a element, postero-lateral view, 96-13-B5, *paratype*, GSC 119145, ×160

Figs. 12–14. *Scolopodus krummi* (Lehnert)

12. e element, inner lateral view, 96-4-K4, GSC 119317, ×65

13. b element, inner lateral view, 96-4-K4, GSC 119318, ×65

14. a element, inner lateral view, 96-4-K4, GSC 119319, ×65

Figs. 15–17. *Scolopodus subrex* Ji and Barnes (early form)

15. a element, inner lateral view, 96-13-B8, GSC 119284, ×62

16. a element, inner lateral view, 96-13-B8, GSC 119285, ×62

17. b element, inner lateral view, 96-13-B8, GSC 119286, ×62

Figs. 18–20. *Scolopodus subrex* Ji and Barnes (late form)

18. e element, inner lateral view, 96-13-B14, GSC 119287, ×62

19. a element, inner lateral view, 96-13-B14, GSC 119288, ×62

20. b element, inner lateral view, 96-13-B14, GSC 119289, ×62

Figs. 21, 22. *Semiacontiodus* cf. *S. cornuformis* (Sergeeva)

21, 22. inner lateral views, 96-13-B16, GSC 119226, 119227, ×100

Figs. 23–25. *Semiacontiodus nogamii* Miller

23. e element, inner lateral view, 96-13-A6, GSC 119210, ×130

24. a element, inner lateral view, 96-13-A6, GSC 119208, ×130

25. c element, posterior view, 96-13-A6, GSC 119209, ×130

Figs. 26–29. *Stolodus stola* (Lindström)

26. a element, antero-lateral view, 96-13-B25, GSC 119246, ×130

27. c element, posterior view, 96-13-B25, GSC 119247, ×130

28. a element, inner lateral view, 96-13-B25, GSC 119248, ×100

29. b element, inner lateral view, 96-13-B25, GSC 119249 ×100

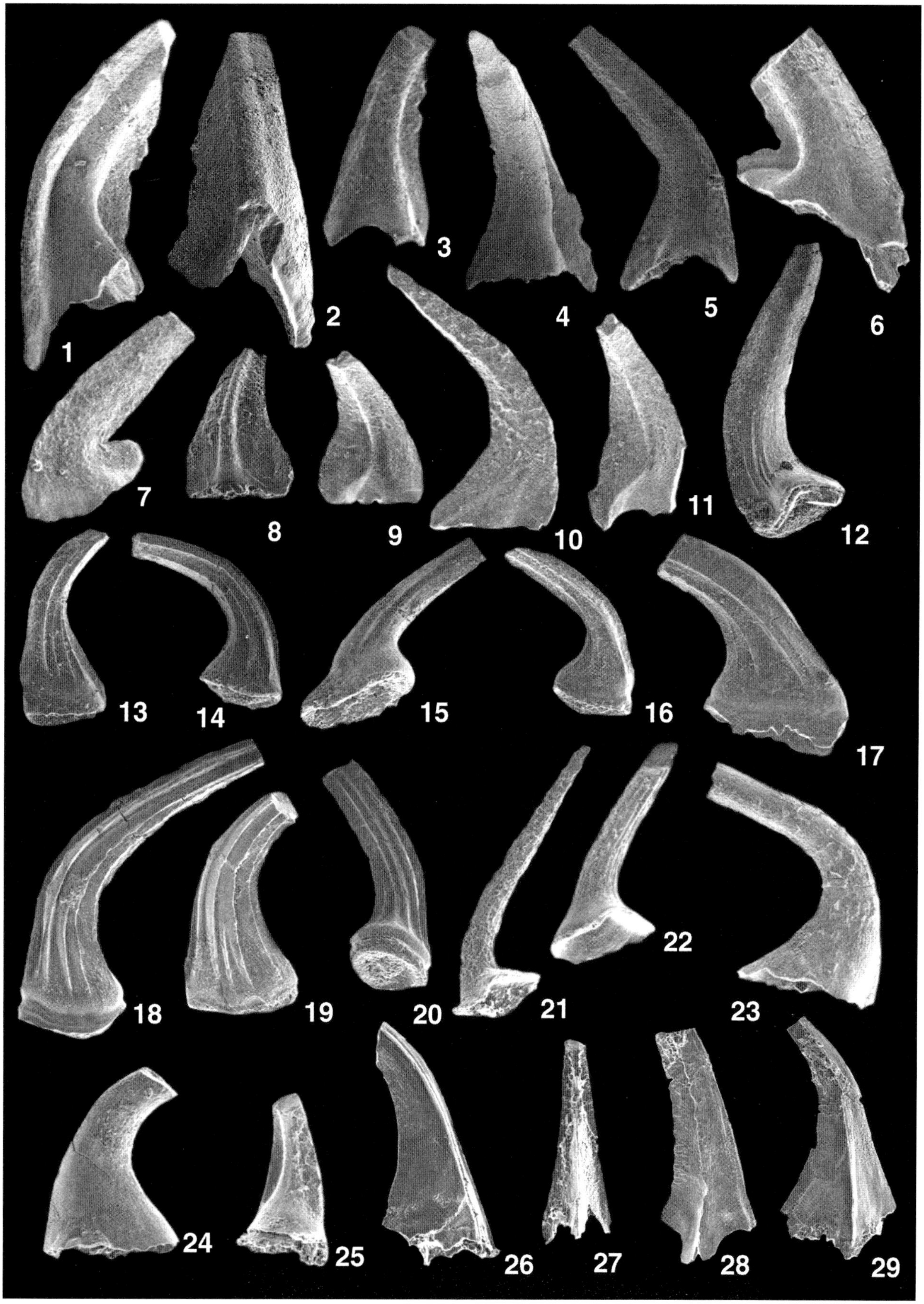
1
2
3
4
5
6
7
8
9
10
11
12
13
14
15
16
17
18
19
20
21
22
23
24
25
26
27
28
29

Plate 15

Figs. 1–4. *Striatodontus prolificus* Ji and Barnes

1. a element, inner lateral view, 96-13-B5, GSC 119171, ×105
2. b element, inner lateral view, 96-13-B5, GSC 119172, ×105
3. c element, posterior view, 96-13-B5, GSC 119173, ×105
4. e element, inner lateral view, 96-13-B5, GSC 119174, ×105

Figs. 5–9. *Striatodontus strigatus* n. sp.

5, 6. a element, inner and outer lateral views, 96-1-9, *holotype*, GSC 119167, ×54

7. b element, inner lateral view, 96-1-9, *paratype*, GSC 119168, ×80
8. c element, inner lateral view, 96-1-9, *paratype*, GSC 119169, ×80
9. e element, inner lateral view, 96-1-9, *paratype*, GSC 119170, ×80

Figs. 10, 11. *Stultodontus costatus* (Ethington and Brand)

10. e element, inner lateral view, 96-13-B5, GSC 119189, ×180
11. a element, inner lateral view, 96-13-B5, GSC 119190, ×180

Figs. 15, 16. ?*Stultodontus* sp.

15. a element, inner lateral view, GP-94-4-5, GSC 119409, ×95
16. e element, inner lateral view, GP-94-4-5, GSC 119410, ×95

Figs. 12–14. *Teridontus gracillimus* Nowlan

12. a element, inner lateral view, 96-13-A7, GSC 119201, ×180
13. c element, posterior view, 96-13-A7, GSC 119202, ×180
14. e element, inner lateral view, 96-13-A7, GSC 119203, ×180

Figs. 17–19. *Teridontus nakamurai* (Nogami)

17. a element, inner lateral view, 96-13-A6, GSC 119198, ×130
18. c element, posterior view, 96-13-A6, GSC 119199, ×130
19. e element, inner lateral view, 96-13-A7, GSC 119200, ×130

Figs. 20–22. *Teridontus obesus* Ji and Barnes

20. a element, inner lateral view, 96-13-A7, GSC 119204, ×130
21. c element, posterior view, 96-13-A7, GSC 119205, ×170
22. b element, inner lateral view, 96-13-A7, GSC 119206, ×170

Figs. 23–26. *Tricostatus* cf. *T. glyptus* Ji and Barnes

23. a element, posterior view, 96-13-B4, GSC 119148, ×125
24. c element, posterior view, 96-13-B4, GSC 119147, ×125
25. e element, inner lateral view, 96-13-B4, GSC 119149, ×125
26. a element, inner lateral view, 96-13-B4, GSC 119150, ×125

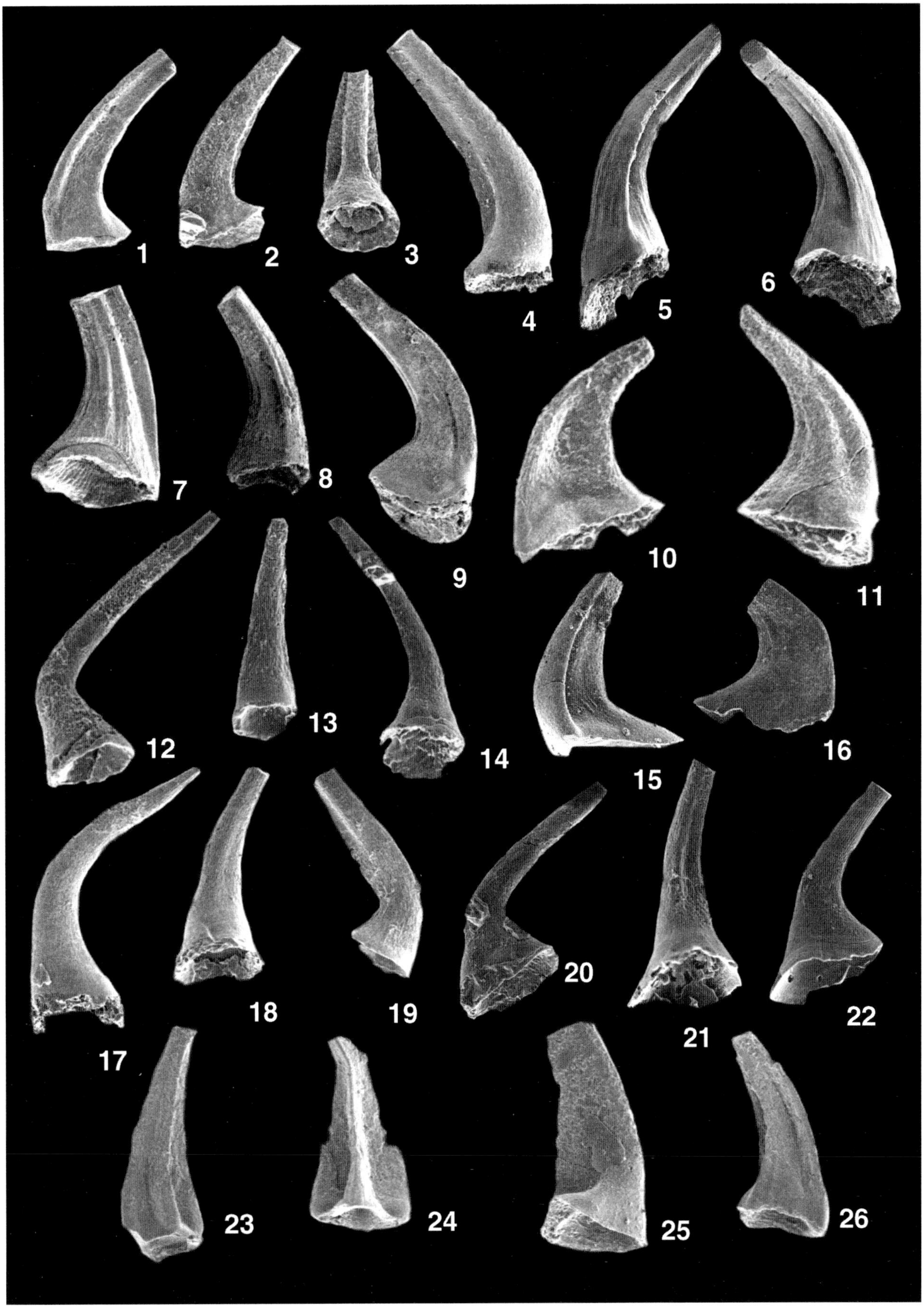
1
2
3
4
5
6
7
8
9
10
11
12
13
14
15
16
17
18
19
20
21
22
23
24
25
26

Plate 16

Figs. 1–4. *Tricostatus infundibulum* n. sp.

1. e element, inner lateral view, 96-1-4, *paratype*, GSC 119141, ×96
2. c element, posterior view, 96-1-4, *paratype*, GSC 119140, ×96

3, 4. a element, posterior and anterior views, 96-1-4, *holotype*, GSC 119142, ×96

Figs. 5–9. *Tricostatus terilinguis* n. sp.

5. e element, outer lateral view, 96-13-B4, *paratype*, GSC 119155, ×80

6, 7. c element, posterior and anterior views, 96-13-B4, *holotype*, GSC 119153, ×80

8. a element, postero-lateral view, 96-13-B4, *paratype*, GSC 119154, ×80
9. b element, inner lateral view, 96-13-B4, *paratype*, GSC 119156, ×80

Figs. 10–13. *Tropodus australis* (Serpagli)

10. Sd element, inner lateral view, 96-13-B23, GSC 119313, ×84
11. Sb element, inner lateral view, 96-13-B23, GSC 119314, ×84
12. P element, inner lateral view, 96-13-B23, GSC 119315, ×84
13. M element, inner lateral view, 96-13-B23, GSC 119316, ×84

Figs. 14–16. *Tropodus comptus* (Branson and Mehl)

14, 15. S elements, postero-lateral views, 96-13-B8, GSC 119215, 119216, ×150

16. S element, antero-lateral view, 96-13-B8, GSC 119217, ×150

Figs. 17–23. *Utahconus utahensis* (Miller)

17, 18, 23. e elements, inner lateral views, 96-1-10, GSC 119105, 119106, 119107, ×100

19, 22. a elements, inner lateral views, 96-1-10, GSC 119108, 119109, ×100

20. b element, inner lateral view, 96-1-10, GSC 119111, ×100
21. c element, postero-lateral view, 96-1-10, GSC 119110, ×100

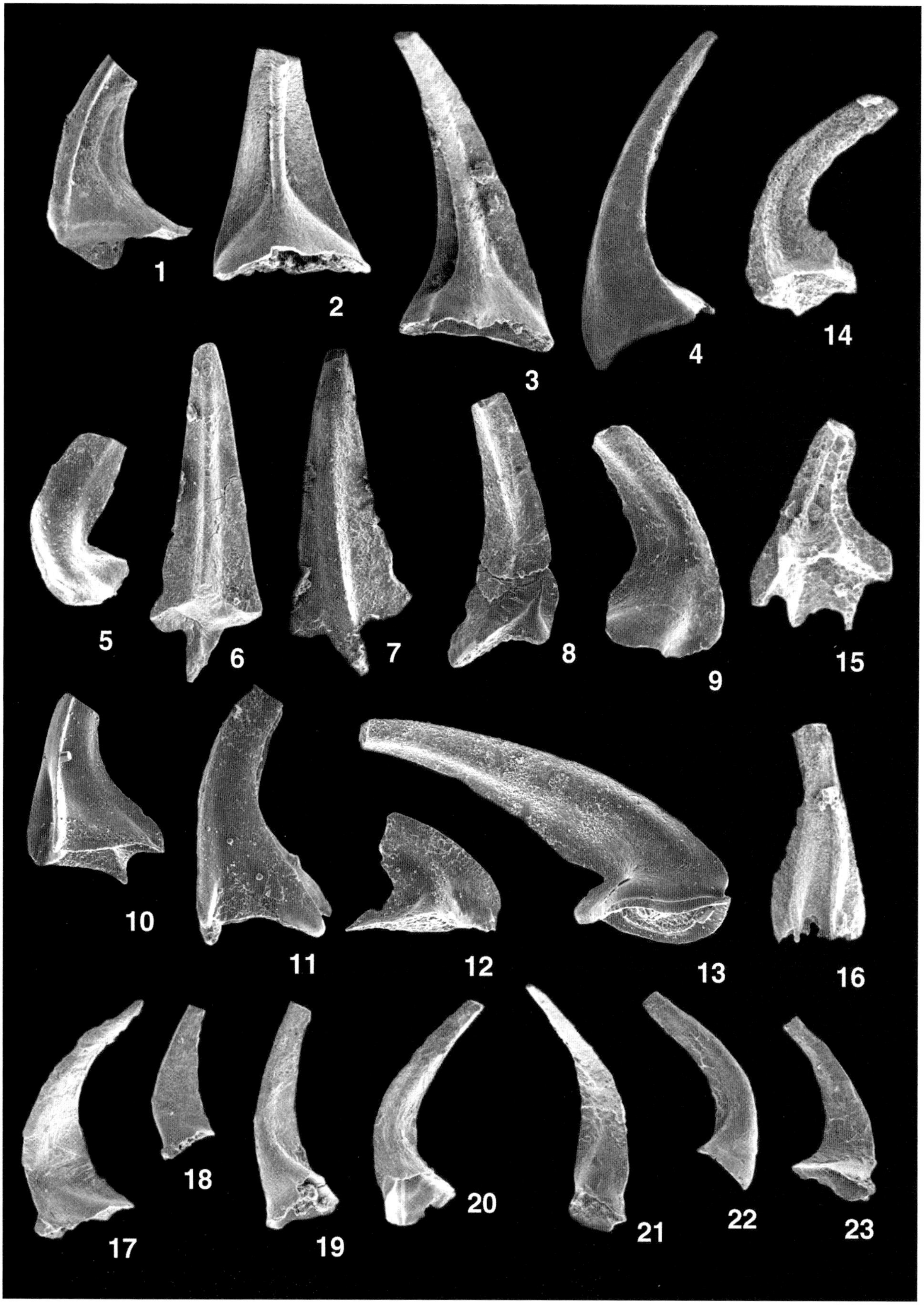
1
2
3
4
14
5
6
7
8
9
15
10
11
12
13
16
17
18
19
20
21
22
23

Plate 17

Figs. 1–3. *Utahconus longipinnatus* Ji and Barnes

1. a element, inner lateral view, 96-1-1, GSC 119164, ×80
2. c element, posterior view, 96-1-1, GSC 119165, ×80
3. e element, inner lateral view, 96-1-1, GSC 119166, ×80

Fig. 4. *Westergaardodina* sp.

4. 96-1-3, GSC 120666, ×100

Figs. 5–8. *Variabiloconus bassleri* (Furnish) (early form)

5. a element, inner lateral view, 96-13-A17, GSC 119179, ×65
6. b element, inner lateral view, 96-13-A17, GSC 119180, ×65
7. c element, posterior view, 96-13-A17, GSC 119181, ×60
8. e element, inner lateral view, 96-13-A17, GSC 119182, ×65

Figs. 9–12. *Variabiloconus bassleri* (Furnish) (late form)

9. a element, inner lateral view, 96-13-B6, GSC 119185, ×110
10. c element, posterior view, 96-13-B6, GSC 119186, ×104
11. b element, inner lateral view, 96-13-B6, GSC 119187, ×110
12. e element, inner lateral view, 96-13-B6, GSC 119188, ×110

Figs. 13–16. *Ventricodus spurius* (Ethington and Clark)

13. c element, inner lateral view, 96-13-A17, GSC 119175, ×110
14. a element, inner lateral view, 96-13-A17, GSC 119176, ×110
15. b element, inner lateral view, 96-13-A17, GSC 119177, ×110
16. e element, inner lateral view, 96-13-A17, GSC 119178, ×110

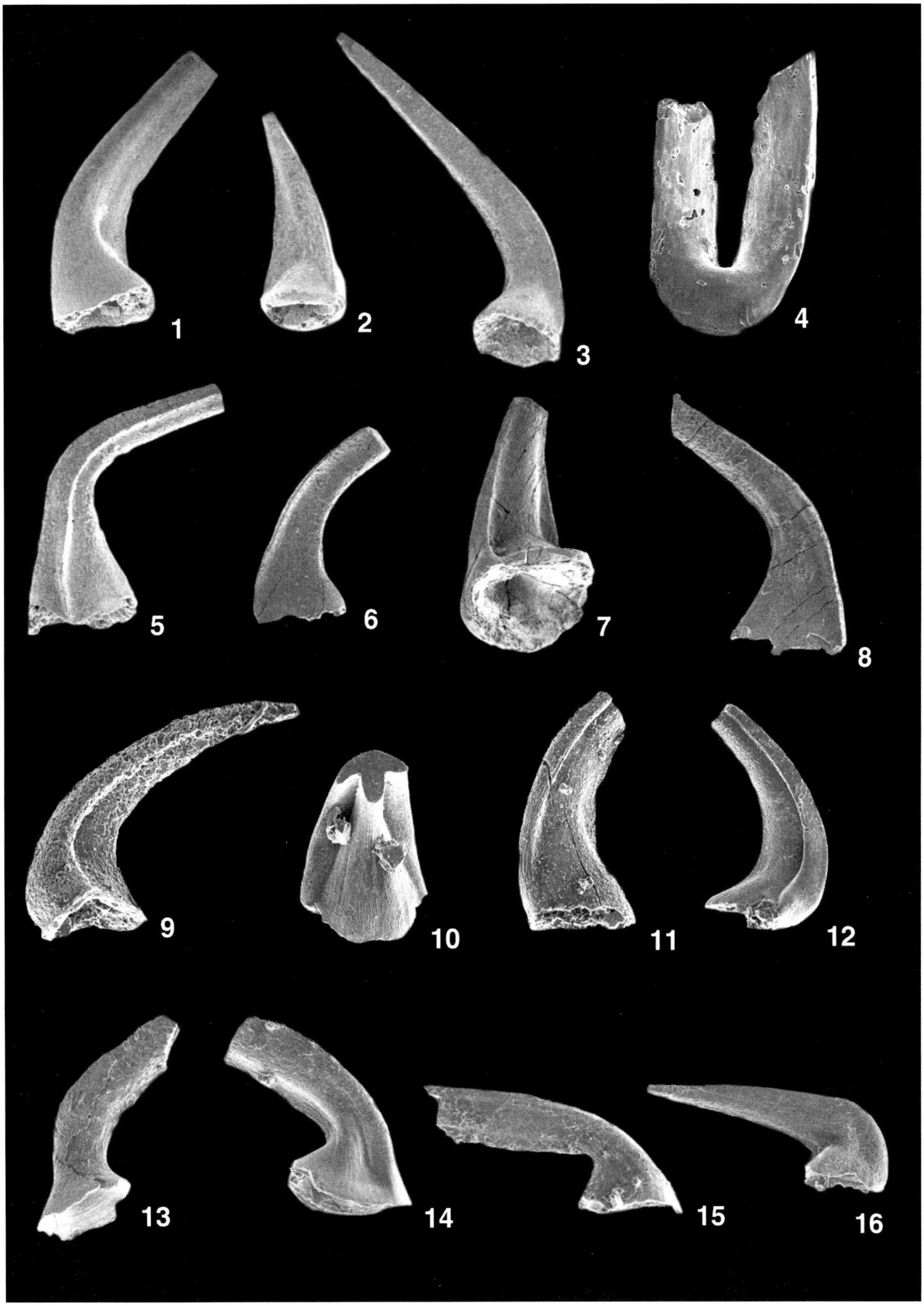
1
2
3
4
5
6
7
8
9
10
11
12
13
14
15
16

Plate 18

Fig. 1. Coniform indet. A

1. inner lateral view, 96-13-B5, GSC 119104, ×100

Figs. 2, 3. Coniform indet. B

2. a element, lateral view, 96-1-6, GSC 119151, ×105
3. e element, lateral view, 96-1-6, GSC 119152, ×105

Figs. 4, 5. Coniform indet. C

4, 5. outer and inner lateral views, 96-1-9, GSC 119157, 119158, ×108

Fig. 6. Coniform indet. D

6. e element, inner lateral view, 96-13-B2, GSC 119207, ×54

Figs. 7–9. Coniform indet. E

7–9. inner lateral views, 96-13-A4, GSC 119218, 119219, 119220, ×115

Fig. 10. Coniform indet. F

10. e element, postero-lateral view, 96-13-B1, GSC 119221, ×84

Fig. 11. Coniform indet. G

11. inner lateral view, 96-13-B25, GSC 119272, ×130

Figs. 12–14. Coniform indet. H

12, 13. inner lateral views, 96-13-A7, GSC 119281, 119282, ×130

14. outer lateral view, 96-13-A7, GSC 119283, ×130

Figs. 15, 16. Coniform indet. I

15, 16. inner lateral views, GP-94-3-8, GSC 119308, 120665, ×50

Figs. 17–21. New genus A n. sp.

17. S element, inner lateral view, GP-94-3-2, GSC 119404, ×60

18, 19. P elements, inner lateral views, GP-94-3-2, GSC 119405, 119406, ×60

20. S element, inner lateral view, GP-94-3-2, GSC 119407, ×60

21. M element, inner lateral view, GP-94-3-2, GSC 119408, ×60

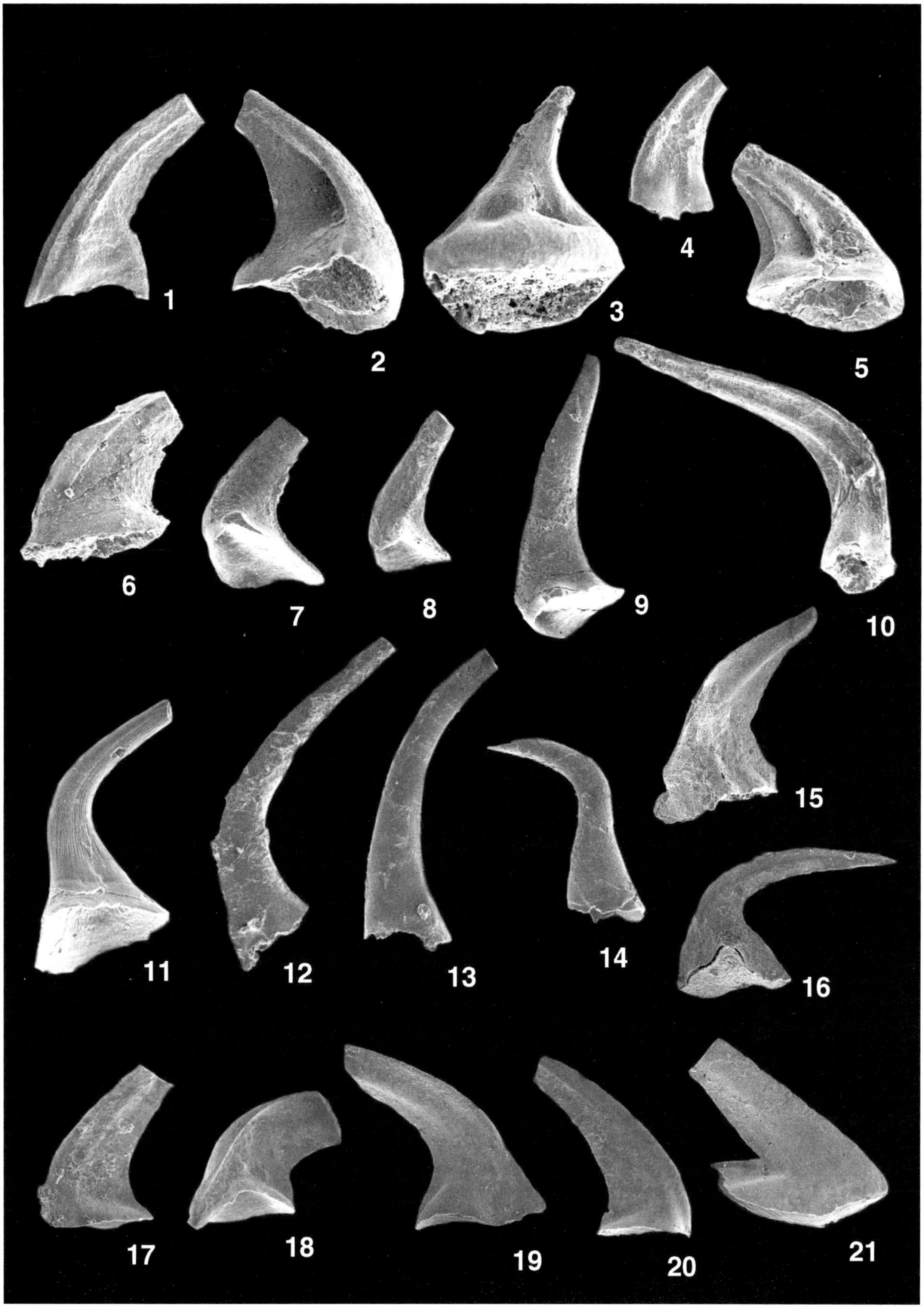
1
2
3
4
5
6
7
8
9
10
11
12
13
14
15
16
17
18
19
20
21

Plate 19

Figs. 1, 2. *Ansella jemtlandica* (Löfgren)

1. P element, inner lateral view, 96-4-3, GSC 119453, ×80
2. M element, inner lateral view 96-4-3, GSC 119454, ×80

Figs. 3–6. *Distomodus staurognathoides* (Walliser)

3. M element, inner lateral view, 96-4-6, GSC 119484, ×44
4. S element, postero-lateral view, 96-4-6, GSC 119485, ×44

5, 6. P elements, oral views, 96-4-6, GSC 119486, 119487, ×44

Figs. 7–9. *Drepanoistodus bellburnensis* Stouge

7. e element, inner lateral view, GP-94-1-20, GSC 119417, ×38
8. c element, inner lateral view, GP-94-1-20, GSC 119418, ×46
9. a element, inner lateral view, GP-94-1-20, GSC 119419, ×53

Figs. 10–12. *Drepanoistodus forceps* Lindström

10. e element, inner lateral view, GP-94-1-20, GSC 119411, ×50
11. c element, inner lateral view, GP-94-1-20, GSC 119412, ×50
12. a element, inner lateral view, GP-94-1-20, GSC 119413, ×55

Figs. 13–16. *Drepanoistodus* cf. *D. pitjanti* Cooper

13. e element, inner lateral view, 96-4-3, GSC 119420, ×70
14. c element, inner lateral view, 96-4-3, GSC 119421, ×70
15. a element, inner lateral view, 96-4-3, GSC 119422, ×70
16. a element, inner lateral view, 96-4-3, GSC 119423, ×70

Figs. 17–19. *Drepanoistodus suberectus* (Branson and Mehl)

17. c element, inner lateral view, GP-94-1-20, GSC 119415, ×45
18. a element, inner lateral view, GP-94-1-20, GSC 119416, ×45
19. e element, inner lateral view, GP-94-1-20, GSC 119414, ×45

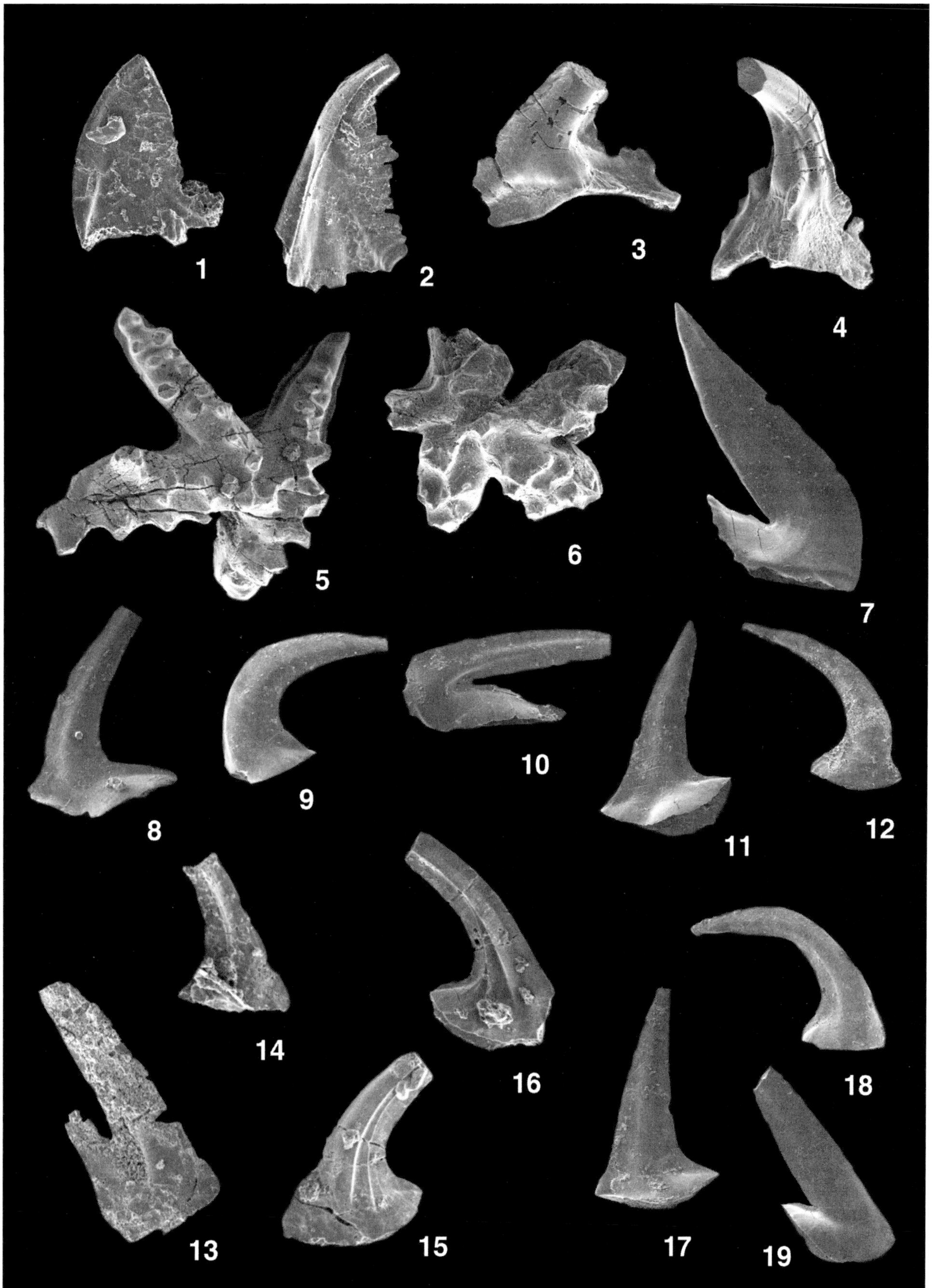
1
2
3
4
5
6
7
8
9
10
11
12
13
14
15
16
17
18
19

Plate 20

Figs. 1–7. *Dzikodus tablepointensis* (Stouge)

1. Sb element, antero-lateral view, 96-4-3, GSC 119438, ×115
2. Sa element, posterior view, 96-4-3, GSC 119439, ×80
3. Sb element, antero-lateral view, 96-4-3, GSC 119440, ×84
4. Sb element, antero-lateral view, 96-4-3, GSC 119441, ×84
5. Pb element, oral view, 96-4-3, GSC 119442, ×80
6. Pa element, lateral view, 96-4-3, GSC 119444, ×85
7. Pa element, oral view, 96-4-3, GSC 119443, ×85

Figs. 8, 9. *Erraticodon balticus* Dzik

8. P element, inner lateral view, GP-94-1-23, GSC 119436, ×84
9. S element, inner lateral view, GP-94-1-23, GSC 119437, ×84

Figs. 10, 11. *Multioistodus subdentatus* Cullison

10. M element, inner lateral view, GP-94-1-24, GSC 119426, ×45
11. S element, posterior view, GP-94-1-24, GSC 119427, ×45

Figs. 12, 13. *Oulodus* sp. A

12. Sa elements, posterior view, 96-4-6, GSC 119490, ×85
13. Sc element, outer lateral view, 96-4-6, GSC 119491, ×85

Fig. 14. *Oulodus* sp. B

14. Sc element, inner lateral view, RR-96-4, GSC 119492, ×70

Figs. 15, 16. *Panderodus unicostatus* Branson and Mehl

15. arcuatiform element, inner lateral view, 96-4-6, GSC 119488, ×40
16. graciliform element, inner lateral view, 96-4-6, GSC 119489, ×40

Figs. 17–20. *Paroistodus horridus* (Barnes and Poplawski)

17. S element, inner lateral view, 96-4-3, GSC 119455, ×64
18. M element, inner lateral view, 96-4-3, GSC 119456, ×50
19. P element, inner lateral view, 96-4-3, GSC 119457, ×50
20. P element, inner lateral view, 96-4-3, GSC 119458, ×50

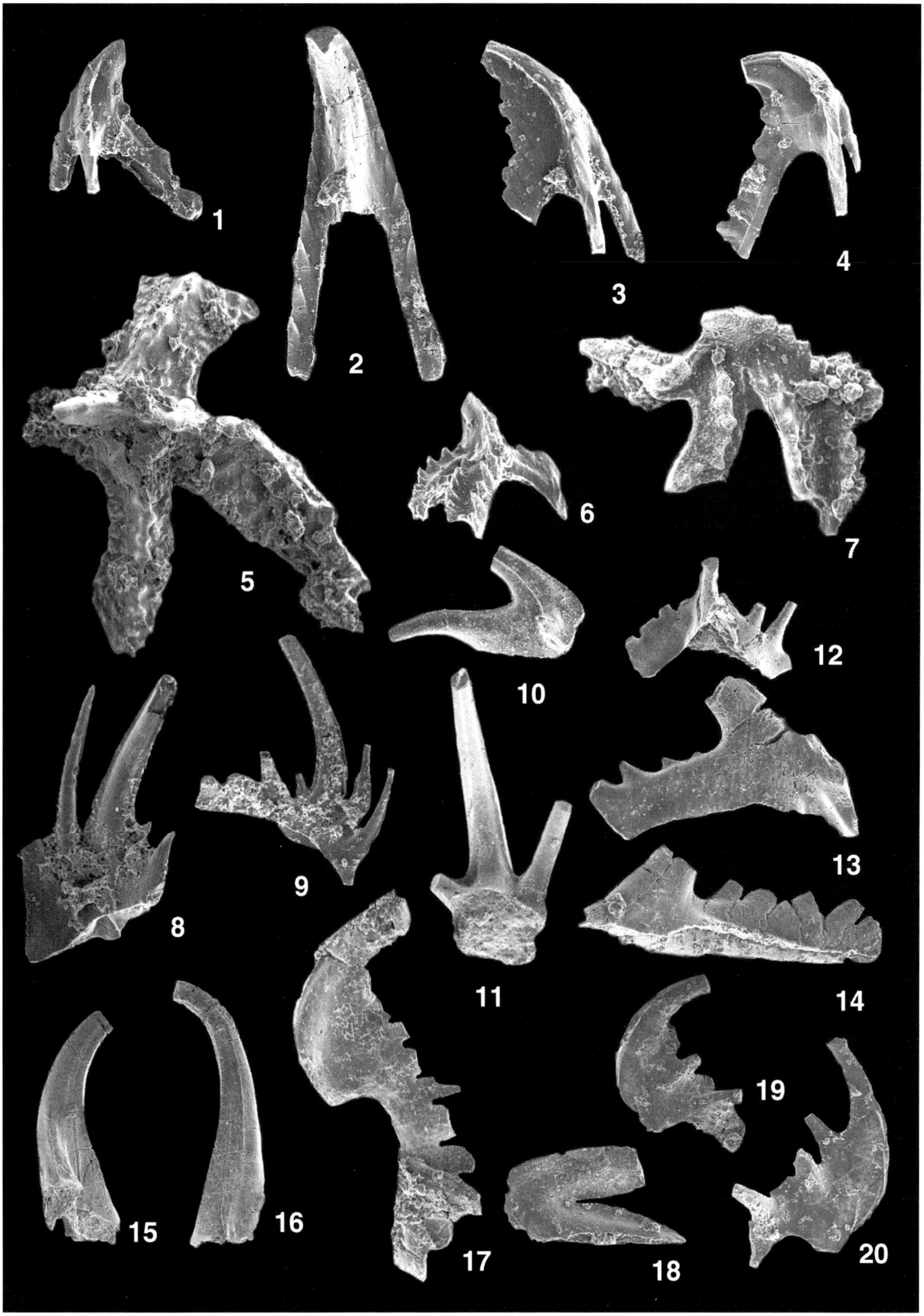
1
2
3
4
5
6
7
8
9
10
11
12
13
14
15
16
17
18
19
20

Plate 21

Figs. 1–3. *Paroistodus originalis* (Sergeeva)

1. M element, inner lateral view, GP-94-1-20, GSC 119465, ×60
2. P element, inner lateral view, GP-94-1-20, GSC 119466, ×60
3. S element, inner lateral view, GP-94-1-20, GSC 119467, ×60

Figs. 4–6. *Paroistodus parallelus* (Pander)

4. S element, inner lateral view, GP-94-1-16, GSC 119414, ×88
5. M element, inner lateral view, GP-94-1-16, GSC 119415, ×82
6. P element, inner lateral view, GP-94-1-16, GSC 119416, ×88

Figs. 7–9. *Periodon aculeatus* Hadding

7. M element, inner lateral view, GP-94-1-20, GSC 119433, ×60
8. Sb element, inner lateral view, GP-94-1-20, GSC 119434, ×60
9. Pa element, inner lateral view, GP-94-1-20, GSC 119435, ×60

Figs. 10–12. *Protopanderodus calceatus* Bagnoli and Stouge

10. a element, inner lateral view, 96-4-3, GSC 119481, ×100
11. c element, inner lateral view, 96-4-3, GSC 119482, ×84
12. e element, inner lateral view, 96-4-3, GSC 119483, ×84

Figs. 13, 14. *Protopanderodus cooperi* (Sweet and Bergström)

13. a/b element, inner lateral view, 96-4-3, GSC 119479, ×90
14. e element, inner lateral view, 96-4-3, GSC 119480, ×90

Figs. 15–17. *Protopanderodus gradatus* Serpagli

15. e element, inner lateral view, GP-94-1-20, GSC 119477, ×50
16. c element, inner lateral view, GP-94-1-20, GSC 119476, ×50
17. a element, inner lateral view, GP-94-1-20, GSC 119475, ×50

Fig. 18. *Pseudooneotodus beckmanni* Bischoff and Sannemann

18. inner lateral view, 96-4-6, GSC 119493, ×65

Figs. 19–22. *Protopanderodus robustus* (Hadding)

19. e element, inner lateral view, GP-94-1-20, GSC 119471, ×30

20, 21. a elements, inner lateral views, GP-94-1-20, GSC 119472, 119473, ×30

22. c element, inner lateral view, GP-94-1-20, GSC 119474, ×30

Fig. 23. *Protopanderodus varicostatus* (Sweet and Bergström)

23. a/b element, inner lateral view, GP-94-1-20, GSC 119478, ×45

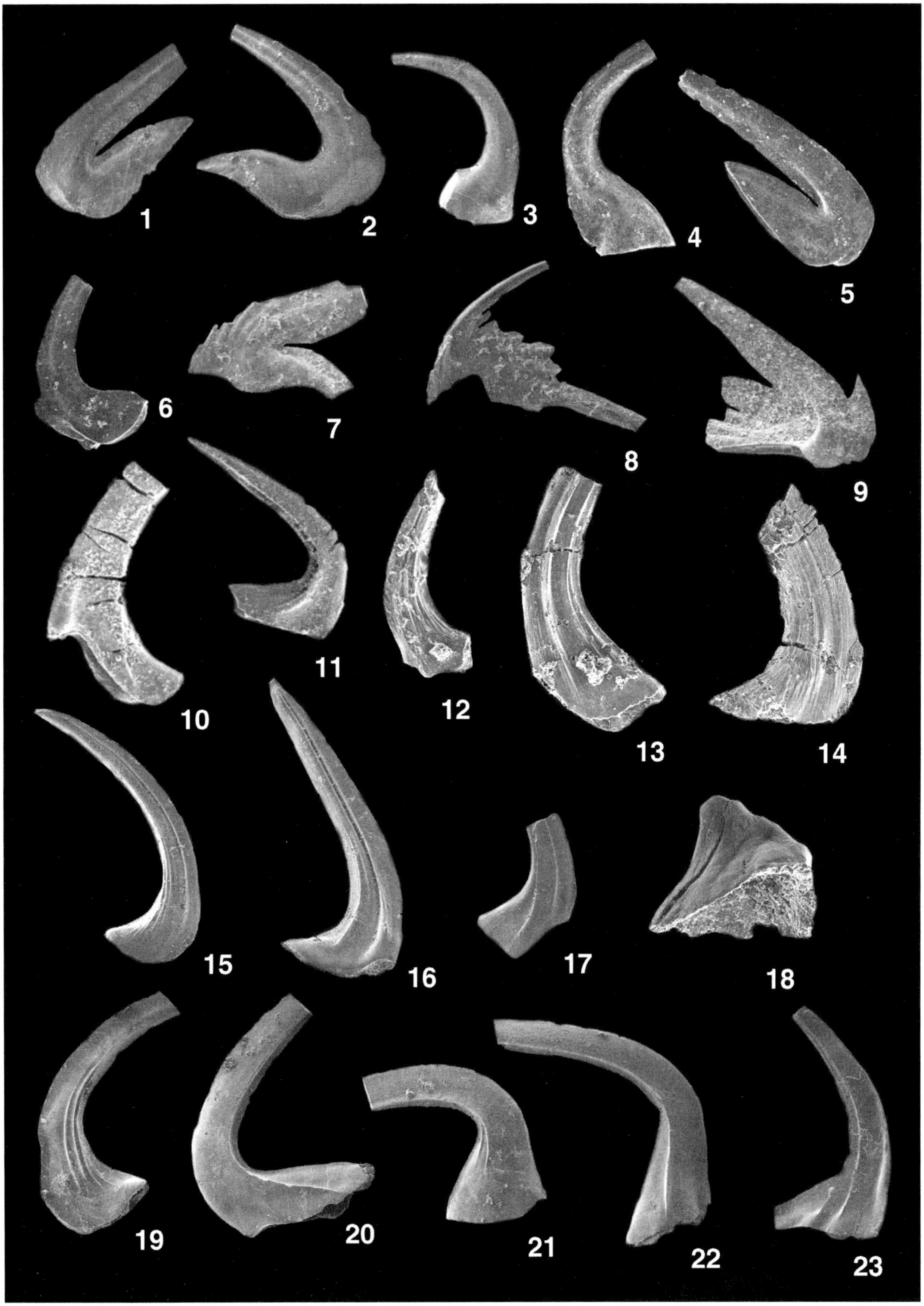
1
2
3
4
5
6
7
8
9
10
11
12
13
14
15
16
17
18
19
20
21
22
23

Plate 22

Figs. 1–5. *Protoprioniodus aranda* Cooper

1. P element, inner lateral view, GP-94-1-19, GSC 119428, ×93
2. S element, inner lateral view, GP-94-1-19, GSC 119429, ×93
3. M element, inner lateral view, GP-94-1-19, GSC 119430, ×93
4. S element, inner lateral view, GP-94-1-19, GSC 119431, ×93
5. P element, inner lateral view, GP-94-1-19, GSC 119432, ×93

Figs. 6, 7. *Spinodus spinatus* Hadding

6. Sc element, inner lateral view, GP-94-1-20, GSC 119424, ×70
7. Sb element, inner lateral view, GP-94-1-20, GSC 119425, ×70

Figs. 8–16. *Triangulodus akiensis* n. sp.

8. Sc element, inner lateral view, 96-4-3, *paratype*, GSC 119445, ×60
9. Sa element, antero-lateral view, 96-4-3, *paratype*, GSC 119446, ×60

10, 11. Sa element, inner and outer lateral views, 96-4-3, *holotype*, GSC 119447, ×60

12. Sd element, inner lateral view, 96-4-3, *paratype*, GSC 119448, ×60
13. M element, inner lateral view, 96-4-3, *paratype*, GSC 119449, ×60
14. Sb element, antero-lateral view, 96-4-3, *paratype*, GSC 119450, ×60
15. Pa element, inner lateral view, 96-4-3, *paratype*, GSC 119451, ×60
16. Pb element, inner lateral view, 96-4-3, *paratype*, GSC 119452, ×60

Fig. 17. *Walliserodus costatus* Dzik

17. S element, inner lateral view, 96-4-3, GSC 119464, ×100

Fig. 18. ?*Walliserodus nakholmensis* (Hamar)

18. P element, inner lateral view, 96-4-3, GSC 119459, ×40

Figs. 19–22. *Walliserodus ethingtoni* (Fåhræus)

19. P element, inner lateral view, 96-4-3, GSC 119460, ×70
20. S element, inner lateral view, 96-4-3, GSC 119461, ×70
21. S element, inner lateral view, 96-4-3, GSC 119462, ×70
22. S element, inner lateral view, 96-4-3, GSC 119463, ×70

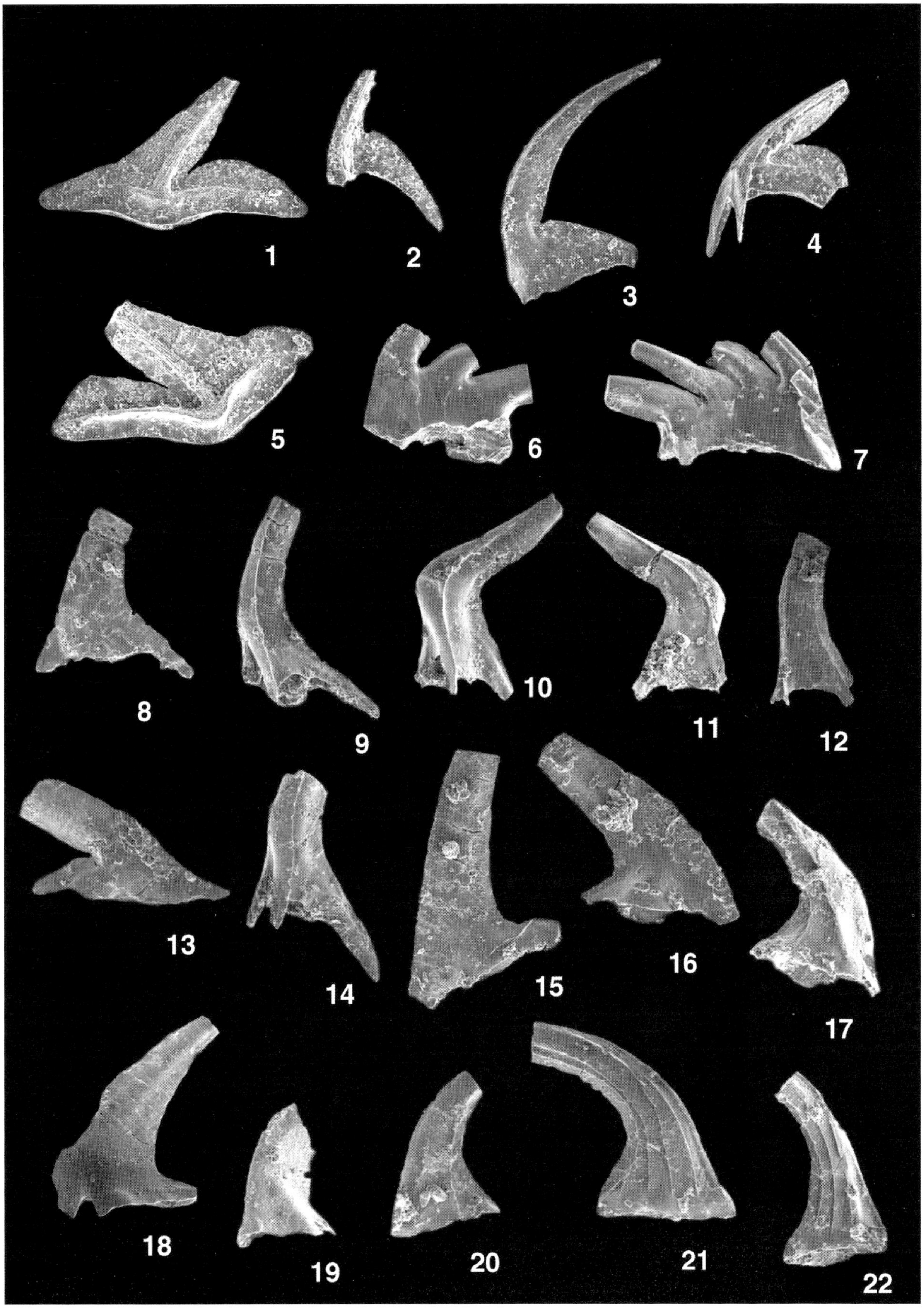
1
2
3
4
5
6
7
8
9
10
11
12
13
14
15
16
17
18
19
20
21
22

Plate 23

Fig. 1. *Anodontus longus* Stouge and Bagnoli

1. a element, inner lateral view, 96-5-17, GSC 119511, ×120

Figs. 2–4. *Ansella longicuspica* Zhang

2. S element, inner lateral view, 96-5-17, GSC 119613, ×110
3. P element, inner lateral view, 96-5-17, GSC 119614, ×100
4. M element, inner lateral view, 96-5-17, GSC 119615, ×100

Figs. 5–7. *Bergstroemognathus extensus* (Graves and Ellison)

5. M element, inner lateral view, 96-5-15A, GSC 119508, ×80
6. Sc element, inner lateral view, 96-5-11, GSC 119506, ×50
7. Sa element, posterior view, 96-5-11, GSC 119507, ×44

Figs. 8. 9. *Drepanoistodus basiovalis* (Sergeeva)

8. a element, inner lateral view, 96-5-12, GSC 119571, ×90
9. c element, inner lateral view, 96-5-12, GSC 119572, ×90

Fig. 10. *Drepanoistodus* n. sp. B

10. c element, inner lateral view, 96-13-B33, GSC 119573, ×88

Figs. 11–13. *Drepanoistodus* n. sp. C

11. a element, inner lateral view, 96-13-B34, GSC 119574, ×44
12. c element, inner lateral view, 96-13-B34, GSC 119575, ×44
13. e element, inner lateral view, 96-13-B34, GSC 119576, ×44

Figs. 14–16. *Drepanoistodus* n. sp. D

14. e element, inner lateral view, 96-5-16A, GSC 119577, ×40

15, 16. a elements, inner lateral views, 96-5-16A, GSC 119578, 119579, ×40

Figs. 17–19. *Drepanoistodus* n. sp. E

17. a element, inner lateral view, 96-5-16, GSC 119580, ×66

18, 19. e elements, inner lateral views, 96-5-16, GSC 119581, 119582, ×66

Fig. 20. *Eucharodus parallelus* (Branson and Mehl)

20. a element, inner lateral view, 96-5-12, GSC 119510, ×80

Figs. 21–24. *Fahraeusodus marathonensis* (Bradshaw)

21. P element, inner lateral view, 96-13-B31, GSC 119597, ×63

22, 23. S elements, inner lateral views, 96-13-B31, GSC 119598, 119599, ×63

24. M element, inner lateral view, 96-13-B31, GSC 119600, ×72

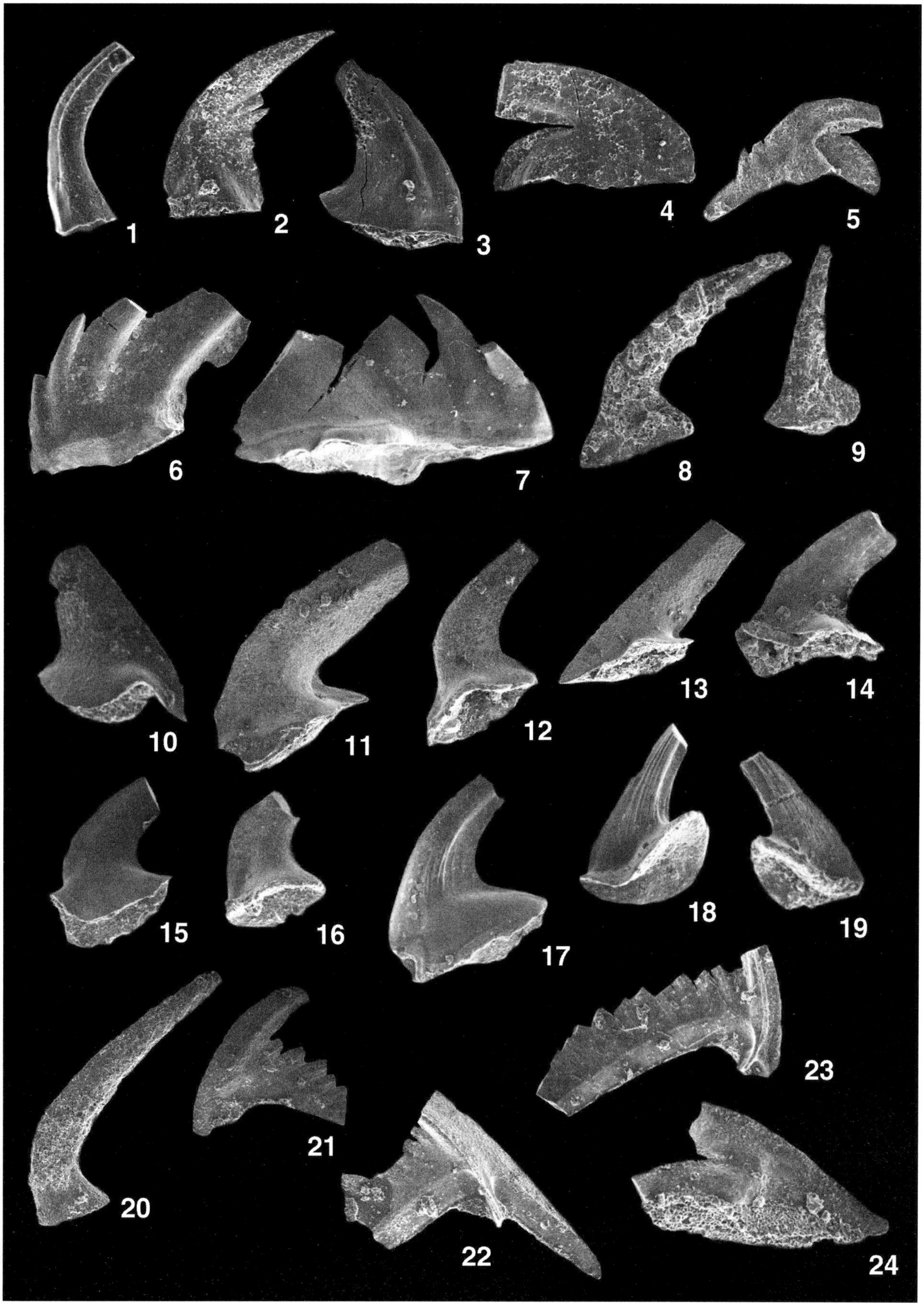
1
2
3
4
5
6
7
8
9
10
11
12
13
14
15
16
17
18
19
20
21
22
23
24

Plate 24

Figs. 1–4. *Juanognathus variabilis* Serpagli

1. a element, inner lateral view, 96-5-16, GSC 119552, ×70
2. c element, posterior view, 96-5-16, GSC 119553, ×70
3. b element, inner lateral view, 96-5-16, GSC 119554, ×70
4. e element, inner lateral view, 96-5-16, GSC 119555, ×70

Figs. 5–8. *Juanognathus jaanussoni* Serpagli

5. a element, inner lateral view, 96-5-16, GSC 119556, ×100
6. c element, inner lateral view, 96-5-16, GSC 119557, ×100
7. b element, postero-lateral view, 96-5-16, GSC 119558, ×100
8. e element, inner lateral view, 96-5-16, GSC 119559, ×100

Figs. 9, 10. *Jumudontus gananda* Cooper

9. P element, outer lateral view, 96-5-15A, GSC 119622, ×110
10. P element, outer lateral view, 96-5-15, GSC 119623, ×70

Fig. 11. *Kallidontus* cf. *K. nodosus* n. gen. n. sp.

11. Pb element, posterior view, 96-5-6, GSC 119509, ×50

Figs. 12–14. *Microzarkodina* n. sp.

12, 14. P elements, inner and outer lateral views, 96-5-15A, GSC 119543, ×125

13. M element, inner lateral view, 96-5-15A, GSC 119544, ×125

Fig. 15. ?*Multioistodus compressus* Harris and Harris

15. Sa element, posterior view, 96-5-18, GSC 119601, ×115

Figs. 16–19. *Oepikodus communis* (Ethington and Clark)

16. M element, inner lateral view, 96-5-16, GSC 119494, ×100
17. Sc element, inner lateral view, 96-5-16, GSC 119495, ×100
18. Sb element, inner lateral view, 96-5-16, GSC 119496, ×100
19. Pa element, inner lateral view, 96-5-16, GSC 119497, ×100

Figs. 20, 21. *Oepikodus* n. sp.

20. P element, inner lateral view, 96-5-16, GSC 119498, ×110
21. P element, inner lateral view, 96-5-16, GSC 119499, ×110

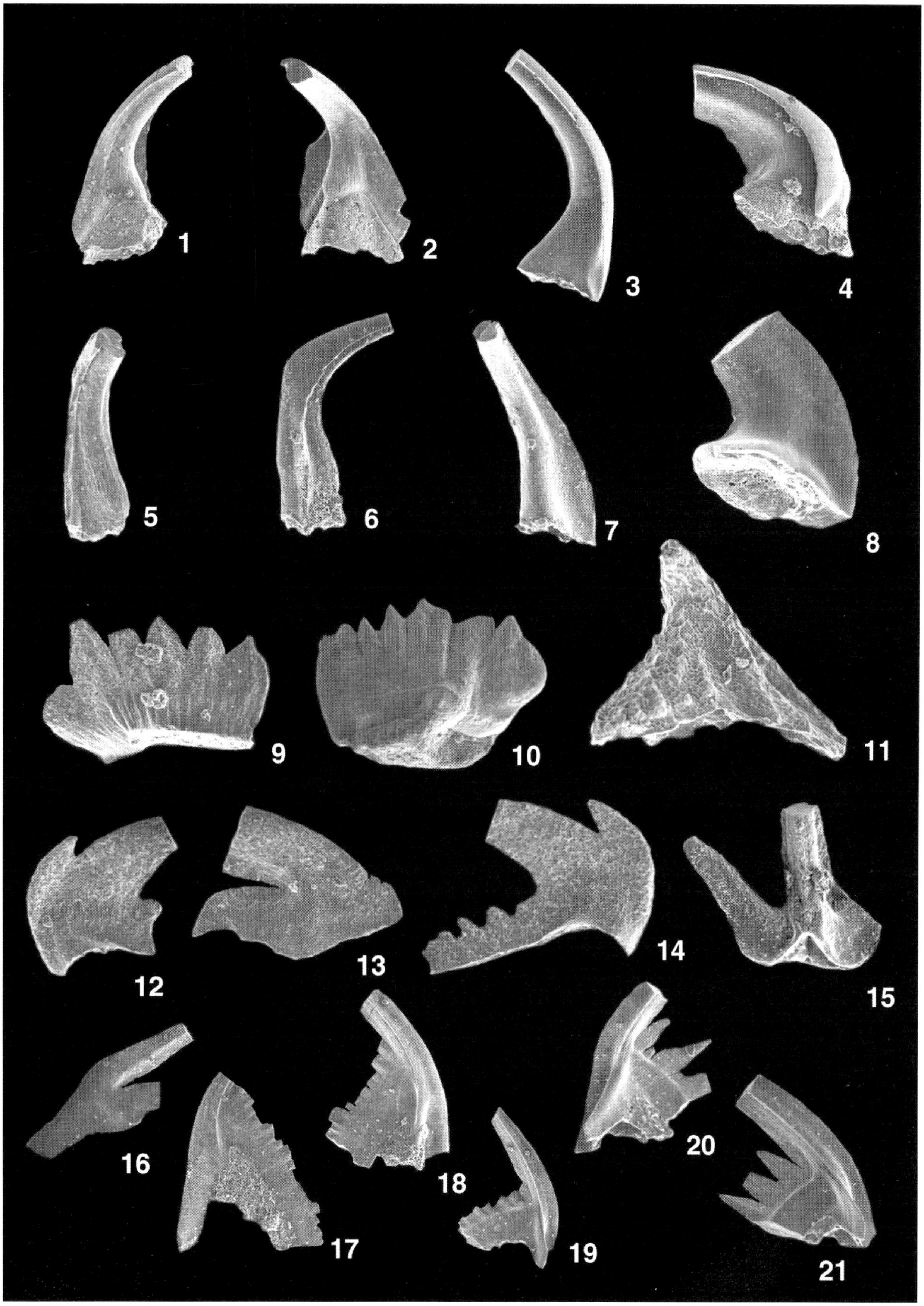
1
2
3
4
5
6
7
8
9
10
11
12
13
14
15
16
17
18
19
20
21

Plate 25

Figs. 1–3. *Oepikodus evae* (Lindström)

1. P element, antero-lateral view, 96-13-B34, GSC 119503, ×86
2. M element, inner lateral view, 96-13-B34, GSC 119505, ×86
3. S element, inner lateral view, 96-13-B34, GSC 119504, ×86

Figs. 4–6. *Oepikodus intermedius* (Serpagli)

4. M element, inner lateral view, 96-5-10, GSC 119500, ×67
5. S element, inner lateral view, 96-5-10, GSC 119501, ×67
6. P element, inner lateral view, 96-5-10, GSC 119502, ×67

Fig. 7. *Oistodus bransoni* Ethington and Clark

7. S element, inner lateral view, 96-13-B30, GSC 119529, ×50

Figs. 8–11. *Oistodus lanceolatus* Pander

8. P element, inner lateral view, 96-5-15A, GSC 119530, ×77
9. Sa element, posterior view, 96-5-15A, GSC 119531, ×110
10. S element, inner lateral view, 96-5-15A, GSC 119532, ×77
11. M element, inner lateral view, 96-5-15A, GSC 119533, ×55

Figs. 12–16. *Oistodus multicorrugatus* Harris

12, 15. S elements, inner lateral views, 96-5-16A, GSC 119524, 119525, ×40
13. P element, inner lateral view, 96-5-16A, GSC 119526, ×40
14. Sa element, posterior view, 96-5-16A, GSC 119527, ×60
16. M element, inner lateral view, 96-5-16A, GSC 119528, ×40

Figs. 17–22. *Parapanderodus striatus* (Graves and Ellison)

17, 22. e elements, posterior views, 96-5-13, GSC 119512, 119513, ×88
18, 19. a elements, inner lateral views, 96-5-13, GSC 119514, 119515, ×120
20. c element, posterior view, 96-5-17, GSC 119516, ×120
21. a element, inner lateral view, 96-5-13, GSC 119517, ×120

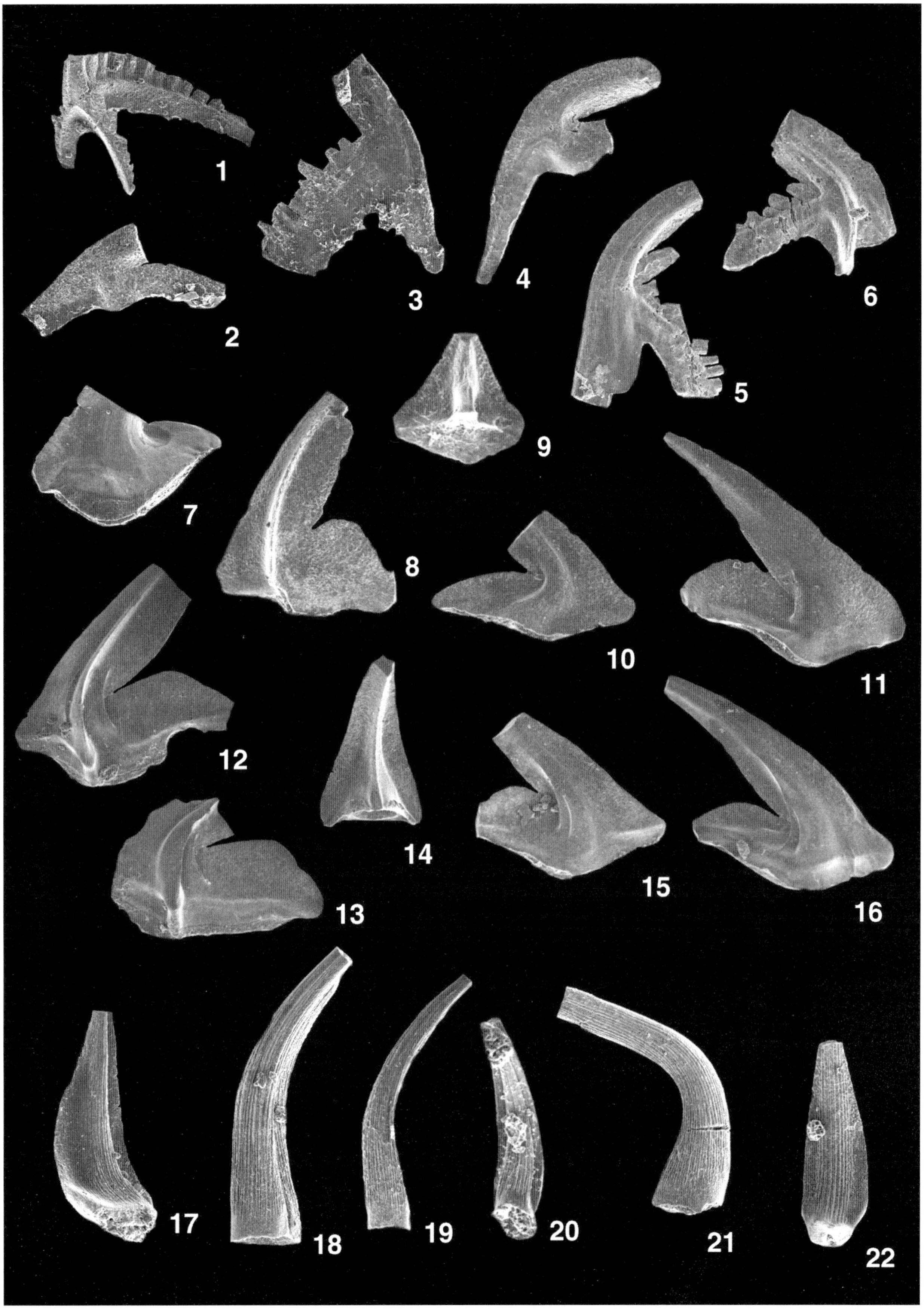
1
2
3
4
5
6
7
8
9
10
11
12
13
14
15
16
17
18
19
20
21
22

Plate 26

Figs. 1, 2. *Paroistodus* n. sp.

1. P element, inner lateral view, 96-5-16, GSC 119560, ×50
2. S element, inner lateral view, 96-5-16, GSC 119561, ×50

Figs. 3–6. *Periodon flabellum* (Lindström)

3. Pa element, inner lateral view, 96-5-16A, GSC 119618, ×85
4. Sc element, inner lateral view, 96-5-16A, GSC 119619, ×85
5. Sa element, postero-lateral view, 96-5-16A, GSC 119620, ×85
6. M element, inner lateral view, 96-5-16A, GSC 119621, ×85

Figs. 7, 8. *Periodon selenopsis* (Serpagli)

7. Pa element, inner lateral view, 96-5-15A, GSC 119616, ×110
8. M element, inner lateral view, 96-5-15A, GSC 119617, ×110

Figs. 9–13. *Planusodus gradus* n. gen. n. sp. (early form)

9, 10. a element, inner and outer lateral views, 96-5-15A, *paratype*, GSC 119590, ×70

11. c element, inner lateral view, 96-5-15A, *paratype*, GSC 119591, ×70
12. b element, inner lateral view, 96-5-15A, *paratype*, GSC 119592, ×70
13. e element, inner lateral view, 96-5-15A, *paratype*, GSC 119593, ×100

Figs. 14–18. *Planusodus gradus* n. gen. n. sp. (late form)

14, 15. a element, inner and outer lateral views, 96-5-16, *holotype*, GSC 119594, ×75

16. c element, inner lateral view, 96-5-16A, *paratype*, GSC 119595, ×70

17, 18. e element, inner and outer lateral views, 96-5-16A, *paratype*, GSC 119596, ×70

Figs. 19–21. *Prioniodus adami* Stouge and Bagnoli

19, 20. S elements, inner lateral views, 96-5-6, GSC 119564, ×70, GSC 119565, ×55

21. P element, inner lateral view, 96-5-6, GSC 119566, ×55

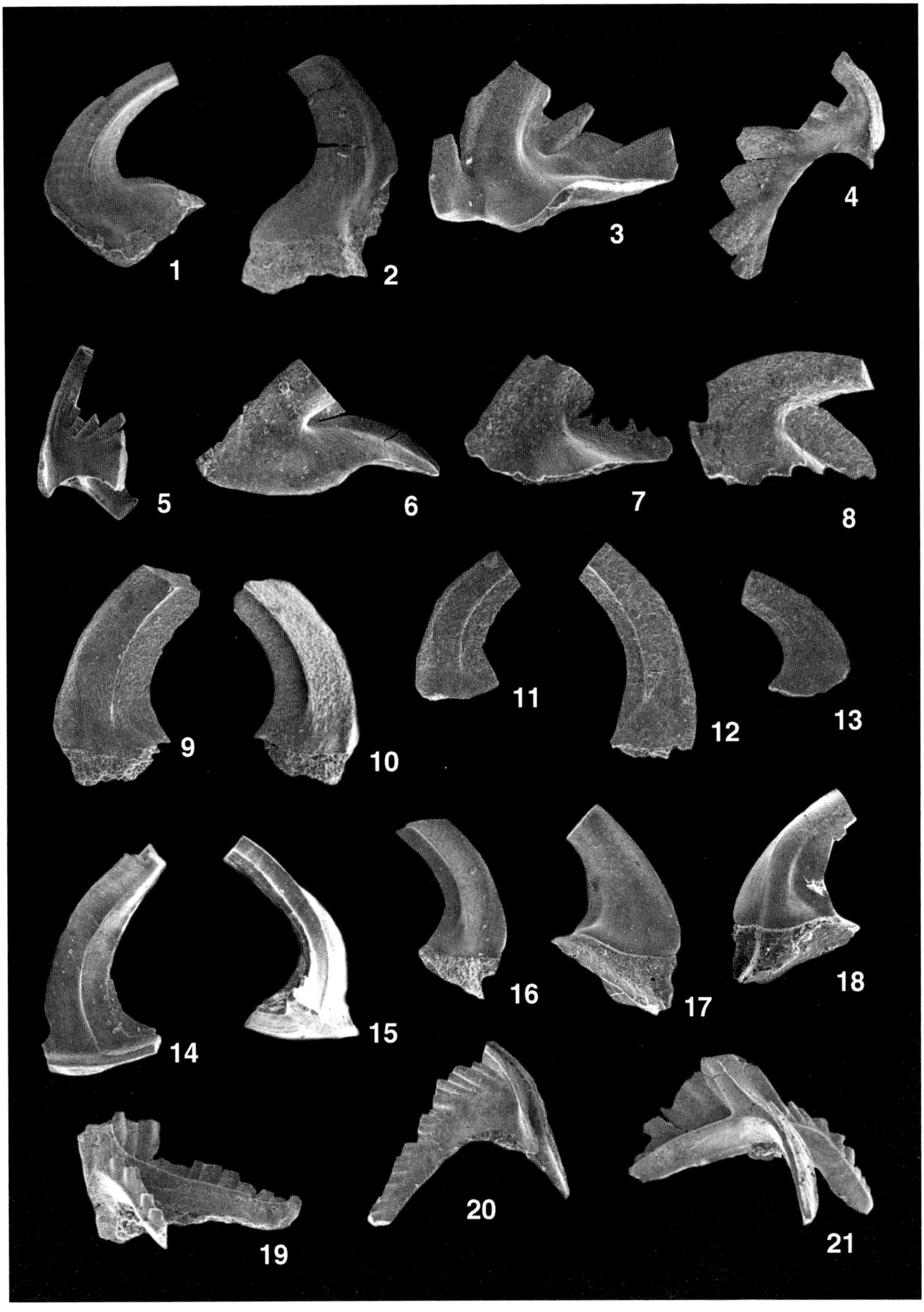
1
2
3
4
5
6
7
8
9
10
11
12
13
14
15
16
17
18
19
20
21

Plate 27

Figs. 1–4. ?*Prioniodus* n. sp.

1, 2. ?Sb elements, inner lateral views, 96-5-6, GSC 119567, 119568, ×85

3, 4. ?Sd elements, inner lateral views, 96-5-6, GSC 119569, 119570, ×85

Figs. 5–7. *Protopanderodus leonardii* Serpagli

5. a/b element, inner lateral view, 96-5-16A, GSC 119624, ×60
6. c element, inner lateral view, 96-5-16A, GSC 119625, ×60
7. e element, inner lateral view, 96-5-16A, GSC 119626, ×60

Fig. 8. *Protopanderodus rectus* (Lindström)

8. a/b element, inner lateral view, 96-5-14, GSC 119627, ×50

Figs. 9, 10. *Protoprioniodus* n. sp.

9. S element, inner lateral view, 96-5-15A, GSC 119638, ×80
10. P element, inner lateral view, 96-5-15A, GSC 119639, ×80

Figs. 11–13. *Protoprioniodus cowheadensis* Stouge and Bagnoli

11, 13. S elements, inner lateral views, 96-5-12, GSC 119635, 119637, ×105

12. Sa element, posterior view, 96-5-12, GSC 119636, ×105

Figs. 14–20. *Protoprioniodus nyinti* Cooper

14. M element, inner lateral view, 96-5-16A, GSC 119628, ×110

15, 17. S elements, antero-lateral and lateral views, 96-5-16A, GSC 119629, 119631, ×84

16. Sa element, posterior view, 96-5-16A, GSC 119630, ×110

18, 19. P elements, outer and inner lateral views, 96-5-16A, GSC 119633, 119634, ×110

20. P element, outer lateral view, 96-5-16A, GSC 119632, ×72

Figs. 21, 22. *Pteracontiodus cryptodens* (Mound)

21. Sd element, inner lateral view, 96-5-18, GSC 119584, ×40
22. Sc element, inner lateral view, 96-5-18, GSC 119585, ×60

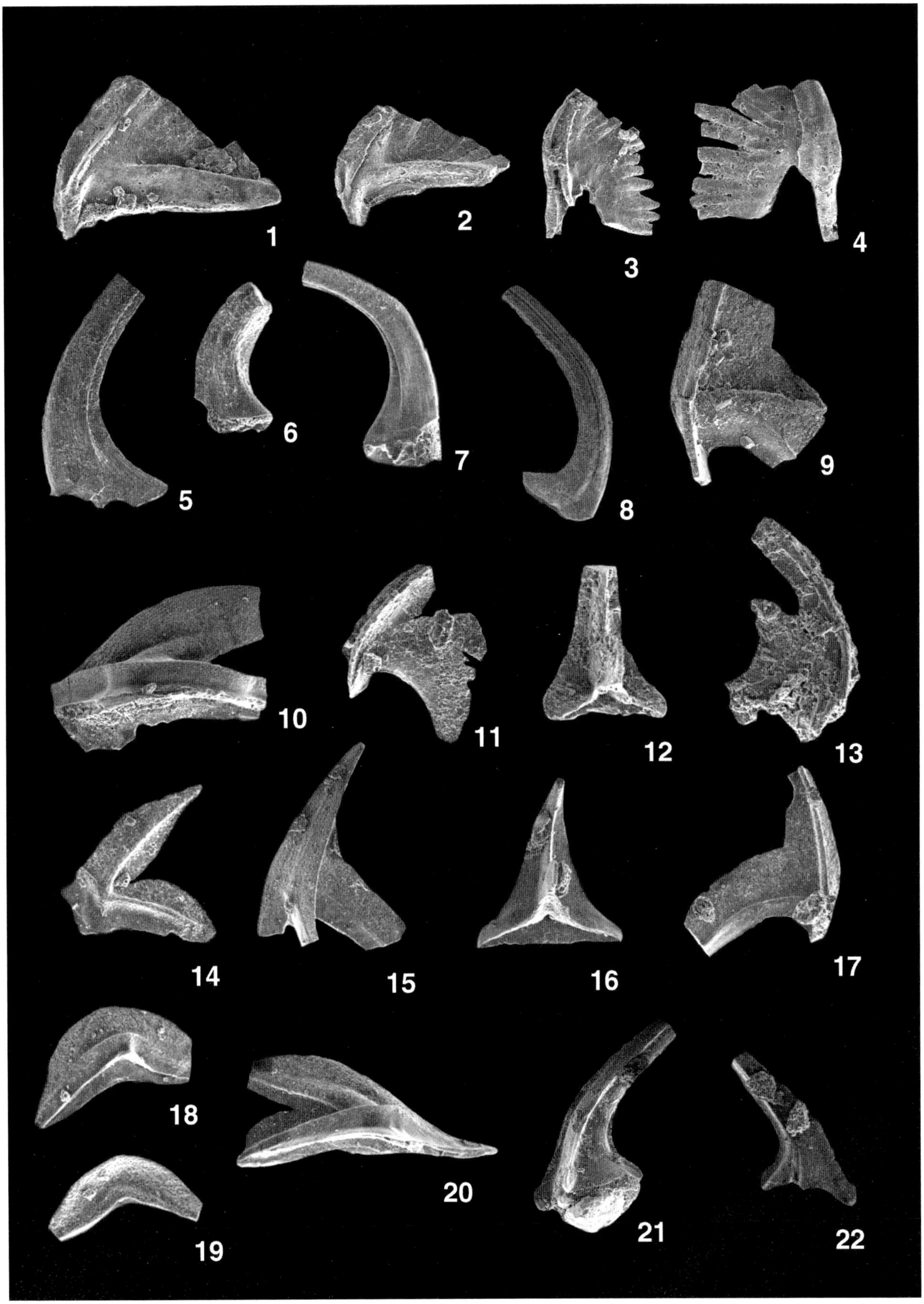
1
2
3
4
5
6
7
8
9
10
11
12
13
14
15
16
17
18
19
20
21
22

Plate 28

Figs. 1–7. *Scolopodus amplus* n. sp.

1. a element, inner lateral view, 96-5-6, *paratype*, GSC 119518, ×40
2. c element, posterior view, 96-5-6, *paratype*, GSC 119519, ×40
3. b element, outer lateral view, 96-5-7, *paratype*, GSC 119520, ×40

4, 5. a element, outer and inner lateral views, 96-5-7, *holotype*, GSC 119522, ×40

6. ?f element, postero-lateral view, 96-5-7, *paratype*, GSC 119521, ×4
7. e element, inner lateral view, 96-5-7, *paratype*, GSC 119523, ×40

Fig. 8. *Tripodus laevis* Bradshaw

8. Sd element, inner lateral view, 96-5-16A, GSC 119583, ×60

Figs. 9–17. *Tropodus sweeti* (Serpagli)

9. Sc element, juvenile, postero-lateral view, 96-13-B34, GSC 119541, ×100
10. Sd element, juvenile, inner lateral view, 96-13-B34, GSC 119542, ×100
11. Sc element, inner lateral view, 96-13-B34, GSC 119534, ×35
12. Sb element, inner lateral view, 96-13-B34, GSC 119535, ×35
13. Sd element, inner lateral view, 96-13-B34, GSC 119536, ×35
14. Sa element, posterior view, 96-13-B34, GSC 119537, ×35
15. M element, inner lateral view, 96-13-B34, GSC 119538, ×35
16. Pb element, inner lateral view, 96-13-B34, GSC 119539, ×35
17. Pa element, inner lateral view, 96-13-B34, GSC 119540 ×35

Figs. 18, 19. Coniform indet. J

18, 19. inner lateral views, 96-5-16A, GSC 119550, 119551, ×115

Fig. 20. Coniform indet. L

20. inner lateral view, 96- 5-16, GSC 119563, ×85

Fig. 21. Coniform indet. K

21. inner lateral view, 96-13-B33, GSC 119562, ×96

Figs. 22, 23. Platform indet.

22. P element, antero-lateral view, 96-13-B34, GSC 119605, ×100
23. P element, oral view, 96-13-B34, GSC 119606, ×100

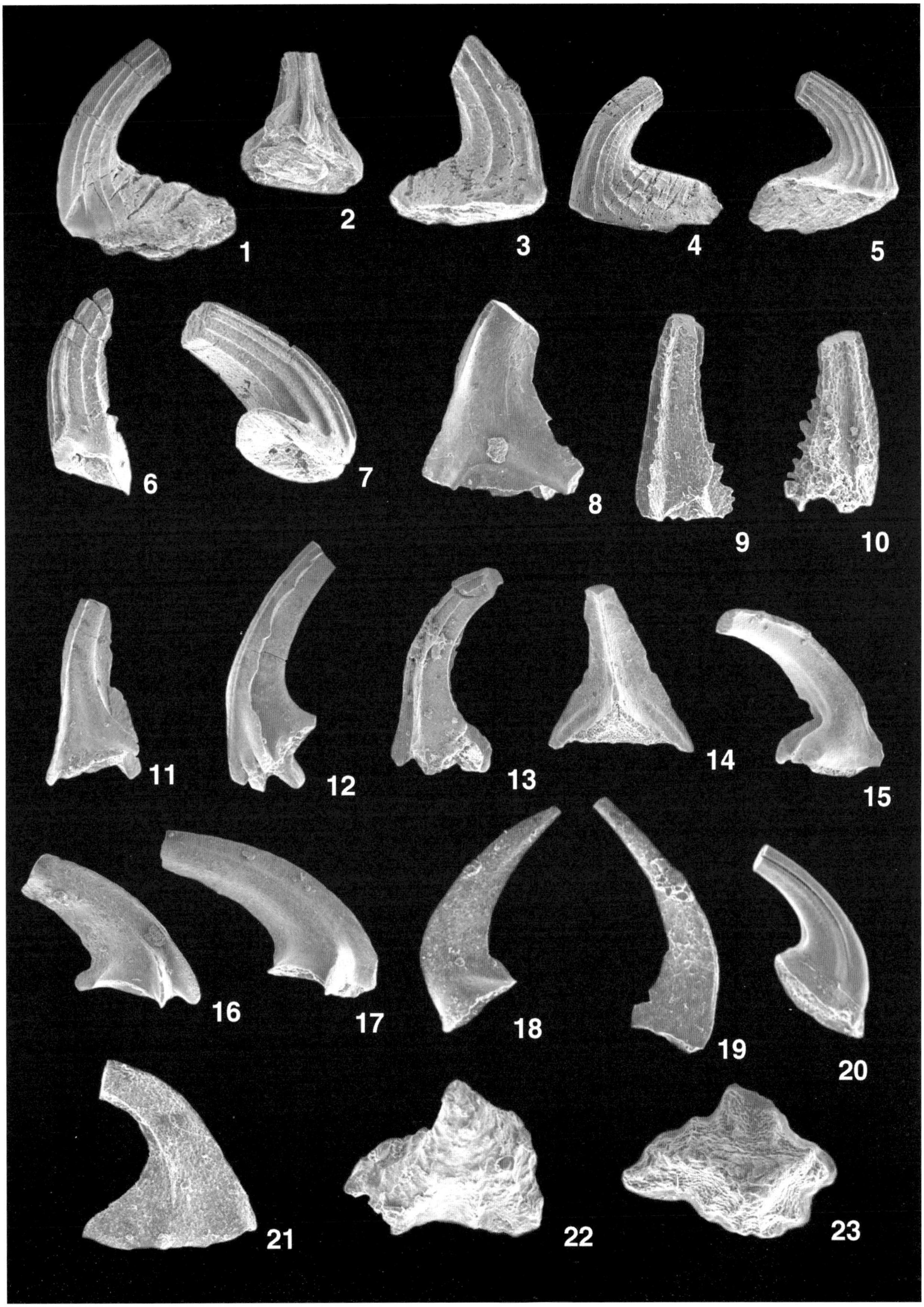
1
2
3
4
5
6
7
8
9
10
11
12
13
14
15
16
17
18
19
20
21
22
23

Plate 29

Figs. 1–3. Coniform indet. M

1. a element, outer lateral view, 96-5-17, GSC 119602, ×115
2. c element, posterior view, 96-5-17, GSC 119603, ×115
3. a element, anterior view, 96-5-17, GSC 119604, ×115

Figs. 4–6. Coniform indet. N

4. e element, inner lateral view, 96-5-18, GSC 119607, ×48
5. a element, inner lateral view, 96-5-18, GSC 119609, ×84
6. c element, posterior view, 96-5-18, GSC 119608, ×84

Figs. 7–9. Coniform indet. O

7. a element, inner lateral view, 96-5-18, GSC 119610, ×84
8. c element, posterior view, 96-5-18, GSC 119611, ×84
9. b element, inner lateral view, 96-5-18, GSC 119612, ×84

Figs. 10–13. New genus B n. sp.

10. e element, inner lateral view, 96-5-6, GSC 119546, ×77
11. belement, inner lateral view, 96-5-6, GSC 119547, ×77
12. a element, antero-lateral view, 96-5-6, GSC 119548, ×77
13. a element, inner lateral view, 96-5-6, GSC 119549, ×77

Figs. 14–17. New genus C n. sp.

14, 15, 17. Selements, inner lateral views, 96-5-16A, GSC 119586, 119587, 119588, ×105

16. P element, anterior view, 96-5-16A, GSC 119589, ×105

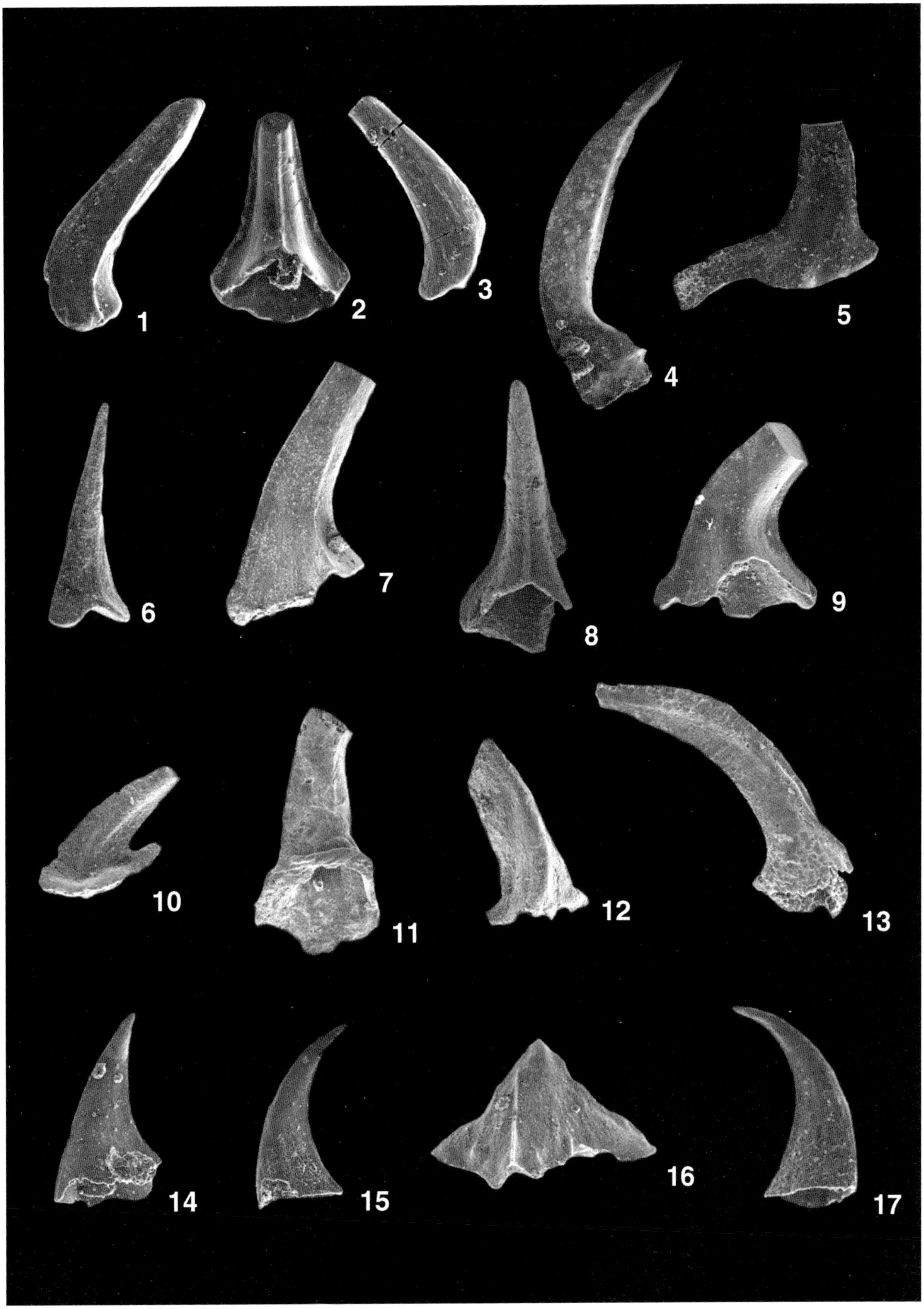
1
2
3
4
5
6
7
8
9
10
11
12
13
14
15
16
17